Erd - E - II - 21
(Erd - B - I)
(TC - III)

Springer

Berlin
Heidelberg
New York
Barcelona
Hong Kong
London
Milan
Paris
Singapore
Tokyo

Monique Mainguet

Aridity
Droughts and Human Development

With 76 Figures and 25 Tables

 Springer

Author
Prof. Dr. Monique Mainguet
Member of the "Institut Universitaire de France"
Université de Reims Champagne-Ardenne
Laboratoire de Géographie Zonale pour le Développement
57 rue Pierre Taittinger
51 100 Reims
France

Translator
Dr. Thomas O. E. Reimer
81 Thabu Nchu, Elemanor
Glenvista 2058
South Africa

The original French edition *L' Homme et la Sécheresse* was awarded the prize "Victor-Amédée du Bocage (1895)" from the Société Géographique de France in 1996.

The translation has been helped by the French Ministry in charge of Culture.

ISBN 3-540-63342-1 Springer-Verlag Berlin Heidelberg New York

Library of Congress Cataloging-in-Publication Data
Mainguet, Monique, 1937- [L' homme et la Sécheresse. English] Aridity : droughts and human development / Monique Mainguet ; [Translator, Thomas O.E. Reimer].
p. cm. The original French edition, L' Homme et la Sécheresse, was awarded the prize "Victor-Amédée du Bocage (1895)" from the Société Géographique de France in 1996–T.p. verso. The translation has been helped by the French Ministry in charge of culture. Includes bibliographical references and index.
ISBN 3-540-63342-1 (alk. paper) 1. Desertification. 2. Nature–Effect on human beings on. I. Title.
GB612.M3413 1998 333.73'6–dc21 98-30231

This work is subject to copyright. All rights are reserved, whether the whole or part of the material is concerned, specifically the rights of translation, reprinting, re-use of illustrations, recitation, broadcasting, reproduction on microfilms or in any other way, and storage in data banks. Duplication of this publication or parts thereof is permitted only under the provisions of the German Copyright Law of September 9, 1965, in its current version, and permission for use must always be obtained from Springer-Verlag. Violations are liable for prosecution under the German Copyright Law.

© Springer-Verlag Berlin Heidelberg 1999
Printed in Germany

The use of general descriptive names, registered names, trademarks, etc. in this publication does not imply, even in the absence of a specific statement, that such names are exempt from the relevant protective laws and regulations and therefore free for general use.

Coverdesign: Erich Kirchner, Heidelberg
Dataconversion: Büro Stasch, Bayreuth

SPIN: 10538746   32/3020 – 5 4 3 2 1 0 – Printed on acid-free paper

Preface

The great beauty of landscape in the arid ecosystems, the value of this natural capital for breeders and farmers, their lessons in the science of survival, and the repeated failure of contemporary development plans are the basis of this book. Two successive expeditions to Mauritania, in 1997 and 1998, revealed an acceleration in degradation which, 3 years after the French edition, justifies this English edition, in which many questions still remain open. Some of these are:

- Aridity: can we believe the theory of a general present-day tendency for dry ecosystems to have less water available now, so that aridity is increasing, when 20 000 years ago Saharo-Sahalian Africa was more arid than in our time?
- Droughts, unpredictable and recurrent: are they becoming more virulent, or is it only that their fury is felt more severely in arid ecosystems than elsewhere, now at the end of this otherwise so successful 20th century, like an acute disease in a weakened body?
- Aridity and drought: the variability of these in time and space, their fluctuating combinations that lead to never-ending modifications of living conditions, all these demand permanent efforts at adaptation. But is mankind confronted with environmental changes caused, in fact, not only by the double action of aridity and drought, but also by our own actions, which are forcing one precarious equilibrium after another?
- Are there specific natural risks to the dry ecosystems – flood, drought, erosion – of the dangerous dynamic forces inherent in these ecosystems becoming accelerated? Are these risks, in fact, greater, ore are the arid ecosystems more vulnerable because we are increasing the risks by our actions in these areas?
- Why do we find the most severe global degradation along the diagonals of the dry lands, especially along the largest of these, from the Sahara to central Asia; and why do we find that the majority of the countries must still be developed within the arid ecosystems? Are the coincidences between dry environments, underdevelopment, severe degradation, and the fight for survival arbitrary?
- For decades we have been filled with pride at our increasing mastery of nature, however, despite enormous progress (mostly material and technical), neither the standard nor the quality of development in arid ecosystems has benefited. Why is this? Must we conclude that it is more difficult to achieve development in these ecosystems than elsewhere?
- How can we explain that the considerable importance of arid ecosystems in global development has, in our days, become a burden? How is it possible that these

lands, the cradle of highly esteemed hydraulic civilizations, still coveted in the 20th century by the great colonial powers, areas with great mineral resources, can become disaster areas, which to the present day have been unable to obtain external aid to achieve their development?
- Why have so many enormous development projects been launched in the dry ecosystems, followed by failures all the more resounding because they were both vast and costly, when, as the proverb has it, "he who can do the most can do the least"? Are we no longer capable of anything except the "most", even when "modest" and "least" seem to be indicated? But why?
- The role played by arid ecosystems in our cultural heritage – in mathematics, art, and things spiritual – should also be investigated: is there a bond between the incontestable beauty of the dry desert lands and this cultural heritage? Can we believe this, and how can it be explained?
- Does the "mastery of survival under extreme conditions" help us to take the decisive step which opens the way to the "more" as represented by an awareness of "beauty", at first perhaps through song and dance, and then by sensing the need to beautify every day life with painting, sculpture, architecture, home decoration, and the arts of carpet weaving and jewellery.
- A last tentative, but not unimportant question: can an arid environment stand the effects of the worldwide and regional demographic explosion, and how can it respond? Can arid ecosystems offer modern man what he needs for his equilibrium? Are they suited for long-lasting development? Must we accept that they do not possess the necessary resources, or should we say, optimistically, that they have not yet learned to utilize their disposable resources fully, and that, in the end, the countries originally classified as "underdeveloped", are merely countries "with delayed development", which will be able to catch up with or even overtake the countries called "developed" up till now?

In the absence of a definitive reply to all these questions, the aim of this book is to illustrate convincingly the primordial importance of the factor "environment" in development, and the necessity of taking it into account, more in the dry ecosystems than in any others. Let us hope that this book awakens curiosity in the reader on the subject of the arid environments, and can encourage fruitful reflection on these regions, which comprise almost 35% of the land of our planet, and 20% of its population.

Monique Mainguet
Reims, October 1998

Contents

General Introduction 1

1 The Spatial Framework, the Concepts of Aridity and Drought: the Soils and the Vegetation 5
1.1 Attempts at Global Localization of Drylands 5
    1.1.1 The Cold Polar Deserts 8
    1.1.2 The Warm Tropical Deserts 10
    1.1.3 The Coastal Deserts 14
    1.1.4 The Lee-side Deserts 16
    1.1.5 The Continental Deserts: The Turan Desert as an Example 17
    1.1.6 Semi-Deserts or Semi-Arid Zones, a Wide Zone of Transition 23
1.2 The Concepts of Aridity and Drought in Drylands 24
    1.2.1 Definition of Degrees of Aridity 25
    1.2.2 The Land Surface Affected by Aridity on the African Continent 26
    1.2.3 The Different Types of Drought 27
    1.2.4 Postglacial Aridification and Advance of the Deserts in Africa? 32
    1.2.5 The Droughts of the 20th Century in North-Equatorial Africa 36
    1.2.6 Study of the Drought of 1984–1985 in Kenya and Its Consequences 41
    1.2.7 What Lessons Do We Learn from the Last Drought in the Sahel? 43
1.3 The Thin Vulnerable Soils of Low Agricultural Potential 46
    1.3.1 The Dryland Soils of North-Equatorial Africa 46
    1.3.2 The Soils of Turan 47
    1.3.3 The Specific Characteristics of Dryland Soils 54
1.4 Vegetation Covers: Economic Potentials and Droughts 56
    1.4.1 Heterogeneity of the Vegetaion Cover in Drylands 57
    1.4.2 Adaptability of Plants to Aridity and Drought 58
    1.4.3 The Vegetation Along the Dryland Diagonal Sahara-China 58
    1.4.4 The Open Vegetation Cover of the Sahel 61
    1.4.5 The Vocation of Grazing the Sahel Steppe 62
    1.4.6 Potential Foraging Value and Carrying Capacity 66
    1.4.7 Weakening of the Vegetation Cover by Human Activities: Different Estimates 69
    1.4.8 Behaviour of the Vegetal Cover During the Drought of 1969–1973 75
1.5 Conclusions 77

2 Resources vs. Hydrological and Aeolian Constraints 79
2.1 Inventory of Resources 79
2.2 Hydrological Constraints in Drylands 81
    2.2.1 Specific Hydrological Features 81
    2.2.2 The Fugacity of the Surface Waters 83
    2.2.3 The Allochthonous Water-Courses 85
    2.2.4 Rill Wash and Runoff in Tropical Drylands 91
    2.2.5 Ponds and Water Bodies:
           The Third Type of Water Reserve in Drylands 98
    2.2.6 The Underground Waters 102
    2.2.7 Degradation of Soils by Water Erosion 107
2.3 The Omnipresent Wind in Dry Ecosystems 112
    2.3.1 Wind – Definition and Basic Principles of Its Dynamics 113
    2.3.2 The Effects of Wind on Vegetation and Soil Humidity 115
    2.3.3 Aeolian Processes in Sand and Loess Drylands 117
    2.3.4 Aeolian Deposits 121
    2.3.5 A Proposal for the Classification of Dunes 122
    2.3.6 The Theory of the Global Aeolian Action System (GAAS) 131
2.4 Conclusions 136

3 Human Genius:
The Search for Water and Its Management – Battle Against the Wind . . 137
3.1 Non-Irrigated Agricultural Systems in Drylands:
    Their Difficulties in the 20th Century 137
    3.1.1 Itinerant Rain-Fed Agriculture and Fallow Periods 138
    3.1.2 The Move Towards Irrigated Agriculture 140
    3.1.3 Livestock Breeding, Nomadism, Pastoralism, and Transhumance 142
    3.1.4 Demise and Mutation of Nomadism 146
3.2 The Geohistory of Water Management: The Hydraulic Civilizations 154
    3.2.1 Aridity, Drought and the Birth of Modern Man 154
    3.2.2 Aridity, Drought and the Birth of Agriculture 155
    3.2.3 Aridity, Drought and the Birth of Hydraulic Societies 157
    3.2.4 The Great Hydraulic Systems of the Drylands 160
3.3 Traditional and Modern Hydraulic Techniques 169
    3.3.1 Modern Search for Water, Luxury Projects 169
    3.3.2 Parsimonious Management of Surface Waters 171
3.4 Traditional and Modern Irrigation 177
    3.4.1 Traditional Irrigation *Bewässerung* 178
    3.4.2 Modern Irrigation 180
    3.4.3 Reservoir Dams 185
3.5 A Response to the Aeolian Constraints
    – The Art of Counteracting the Effects and Damage Done by Wind 192
    3.5.1 Basic Principles for Fixing Mobile Sands and Dunes 192
    3.5.2 Control of Barchans and Linear or Parabolic Dunes 198
    3.5.3 Strategies for the Control of Aeolian Erosion on the Level of the
           Various Units of the Global Aeolian Action System (GAAS) 200
3.6 Conclusions 201

4 From Ingenuity to Decadence: Geohistory of an Actual Decline – Grounds for Hope? ... 203

- 4.1 Proposals for a Definition of Decadence ... 203
- 4.2 Ten Principles Explaining Decadence ... 204
  - 4.2.1 The Naturalist Perspective ... 204
  - 4.2.2 The Neo-Malthusian Perspective ... 204
  - 4.2.3 The Marxist Perspective ... 205
  - 4.2.4 The Anti-Colonialist and Neo-Colonialist Perspective ... 205
  - 4.2.5 The Judicial and Socio-Cultural Perspective ... 205
  - 4.2.6 The Global Perspective ... 206
  - 4.2.7 The Perspective of Non-Adapted Strategies and Technologies ... 206
  - 4.2.8 The Perspective of Underdevelopment as the Cause for the Degradation of the Environment ... 206
  - 4.2.9 The Moral Perspective ... 207
  - 4.2.10 The Perspective of Worldwide Ignorance ... 207
- 4.3 The Oases – Mutation, Decline or Resurgence? ... 208
  - 4.3.1 Description of the Oasis Space: Definition of the Term ... 209
  - 4.3.2 Chronological Framework of Oases According to Their Agrotechniques ... 214
  - 4.3.3 The Oases of the 20th Century: Decadence of a Strategy of Development ... 220
- 4.4 Soil Degradation, Irrigation, Salinization and Decadence ... 234
  - 4.4.1 The Degradation of Dryland Soils by Salinization ... 234
  - 4.4.2 Irrigation and Decadence: Four Case Studies ... 237
- 4.5 Desertification – An Expression of Decadence? ... 248
  - 4.5.1 Difficulties with the Sense of the Word Desertification ... 249
  - 4.5.2 What Is Really Happening? ... 257
- 4.6 A Glimmer of Hope ... 264

General Conclusion ... 269

References ... 275

Geographic Index ... 289

Subject Index ... 295

General Introduction

The English term "drylands" was defined by Barrow (1992) as "environments which are permanently, seasonally or temporarily subjected to a significant deficit in moisture". One third of the emerged land mass – 35–37% of our planet or some 45 million km^2 – is dry, and is inhabited by 15–20% of the world's population. Being azonal, these lands are found in polar, temperate, subtropical or tropical latitudes. For these reasons the subject "Mankind, aridity and droughts in drylands" seems worthy of consideration, further justified by some basic facts:

- Water, the lack or scarcity of which forms the core of this book, paradoxically covers 70% of our planet, but 90% of it is neither available nor amenable for human consumption. This may be because of its salinity (seawater) or its location either in the polar ice caps or in deep groundwaters which also may be saline; the accessible freshwaters are not always directly consumable because of being polluted.
- The treatment of water entails a high consumption of energy and chemicals, the residues of which are difficult to handle.
- The desalination of salt water is costly because of its energy-intensive nature. Modern techniques using salty water for agricultural purposes witness to a new approach and economic progress, the water no longer being desalinated, but used in its saline state for irrigation.
- Tapping deep fossil aquifers by pumping is irreversible on a historic time scale and leads to compaction of the aquifers, subsidence of the soil surface and the ingression of seawater into the continents. Such subsidence is at its worst in Mexico City and in the San Joaquin Valley of California. In Denver (Colorado) pumping has led to settlement with effects similar to those of an earthquake. Bangkok and Venice are other examples that come to mind.
- Storage of water in large dams is expensive as far as investment is concerned and because of the loss due to evaporation which may only be avoided by underground storage which is still more expensive.
- The transport of water is also costly in terms of investment and management and its effects on the environment are difficult to predict.
- Urbanization and other land uses involving deforestation increase the loss of water runoff, flash-floods and inundation, the latter sometimes assuming catastrophic dimensions as in France in 1988 in Nimes, in 1992 in Vaison-la-Romaine, Valréas and Bollène, and also along the Mississippi in 1993.

- Any change in land usage, in the use of soils, i.e. any modification of cultural practices and vegetation (e.g. changes in land tenure) modify the hydrological parameters and affect the local hydrological equilibrium.

So in drylands where climatic fluctuations and repeated droughts are known, management of water reserves should be more parsimonious and based on an integrated approach which takes all activities into account. Such an integrated management exists in the industrialized countries, but it is important that it is transferred to countries on the road to development, especially to those in which water is scarce, and that the economic value of water is recognized. The former Soviet countries in Asia do not subscribe to this even today. It is necessary to adapt such programmes to a stricter consideration of the environmental constraints.

During the preparation of the Earth Summit in Rio de Janeiro (July 1992) the International Conference on Water convened some 300 experts from 113 countries in Dublin who proposed the following plan of action:

- "protect the water resources and the aquatic ecosystems which should be understood in their global nature (water courses, their beds and banks, aquatic flora and fauna, vegetation along the banks, etc.);
- satisfy the material needs of supply of potable water, treatment, recycling and disinfection, and combating pollution. The problem is even more acute in the immense megalopolis complexes but also in the rural areas especially with the necessary development of irrigation;
- assure the protection of populations and goods against natural disasters especially in the form of flooding (retaining barrages, dykes, etc);
- resolve international conflicts in which the control of water occupies a central importance; this is especially the case for the Middle East where it involves Israel, Jordan, Turkey, Syria, Iraq and Greece, but also in Latin America." (Plenum 1992).

Drought as related to man is latent for man throughout our entire planet and inherent in his presence, especially because instead of being a considerate consumer he behaves like a predator of water particularly as his activities and his demography increase, and all the more so when such "development" affects environments which are already "dry" by nature. According to UNEP (United Nations Environmental Programme 1992) some 69% of all water used goes to agricultural uses. The average consumption of a citizen of the United States amounts to 70 times that of a man from Ghana.

Based on present demands, for the year 2000 an annual water consumption of 5000 km^3 and even 20 000 km^3 is predicted. This demand has to be considered against the total minimum water resources which have been estimated as 40 000 km^3 per year (Kundezewicz et al. 1991).

As aridity and droughts cannot themselves be fought directly, their effects must be combatted all the more energetically. Ambitious programmes for the fight against desertification have been produced, but mainly in the documents of large international organizations and less so in the field. However, would it be really realistic to believe that such a combat at such high cost could be carried out on the global scale of our planet?

General Introduction

The three main terms of the title of this book, *Aridity, Droughts and Human Development*, apply to the "*drylands*".

In order to avoid the ambiguity of the word "desert", which is too restrictive, the term "dryland" will be used here in a generic sense, including the deserts proper as well as their semi-arid and subhumid dry margins. These margins develop in a "continuum" from the deserts but human activities do not take such boundaries into account. Although the semi-deserts and their semi-arid border areas do receive more precipitation than the deserts proper, they actually exhibit seasonally, temporarily or accidentally characteristics similar to those of the deserts to which they are adjacent.

Nevertheless, retaining the word desert because of its simplicity and for the driest areas, we shall encounter its complexity when we delineate areas referred to as deserts and when we try to understand the relationship between man, aridity and drought and how these relationships have already been affected by high degrees of development prior to the crisis of the 20th century.

Areas referred to as "dry ecological systems" range from the cold polar deserts which, however, because of their highly specific nature, are excluded from this book, to warm tropical deserts surrounded by tropical semi-arid and subhumid drylands which include the subtropical Mediterranean latitudes. Coastal tropical deserts are a third group. Another group covers the lee-side deserts, and yet another the continental deserts. This book deals less with the original characteristics of the deserts than with the undeniably harsh relationship between man and aridity and the resulting compromises.

The title of the book and its subject are of interest from three points of view:

1. Placing the discussion into a concept of spatial geography which at present is somewhat outdated, i.e. zonal geography.
2. Tying it to "geohistory" which places geographical phenomena into a time frame and forces a scientist to consider the geographical environment before looking at the historic situation.
3. Taking into account the degradation of the environment, the geographical expression for a decline.

It is a question of:

- *Zonal geography*, as according to Pinchemel (1970) "geographic phenomena may be judged from a description and an explanation within the framework of climatic zones". Zonal geography studies especially the relationship between a natural environment, connected to zonal mechanisms, and man.
- *Climatic zones*; the word "zone" (from the Greek word for belt) is used here to describe the terrestrial climatic belts which are discontinuous and of unequal width, arranged along the parallels and subdivided into smaller geographical units, the regions.

Under *geohistory* we understand the evolution of the history of man, of his ways of life and ways of thinking and of his mentality as a function of his geographical framework. One of the most convincing statements on this subject is by Suzuki (1981): "No differentiation would have appeared in humanity without a change in the environment or without a difference in the environment between zones of the planet." However, is this statement not somewhat too "Lamarckist", suggesting that the evolution of civilization took place under the pressure of the environment and its changes?

At the end of this century, the deserts are still fascinating: their landscapes of dunes and bare rocks are abundantly used in advertising and in the movies. What is the reason for this fascination? Could it be that the mineral world – the biotope – creates an impression of apparent simplicity? It is in the connection with biocoenoses – the living world – that complexity appears. A multitude of books have dealt with drylands. We are hoping for a new vision, bringing man, aridity and drought into context, but even here the existing literature appears to be abundant. After looking through the treatises of the last few years a new clarification of this synthesis appears to be justified. The slow-down and even decline of development in the drylands, the degradation of the environment, insecurity in the face of programmes which are put into action without adaptation, or, in one word, a certain form of decadence, have appeared convincing to us because of their concordant appearance.

In the hostile and even restrictive drylands life is precarious, and in particular that of man who is anatomically and physiologically ill adapted to the desert and its solitude where by definition directly available water is rare. This confrontation has forced man to draw upon the better part of his intelligence, his capacity for adaptation and his creative ingenuity. In addition to the damage wraught for his own survival, he has made deserts the cradle of great civilizations. Survival and creative miracles became possible only because of the collective organization of the civilizations in them where furthermore the development of spirituality has permitted man to tame his own individualism. It is there that millennia ago "hydraulic civilizations" were born which lasted up to the 20th century. They were endangered and are still in danger in our times of degrading environments. So despite these major constraints, what can a country, on its way to development until it becomes set back by natural causes, do to avoid degradation of the environment and famine from becoming the ultimate price?

These thoughts have to be treated one after the other:

- The *spatial framework* delineating the drylands and their characteristic features of drought and aridity, their poor soils and the specific nature of their vegetative cover.
- The *environmental constraints*, in particular the scarcity of water and the omnipresent wind.
- The *geohistoric response* showing how the drylands lead to original ways of life, of thought and of religion, and how human genius, inventing techniques for reducing the scarcity of water, has made them the cradle of modern man, of agriculture, irrigation and of sophisticated hydraulic societies.
- The *decline as a fact*: degradation of the environment and economic decline. The analyses of desertification during the 20th century, its causes, mechanisms and consequences will be carried out here. Why is it that the drylands became the most degraded regions of our planet?

CHAPTER 1

The Spatial Framework, the Concepts of Aridity and Drought: the Soils and the Vegetation

Presenting a geographical definition entails among other objectives the study of the relationship of man with the "natural" world and as a first prerequisite the understanding of the spatial position of the phenomena studied. This implies a description of the framework and of its dynamics whatever the degree of human influence on it may be. The global aspects of the localization of drylands will be discussed as well as their particular features on the various continents. An understanding of the couplet man–drought also necessitates an understanding of the climatic data inherent to drylands. The characteristic long-term feature of aridity and the short-term feature of drought will have to be defined and differentiated. Because of their specific nature, the soils and covering vegetation in drylands will have to be taken into account, less as environmental parameters than as the substrate for clarifying the relationship man–aridity–drought.

1.1
Attempts at Global Localization of Drylands

The first attempt at globally localizing drylands goes back to Martonne and Aufrère (1928) with their concept of "arid diagonals" which still retains some interest. However, we shall replace this term here by "dryland diagonals". Although generally attractive, the former term is rather simplifying as the diagonals are discontinuous and heterogeneous in all their climatic data and especially in their degree of aridity.

The Northern Hemisphere

- The *Sahara-Arabia-Thar* diagonal followed north of the tropical world by the central Asian deserts (Turan) and the *deserts* of temperate latitudes in *China*. The aridity of the Sahara, western Asia (Middle East, Arabian Peninsula, Rajastan) may be related to their position between the areas of influence of two main sources of precipitation, viz. the polar front and the equatorial westerly currents. The westerly Mediterranean and equatorial currents are blocked by the Himalayan chain which they cannot surmount because of their insufficient height of 3000–5000 m. The deserts of central Asia and China are continental deserts as well as lee-side deserts at temperate latitudes.
- The dryland diagonal of California, the Sonora and northern Mexico.

The Southern Hemisphere

- The dryland diagonal of the Namib-Kalahari.
- The dryland diagonal of South America, extending from the equator (including the Galapagos) to north of 30° S, crossing the Andes at Antofagasta and running along the eastern foot of the Cordillera down to Patagonia.
- The dryland diagonal of Australia, covering some 5 million km² or about 75% of the continent. The middle part in the central-western portion of the continent is a sand desert. Ollier (1992) stated: "Australia possesses few true deserts but is nevertheless after Antarctica the driest continent of all. This paradox is the result of two causes: Firstly, Australia was more arid in the past and has inherited some of its specific landscapes from its arid paleoclimates and, secondly, the rains here are so erratic that many of the semi-arid parts appear to be arid."

On the five dry diagonals grades of aridity may be distinguished:

- *Hyperarid* (the *Kernwüste* of the German geographers) when the averave rainfall is 10–50 mm/year and the interannual coefficient of variability above 40%. These are the centres of the most extensive continental deserts (central Sahara and central Arabia) and the littoral deserts of Peru/Chile and Namibia. Precipitation here is insufficient for agriculture, except in natural or artificial oases.
- *Arid or desert-like*: average rainfall is 50–100 mm/year and the interannual coefficient of variability 30–40%. Rains are restricted to a few months per year. The regime affects the centres of the continental deserts (southwestern USA, Kalahari, Australia) and the periphery of the most extensive continental deserts. Precipitation is still insufficient for agriculture.
- *Semi-arid or Sahelian:* average rainfall here is 150–500 mm/year and the interannual coefficient of variability 20–30%; rain-fed agriculture is possible but remains hazardous.
- *Subhumid-dry or Sahelian-Sudanian:* average annual precipitation is 500–800 mm spread over a rainy season of 6 months. Aridity here is a recurring seasonal constraint and the interannual coefficient of variability may still reach 20%.

These ecological systems make up a group difficult to delineate, referred to as deserts in this book, which explains why its contents are so varied. More than 40% of the world's deserts are classified as warm deserts, the others exhibit warm summers and cold winters (Parsons and Abrahams 1994).

Twenty-five years after Aufrère and Martonne, at the request of UNESCO, Meigs (1953) presented a second attempt at a global study of deserts by drawing up a locality map of drylands based on the index of available humidity of Thornthwaite (1948). In 1977, UNESCO published a synthetic and very useful map of the distribution of drylands. The degree of aridity here is defined by the bioclimatic index: P/PET in which P = total annual precipitation and PET = potential average evapotranspiration from measurements at 1600 meteorological stations.

Still later, Heathcote (1983) estimated from own data that 37% of the drylands are found on the African continent, 34% in Asia, 13% in Australia, 8% in North America, 6% in South America, and 2% in Spain.

1.1 · Attempts at Global Localization of Drylands

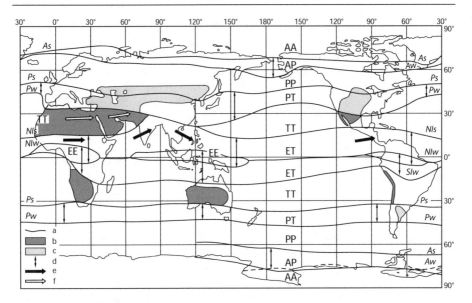

Fig. 1.1. Dynamic classification of climates and distribution of drylands (after Alissov 1954, modified by Suzuki 1981). *A* Arctic or antarctic front; *P* polar front; *NI* northern intertropical convergence; *SI* southern intertropical convergence; *w* position of fronts in winter; *s* position of fronts in summer. *A*, Arctic or antarctic air; *P*, polar air; *T*, tropical air; *E*, equatorial air. Combination of these air masses in summer and winter: *a* Northern and southern position of fronts or intertropical convergence; *b* arid zone beyond the reach of the fronts or by the intertropical convergence; *c* arid zone on the lee-side location to the westerlies; *d* seasonal migration of the fronts or of the intertropical convergence; *e* summer position of the west equatorial circulation; *f* northernmost position of the west equatorial circulation

An attempt at a simplified localization of the drylands (arid and semi-arid) of our planet (Fig. 1.1) was proposed by Suzuki (1981) based on the map of Alisson (1954). The map of Suzuki allows us to locate drylands in one single document by evaluating the position of air masses and fronts. The main air masses of our planet and the fronts separating them from each other are:

- *Air masses*
  - A: arctic or antarctic
  - P: polar
  - T: tropical
  - E: equatorial
- *Fronts*
  - *A*: arctic and antarctic between A and P
  - *P*: polar between P and T
  - *ITC*: intertropical convergence
  - *NITC*: northern intertropical convergence *NI*
  - *SITC*: southern intertropical convergence *SI* (both between T and E), and still referred to as intertropical (*ITF*)
  - *W*: position during boreal winter
  - *S*: position during boreal summer

When the equatorial air mass between air masses T_{north} and T_{south} is missing, the NITC and SITC combine to a single front.

The fronts move north and south with these movements being responsible for the seasons. A region is dry when the front passes along it and it is rainy when the front passes across. Each part of the Earth is thus dominated by one or two air masses in summer or winter which led Suzuki to conclude that on either hemisphere seven combinations are possible: AA, AP, PP, PT, TT, TE and EE. In an exemplary manner he further distinguished the different types of desert as shown in his map: cold polar, tropical warm, coastal, lee-side and continental.

In the tropics, the deserts attain their maximum extent. In those latitudes the air settles down from high pressure systems and maintains a permanent aridity concurrently with three other factors:

- *Continentality*: in winter the colder continental air masses are the site of an anticyclone; despite being the site of low pressure systems in summer, the remoteness of these deserts explains why the moist oceanic air masses cannot reach them. The Sahara and the deserts of central Asia are examples for this situation.
- *North-south running mountain barriers* keep the moist oceanic winds from reaching their eastern foothills; this is the case for the non-tropical deserts of North and South America.
- *Ocean currents* are responsible for the aridity of the lands lying along their eastern shores: the Namib of southern Africa and the Atacama of South America.

The above classification leads to a more difficult regional analysis of the deserts which are highly different from continent to continent depending on their latitude and their degree of continentality, despite the impression of an apparent homogeneity. Their main characteristics will be described, starting with the cold deserts and taking their particular features into account, viz. their low temperatures and their having water present in solid form; these features are more decisive for the absence of life in such deserts than aridity and drought.

1.1.1
The Cold Polar Deserts

The deserts of the polar latitudes are less well known. The map of Alisson places them:

- *North of the line As*, the summer position of the northernmost north polar front reached in summer but not crossed further
- *South of the line Aw*, the winter position of the southernmost south polar front reached during the southern winter but not crossed further.

Pagney (1994) summarized the characteristics of the polar climates under the following headings:"… severity of frost, absence of a summer; extended illumination and polar night; drought and turbulence …". The polar areas may be classified as desert-like because of the scarcity of precipitation and the low temperatures favouring a

1.1 · Attempts at Global Localization of Drylands

Table 1.1. The low precipitation (mm/year) of the Arctic (measurements of Pagney 1994)

Beaufort Sea	104	8 dry months with $P < 10$ mm
Canadian Archipelago: Ellesmere Isl.	146	7 dry months
Central Greenland: Eismitte	109	
Laptev Sea	131	
Eastern Siberia	131	
Wrangel Island	104	
Vicinity of North Pole between May 1958 and May 1959	<200	
Antarctica: Vostok Station	< 50	

low content of water vapour in the air. Furthermore, the skin-like anticyclones stabilize the seasons (Table 1.1). The weak precipitation is snowy and more abundant in the west than in the east of the Arctic ice-cap, in the centre of which it may nevertheless snow for more than 200 days/year. The snow stops in June, disappears in July and returns again in August.

Joly (1957) raised the question of the uniformity of the arid subpolar environments: "One could ask oneself …" whether they are "… really comparable to the drylands of the warm and temperate regions, or in other words, whether the biological influence of the cold does not suppress that of the aridity itself." He distinguished two ecosystems: the *subpolar desert-like*, the domain of permanent ice, and the *subpolar semi-arid*, the domain of annual frost and tundra.

In the Northern Hemisphere, the latter domain forms a zone running from the Bering Straits to Norway along the north coast of the Eurasian continent, covering the coastal fringe of Greenland and of Labrador and from these along the north coast of North America over to Alaska. The ecosystem of the *tundra* grades into the *taiga*, a band of forests separating the former from the arid zones of temperate latitudes.

In the Southern Hemisphere at high latitudes the configuration of the continents is such that the dry semi-arid domain is virtually completely lacking except for some small islands with windy climates which are sometimes even rich in rain. In contrast, the subpolar desert-like regime is well represented by the 14 million km² of inland ice of the Antarctic continent, the most extensive ice fields of our planet, with a thickness of 4.5 km.

The preponderance of cold with a highest average monthly temperature below 10 °C leads to a slow-down of plant growth:

- In the *subpolar desert-like domain* the accumulation of water in the form of ice excludes any vegetation.
- In the *subpolar semi-arid domain* the long period of summer insolation of 3–5 months and the melting of the soil down to 2 m permit the growth of the tundra (Photo 1.1). The *rocky tundra steppe* of the *barren-ground* type in which the plants concentrate in cushions is distinguished from the *closed-floor tundra* rich in woody plants. The latter do not exist in the tropical dry steppe.

Photo 1.1. Pingo and barren tundra in the Great North of Canada. Aerial view of the tundra near Tuktoyaktuk upstream of the Mackenzie delta in the northern part of Canada: Ibyuk *pingo* caused by diapirism by an underlying ice lens. At these latitudes (65° N) close to the Arctic Circle the barren vegetation does not result from the lack of water as one can see, but from the cold and especially the solid state of the groundwater (Photo by M. Mainguet)

Amongst the other consequences of the cold are the precarious nature or impossibility of agricultural activity and the isolation resulting from low population densities. On 30 October 1992 a new state, Nunawont, was established in the Great Canadian North. It was handed by Canada to the 17 000 Eskimos who are spread over a territory of 2 million km^2. It will achieve complete autonomy within a period of 16 years.

After the above information we shall no longer talk about the dry cold domain and concern ourselves only with the dry domains of the tropical and temperate zones.

1.1.2
The Warm Tropical Deserts

We are dealing here with the truly arid (and even hyperarid) areas which are virtually permanently dry and receive less than 100–150 mm of rain per year, the rains falling during the cooler season along the Mediterranean fringe of the Sahara and during the warm season over the Sahelian fringe. Adjacent to these zones may be semi-arid and subhumid drylands which may be seasonally dry, receiving 600–800 mm of annual precipitation.

Intense insolation, elevated temperatures and strong evaporation lead to a low availability of surface water. The annual insolation "is above 3250 h in the Sahel to the south of the Sahara and in central Australia which are both crossed by the tropics

1.1 · Attempts at Global Localization of Drylands

Photo 1.2. Mount Sinai. The impressive beauty of the granite domes of Mount Sinai at dawn in April '74. The rocks highlighted by the rising sun are barren and numerous joints are visible (Photo M. Mainguet)

under which the theoretical insolation reaches 4100 h. It decreases in the direction of the Mediterranean margins to less than 3000 h or not more than about 75% of the theoretical insolation in central Asia". In the Chinese deserts it amounts to 2500 to 3000 h annually. "It decreases still further in the coastal deserts and in particular in those along the west coasts of the continents." (Toupet 1992; Photo 1.2).

"The clarity of the dry atmosphere" and the absence of clouds explain the elevated temperatures and led Viers (1968) to state: "There is no month when one does not record more than 50 °C in the shade at Tamanrasset even though at an elevation of 1376 m and the record values for August were 69.5 and 70.1 °C. Over sand or bare rocks, temperatures from 72 to 78 °C have been recorded. This explains also why despite the nocturnal lows the mean monthly temperatures reach and surpass 35 °C at Aoulef and in the Tunisian Sahara. In contrast, the nocturnal minima in the subtropical countries are low: at Colomb-Béchar (30° N) −6 °C have been recorded and −3 °C at Bilma (Niger, 18° N). At Colomb-Béchar an average of 14 frost days in sheltered parts has been registered and 17 days at Laghouat. It may freeze ... down to Darfour at 12° N. There are thus differences between day and night temperatures ... which exceed 30 °C in the air and 50–60 °C on surface of rocks or sand." In the Chinese deserts the average nycthemeral amplitude varies between 30 and 50 °C and may reach 60 °C.

The location of the areas subject to a lack or scarcity of rains is closely defined in the paper of Suzuki: As the ITF and the polar front are responsible for precipitation, the zones of the earth running along the one or other of these fronts receive rain or snow once or twice yearly. The zones TT of this map which are not reached or traversed annually by these fronts are permanently dry areas, and among these are tropical warm arid, semi-arid and subhumid drylands along the tropics. They are outside the zones which are reached either by the polar front or the intertropical convergence front (Fig. 1.1): they are the great dryland diagonals and grand terrestrial deserts described by Martonne and Aufrère. The Saharan-Arabian desert and central Australia are examples of this type of desert.

1.1.2.1
The Saharan-Arabian Desert

With a surface area of 12×10^6 km^2 it is the largest desert of our planet, covering about 9% of the surface of the continents not under permanent ice, and extends over about 7500 km from the Atlantic to the Persian Gulf (Oberlander 1994, in Abrahams and Parsons 1994). Except for its eastern and western margins where precipitation can climb to 100 mm/year, it is hyperarid over most of its area with precipitation rarely exceeding 10 mm/year in the Libyan and Egyptian sectors. The summers are exceedingly hot and the winters mild, although the temperatures then may fall below freezing.

The landscapes, inherited from pluvial periods, are marked by abundant traces of runoff dating from the last humid phase of the Holocene, Quaternary or upper Tertiary. The relict drainage network is indicated by the deep canyons of the basins bordered by mountainous slopes and by immense pebble fans forming the numerous alluvial regs of the Sahara: The Igharghar, Saoura, Tamanrasset and Tafassasset Wadis of the Sahara and the Ar Rima, Brik and Dawasar of the Arabian Peninsula, which are not at all or only rarely functional in our times, are as large as the largest water courses of the temperate zones. Temporary runoff and local sanding-up give them a fragmented longitudinal profile. In the 1980s El-Baz discovered in Egypt with the aid of images of the Radar-Sar satellite a network of fluvial channels several dozen kilometres wide which had been cut into the bedrock and fossilized under aeolian sands.

McCauley *et al.* (1982) suggested that a palaeo-Nile was connected with Lake Chad prior to becoming cut off from the latter during a tectonic uplift some 15 Ma ago. In order to explain the present course of the Nile they proposed that it captured the upper Nile by regressive erosion induced by a lowering of the level of the Mediterranean because of desiccation some 6 Ma ago. This interesting hypothesis is still debated.

The effects of aeolian mechanisms make themselves felt over vast portions of the Saharan-Arabian desert: here, accumulations of sand cover some 20% of the area and landscapes sculptured by wind to such an extent that the Sahara desert is more rocky than sandy. There are active sand seas arranged in chains, from the north to the south of the global system of aeolian action (to be defined in the second part of this book: *The omnipresent wind* ...) which make up the Sahara-Sahel complex. In contrast, south of 21° N the sand seas are fixed or vegetated and no longer active, although during the last drought of 1968–1985 they lost their vegetative cover and became reactivated, as was the case in the ancient Hausa sand sea of Niger. There are five chains of sand seas of which we shall mention but two:

- *In the western part of the Sahara from north to south*: Grand Erg occidental → Er Raoui sand sea and Iguidi sand sea (both at the same latitude) → Magteir and Ouarane sand seas (both at the same latitude) → Ijafène sand sea → Trarza,

Mreyye and Azamad sand seas (all three at the same latitude) → Telak and Ferlo sand seas (both at the same latitude).
- *In the eastern part of the Sahara from north to south:* Calancho sand sea and Great Sand Sea (both at the same latitude) → the Libyan-Chad and the Sélima sand seas (all at the same latitude) → Mourdi sand sea → Manga and *Goz* of Kordofan sand seas (at the same latitude) → Hausa sand sea.

The two chains start around 30° N and end at Sahelian latitudes. This situation explains why the aeolian dynamics of the Sahara and the Sahel are interdependent. South of the Sahara in the Sahel zone, the Mauretanian sand seas, those of the Manga in Chad-Niger, the Hausa sand sea, and the *goz* of the Sudan consist of material slightly aggregated and coloured red by inherited palaeosols. They are presently the site of gullies, of accelerated reactivation under the effects of the latest climatic variations and the drought of 1968–1985 and of overexploitation of the vegetative cover.

1.1.2.2
The Deserts of Australia

In contrast to the Saharan-Arabian desert, those of Australia and also of central Asia at the border between arid and semi-arid regions are no true deserts: the mean precipitation surpasses 100 mm/year and may even reach 400 mm/year. Situated along tropic of Capricorn they receive the precipitation of the southern summers: the humidity is sufficient to permit a vegetation of bushes and xerophytic grasses and even in the centre of the desert extensive livestock farming is possible. Like the Saharan-Arabian desert, Australia possesses numerous signs of ancient hydrographic systems from more humid periods, but these are older than those of the Sahara. Australian geographers (e.g. Oller 1977; van den Graaf *et al.* 1977) attribute these to the phase of separation of the Australian continent from Antarctica some 75 Ma ago.

In proportion to its surface area Australia possesses the most extensive cover of aeolian sands forming longitudinal dunes up to 300 km long and 20–30 m high arranged in a circular pattern around the central Australian anticyclone with the zonal migration explaining the opening of the dune system to the west. They are evidence for a negative sediment balance and represent sand seas in the state of diminution. Transverse dunes are rare and only small whereas the other types (pyramidal dunes) are missing. The dunes are mostly fixed except for their upper part. The intense red coloration of the sand is a sign for the great age of these sand seas.

Australia is the continent of *lunettes*, sand structures of parabolic plan view, which are concave in wind direction and consist of clayey-sandy salt-rich material. They are formed leeward of temporarily saline lakes which after drying out became areas of deflation. The *lunettes* were used by Bowler (1976, in Oberlander 1994) to establish a stratigraphy: a dry period prior to 45 000 B.P., a humid period between 45 000 and 26 000 B.P. and a long dry period from 17 000 B.P. until the present.[1]

Australia is also the desert in which *duricrusts* (mostly silcretes and laterites) make up the dominant part of many landscapes with the laterites being particularly

---
[1] B.P. = Before Present, the present usually being defined as 1950 A.D.

frequent in the western part of the continent. These crusts are attributed to older more humid climates and represent remnants of an erosion surface formed up to the end of the Tertiary (Oberlander 1994, in Abrahams and Parson 1994).

The Australian continent harbours the most well-known inselbergs of our planet: Ayers Rock and the Olgas. Ayers is a monolith consisting of Cambrian arkoses towering 350 m on three sides over Tertiary pediments (Twidale 1978). The Olgas make up an *Inselgebirge* of some 30 domes shaped out of a polychrome upper Palaeozoic conglomerate (Ollier and Tuddenham 1961).

1.1.3
The Coastal Deserts

The four foggy littoral deserts – Baja California and southwestern Morocco in the Northern Hemisphere, the littoral desert of Peru and the South-West African one in the Southern Hemisphere – do not possess the same geographical characteristics. Those of the Northern Hemisphere in subtropical latitudes do have sunny phases whereas those of the Southern Hemisphere are of elongate shape and foggy virtually throughout the year. Their development in tropical and subtropical latitudes along the eastern shores of the oceans results from the combination of a vast anticyclonic cell, with cold coastal currents originating from higher latitudes, and a phenomenon called *upwelling*, the rise to the surface of colder deep oceanic waters. As a consequence the air masses along the coast are cooled down and stable over a thickness of several hundred metres. The global situation changes according the activity of anticyclones and the cold marine currents.

The persistence of this thermal inversion explains the low monthly average temperatures of 13–22 °C, which are rather low for these latitudes, and the low annual thermal amplitudes of 6–7 °C. At the Namibian part of Walvis Bay (23° S) the mean temperature is 17 °C, the maximum 27 °C, the annual amplitude 6 °C and the mean annual precipitation a mere 25 mm. The climate may be even hyperarid: Antofagasta in the northern part of Chile has registered only two weak rainfalls between 1951 and 1960 and the mean at Lima (12° S) is 36 mm/year, but the town received a downpour of 1513 mm as the result of the "El Niño" phenomenon during which the cold Humboldt current is replaced by a warm current, usually around Christmas.

The scarcity of precipitation – with averages of a few fractions of a millimetre in the Atacama desert of Chile – leads in the lower parts of the relief to a virtually inert morphogenesis as the runoff is insufficient to coalesce and the water cycle does not pass beyond a superficial moistening by fogs and dew. Aeolian action is also low as the air remains stable and thus is little conducive to wind. Haloclastic action is felt prominently over several dozen kilometres from the coast. In Baja California and on the western, Andean slopes of the Peruvian–Chilean desert, which in its middle stages receives mud slides and torrential down-pours endangering human infrastructures in their course, morphogenesis becomes active only with height.

The arid part of the Namib desert extends along the west coast of southern Africa from the Kunene River forming the boundary between Angola and Namibia at about 15° S down to the Orange River at 32° S or over some 1280 km between these two perennial allochthonous rivers (Fig. 1.2). It owes its existence to the cold Benguela current giving rise to a pronounced stability of the air below the thermal inversion at a

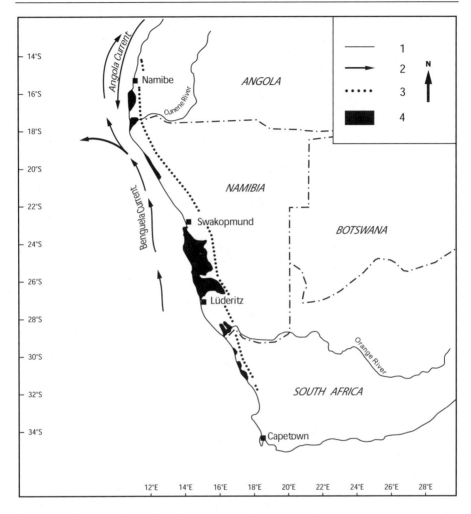

Fig. 1.2. Position of the littoral Namib Desert (after Lageat 1994). *1* Perennial water-courses; *2* ocean currents; *3* continental border of the desert; *4* areas of sand accumulation

height between 500–1000 m and the up-welling which blocks any ascendance and explains the permanent presence of cold waters along the coast. As a result, fog and dew supply a water equivalent of 40.4 mm/year (February 1967–August 1974), whereas the mean annual rain precipitation does not rise above 12.9 mm/year (Nieman et al. 1978, in Lageat 1994). The hidden precipitation favours the growth of lichens. The absolute extreme temperatures at Swakopmund are 37.5 and 3.4 °C (Lageat 1994), but frost is unknown here.

With a mean width of 130 km the desert is bordered along its eastern edge by an escarpment and rises from the coast by about 1000 m to this barrier which causes a föhn effect from the easterly winds. The Namib is divided by the Kuiseb River into a

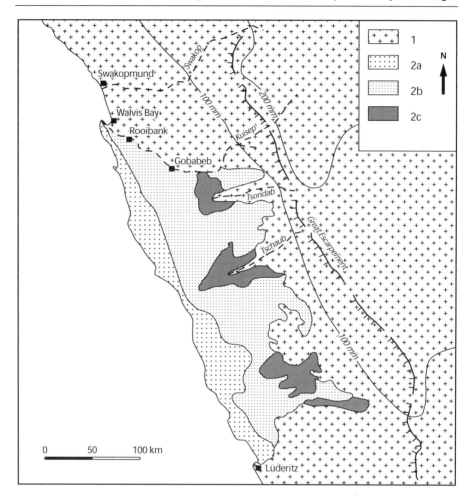

Fig. 1.3. The central Namib and its sand deposits (after Lageat 1994). *1* Rock outcrops; *2a* sand sea with barchanic dunes; *2b* area of linear dunes; *2c* area of pyramid dunes

sandy south and a rocky north where the landscape is covered by a gypsum crust of allochthonous – probably aeolian – derivation which may epigenetically replace limestone crusts (Fig. 1.3).

1.1.4
The Lee-side Deserts

The fourth group of (arid or semi-arid) drylands, the lee-side deserts of temperate latitudes, are areas sheltered from dominant winds by a range of mountains. The winds are west winds – or the *westerlies* – which emanate from cyclones. Such winds encounter the Rocky Mountains and the Andes which block them in such a way, that their lee-side eastern piedmonts experience a föhn effect which increases the lee-side effect and the dryness. This is the *chinook* of the North American Great Plains or the *zonda* of the cold Patagonian deserts. At the eastern foot of the Rockies the desert is semi-arid and at the Argentinian side of the Andes even truly arid. Lee-side deserts are also found in intramontane basins where humid currents again develop into föhns as in the inner basin of the Rockies, the high plateaus of the Andes, the high plateaus of Tibet on the northern side of the Himalaya mountain range, which shelters them from the monsoons, as well as the Iranian (Lut) and the Afghan plateaus.

The Nordeste of Brazil, paradoxically a desert of equatorial latitude, is semi-arid according to its mean precipitation (500–600 mm/year), but has 300 mm/year at its centre. Two mechanisms are responsible for the rains in this area: temperate cyclones coming from the south, for which the Nordeste is the northernmost area of influence, and the intertropical front (ITF) which arrives from the north in winter. Both, however, may fail to arrive over several consecutive years. The Nordeste is furthermore on the lee-side of the coastal Borborema Range (900 m) which blocks the trade winds and leads to a subsidence effect on their western piedmont (Demangeot 1972). It has only temporary runoff and underground water reserves are scarce.

Pélisier (1989) estimated that the xerophitic *caatinga* of the Brazilian Nordeste, made up of crassulescent cacti and woody deciduous plants, possesses a denser and higher bushy and spiny aerial biomass than the vegetation of the Sahel and thereby acts as a better protection for the soil and is more resistant. This explains why despite the aggressive nature of the (violent) down-pours and the steep slopes of a landscape which is more varied than the Sahel, water erosion is less damaging in the Nordeste than in the Sahel.

1.1.5
The Continental Deserts: The Turan Desert as an Example

With cold winters and searing summers, this desert of temperate latitudes features the highest thermal amplitudes. Its continental nature combines with its lee-side location, and perturbations from the west arrive only in an attenuated form. The climate has by no means the paradise-like summers reported in the legends about the oases of Samarkand and Bukhara.

The Turan region, consisting of 80% plains and 20% mountains, is subdivided into four areas:

- In the north the sub-boreal deserts of Kazakhstan with their monotonous topography containing, in the immediate vicinity of Lake Aral, the Barsouki sand sea.
- In the south the subtropical deserts of the central Asian type which themselves are subdivided into two large sand seas: the Karakum (black sand) in the southwest and the Kyzylkum (red sand) in the east.
- In the west the Ust-Urt plateau between the Caspian Sea, the Ural Mountains and Lake Aral.
- Finally, the alluvial plains and deltas of the Amu and Syr Daria in the centre which are superimposed onto various types of deserts and represent, like the Nile in Egypt, an ecological "anomaly".

1.1.5.1
The Degree of Aridity

The Turan (Fig. 2.3), the vast endoreic basin of Lake Aral, measures some 3.5 million km^2 and is locked to the south-west by the Kopet Dag, to the east by the mountain ranges of the Hindukush and the Pamir (7495 m) and, farther to the north by the Tienshan (7440 m). It opens to the north into Siberia where the boundary of the hydrographic basin becomes poorly defined at an altitude below 200 m. In the south, the mountain ranges between Iran and Afghanistan drop down to 2000 m and through this passageway the Indian Ocean monsoons which are partly blocked by the high altitude of the Iranian plateaus and of the Pamir-Hindukush complex may intrude to bring rain to the foot of the Tienshan and the Altai. These mild föhn winds coming from the mountain ranges may also reach the Aral region. The centre of the basin lies during the summer under a thermal low pressure system which is responsible for the advent of these winds.

With its semi-arid to dry steppe nature, the Turan is comparable to both the south-Saharan Sahel and the sub-Atlas fringe. Because of its continental location and mountainous borders the Turan is subjected in winter to the Siberian anticyclone, whereas during summer cool and humid air from the North Atlantic brings humidity to its western part via the Mediterranean and the Black Sea and additionally via the Caspian Sea and the wide gap between the Urals and the Caucasus. Its desert nature is defined climatically by precipitation below 100–150 mm/year.

The aridity is less pronounced than in the Sahara because the potential evaporation does not exceed an average of 1000 mm/year. Evaporation may reach, however, 2300 mm/year at Repetek some 100 km SW of Tachaouz in the Karakum Desert. Actual evaporation over the waters of Lake Aral is around 1000 mm/year, ranging from 1450 mm/year in the north, 1010 mm/year in the south to 950 mm/year over Lake Sary-Kamish to the SE of Lake Aral, and depends on temperature, wind and salinity of the waters, as salt water evaporates less readily than freshwater. In comparison to the Sahara, the Turan is located farther north, has lower mean temperatures and a higher humidity of the air. Because of this, this region is less arid than the Sahara, and possesses a vegetation that is less patchy and still frequented by nomads.

1.1.5.2
The Winds and the Drought

The low pressure systems spin off cyclones which contribute to the aeolian transport of particles from January to April especially to the southeast and east of Lake Aral. The dust storms and dry "fogs" have been described by earlier travellers. Their annual frequency is above 25 days and at Ashkabad a dry "downpour" over 8 h deposited some 20–30 m^3 of dust per hectare. Nukus in the upper part of the delta receives an average of 35 dust storms per year. According to Bugaev (1957), these dry dust storms were so violent in 1882 that the transported particles abraded the copper wires of the telegraph lines along the southern Transcaspian.

Due to its effect as a local climatic regulator, Lake Aral creates breezes which freshen the climate beyond its immediate surroundings. The thickness of this cushion of humid air was once up to 9 km and made itself felt over several hundred kilo-

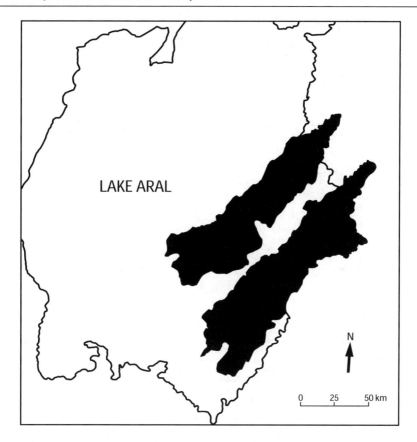

Fig. 1.4. NE–SW oriented dust clouds over Lake Aral (after satellite images of IS3/Meteor-Priroda of May 18th 1975)

metres to the south of the lake. With the drying of the lake this air cushion has shrunk to becoming virtually non-existent and the most frequent dust storms now progress from NE to SW (Fig. 1.4).

The global aeolian circulation over the Asian deserts during summer was described by Akio-Kitoh (1983). Figure 1.5 presents for the northern summer months (June–August) the aeolian circulation between the tropical (10° N) and the temperate (70° N) latitudes and between 20° E–120° E longitudes, i.e. between the central Mediterranean in the west to the Gulf of Tche Li of the Pacific in the east:

- Over temperate latitudes the surface winds form a flow pattern from the west which becomes deflected to the south from north of the Mediterranean and the Black Sea.
- Over the Caspian Sea this deflection causes the winds to turn back against their original direction and to circulate east to west.

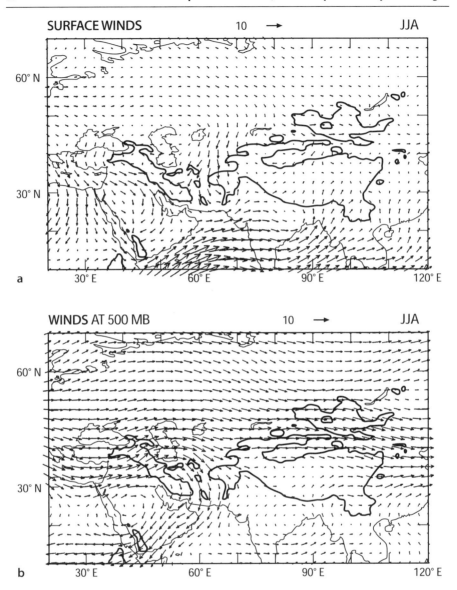

Fig. 1.5 a,b. Aeolian circulation over central Asia during June–August (after Akio Kitoh 1993). a Winds at the soil–atmosphere interface; b winds at 500 mbar (equivalent to an altitude of 5.6 km)

- The Caucasus-Taurus-Zagros-Elbrus complex forms a barrier which causes a major deflection over Arabia where surface winds form a circulation pattern in a clockwise direction.
- Lake Aral, to the north and directly opposite the topographic low separating the Hindukush from the southern end of the Zagros, occupies a particular pivoting

position around which the western circulation turns to become north-south down to the shores of the Gulf of Oman where it encounters the circulation of the Indian Ocean monsoons. This situation shows that, contrary to generally held opinion, the loess deposits on the northern foothills of the mountain ranges bordering the basin to the south were derived from the deflation of the Karakum sand seas, of the deposits of the Amu and Syr Daria and of palaeo-courses of water in these areas.

- At 850 mbar or 2.5 km altitude the air circulation is still the same as at ground level; at 700 mbar or 3 km altitude it remains the same except that the southern circulation around Lake Aral extends still farther south into the Gulf of Oman. At 500 mbar or 5.6 km altitude Lake Aral no longer forms a pivoting point of circulation; the basin is crossed in its northern two-thirds by a westerly current. It is only in the southernmost third that this western flow turns to become NW-SE. Above 300 mbar or 9.2 km altitude the western circulation is re-established without any influence of topographic obstacles.

1.1.5.3
Precipitation

The annual precipitation on average amounts to 90–120 mm over the plains and to 400–500 mm over the piedmont, but rises to 2000 mm over the western slopes of the Tienshan. It drops to 200 mm over the peripheral northern areas and to below 30 mm in the Faim Steppe southeast of Tashkent. Over the direct vicinity of Lake Aral it is around 100 mm/year with a rather high air moisture because of the intense evaporation over the surface of the lake. Prior to the 1960s Lake Aral raised the surface humidity and precipitation by about 10 mm/year.

The rains are irregular as in all deserts (Table 1.2). It also snows in the Aral basin; but when evaluating these data, it has to be kept in mind that the thickness of snow cover is ten times the thickness of the same amount of water: the mean snow thickness is 6 cm/year in southern Kazakhstan, 2.5 cm/year at Nukus in Karakalpakstan and below 1 mm/year in Chardzu. At Kyzyl-Arwart, 250 km northwest of Ashkabad, it snows every ten years, at Kerki on the Amu-Daria every 50 year and more frequently travellers of old reported that their caravans were enveloped in eddies of snow. The at Ust-Urt where the mean annual number of days with snow is 70 at Aralst north of Lake Aral, 37 at Tashkent, but not more than 4 at Bayram-Ali.

Table 1.2. Turan: precipitation for the period 1910–1955 in mm/year (Létolle and Mainguet 1993)

	Spring	Summer	Autumn	Winter	Total
Lake Aral	26	25	31	23	105
Kyzyl-Orda	37.5	15	22.5	31	107.5
Bayram-Ali	59	2	14	47.5	122.5
Kerki	70	1	17	75	162.5
Ashkabad	80	10	50	90	230
Turtkul					95
Nukus					77.5

1.1.5.4
The Contrasting Temperatures

Taking into account the latitude and the generally clear skies, the area receives a large quantity of solar energy, i.e. 120–160 kcal/cm^2, of which a large part is reflected into the atmosphere so that the air is frequently overheated. In sheltered areas of the Kyzylkum Desert over 50 °C was recorded and the maxima are nearly always above 40 °C. The temperature of the ground surface surpasses 70 °C and, even in winter, that of the sand is above 50 °C. Nevertheless, at a depth of 10 cm, the temperature is 20–30 °C lower. Cooling at night is intense: the temperature drops from 40 to −5 °C during the nycthemeron with the values smoothed by irrigation and plant cover, as e.g. cotton absorbs some 15% of the solar energy received.

The temperature extremes are presented in Table 1.3. Lake Aral reduces the severity of the differences. Prior to the activities of the 1960s the mean temperature over the centre of the lake in January at −5 °C was 2 °C higher than over its south coast and 6 °C higher than over the north coast. Because of the low winter temperatures, Lake Aral was frozen over for at least 5 months/year and shipping was interrupted.

The duration of frost is 140 days in the south of the Karakum and 210 days in the north. The cultivation of cotton requires a mean daily temperature above 14 °C, a condition that is met in the south only during 6–7 months/year and still less in the north towards Lake Aral. Northeast of the Syr-Daria the warm season is too short for growing cotton. Frost penetrates beyond a depth of 2 m in the soils of the Turan, forcing the desert fauna to dig underground during this period. The lower reaches of the Syr-Daria are frozen from December to the end of March and those of the Amu Daria below Nukus for 2–3 months. Upstream the Amu Daria freezes only along its banks.
In conclusion, it can be stated that the diversity of the deserts is impressive. Their common factor is their present aridity which, however, masks a palaeoclimatic instability caused by the superimposition of landscapes inherited from humid and dry phases. The dry periods of the pre-Quaternary, their durations and their different degrees of dryness has led to such diversity of landscapes.

Table 1.3. Turan: some temperature statistics in °C (Létolle and Mainguet 1993)

	Karzalinsk	Bayram-Ali
Extreme temperatures (°C)	42.5 − 32.8	45.2 − 25.5
Duration of frost (days)	172	215
Duration of life cycle of annual plants (days)	204	208
Daily variation (Karzalinsk) April, Temp. (°C)	6 a.m: 3 9 a.m.: 20 1 p.m.: 28 6 p.m.: 21 9 p.m.: 10	

	Temperature difference soil/air (°C)	
	Air	Soil
Karakum	33.5	64
Repetek (near Tashkent)	42	79.3 (20.6.1915)

1.1 · Attempts at Global Localization of Drylands

Among the drylands in which sandy expanses dominate, i.e. the sand seas of the Saharan-Arabian deserts, Australia, the Namib, central Asia with the Karakum and the Kyzylkum and in the Chinese sand seas enclosed by mountain ranges like the Taklamakan, two major families may be distinguished:

- *The open Saharan-Arabian sand seas* of the Namib and central Asia in which longitudinal dunes predominate as an indication of a negative sedimentary balance.
- *The closed sand seas*, like most of the Chinese sand seas, are dominated by transverse dunes as an indication of a positive sedimentary balance. The different types of dunes and the concept of the sedimentary balance will be outlined in the second part of this book, *The Omnipresent Wind*.

In the open sand seas deflation transports the fine-grained particles and notably the loess in the downwind direction. When deposited they form fertile grounds and as a result, the deserts from which they are derived consist of material depleted in fine-grained components at least on the surface and are areas of low fertility, the utilization of which is difficult: the Ténéré of Niger as well as the Karakum and the Kyzylkum of the Aral basin may serve as examples for this situation.

In closed sand seas, from which nothing can escape, the fine particles constitute a matrix, which, apart from phases of lower value, will give richter soils: Taklamakan in China is the best example of this.

This situation controls texture and structure of the soils and forms the foundation upon which agricultural development programmes should be based. The diversity of the true deserts is enhanced by an entire array of degrees of aridity.

1.1.6
Semi-Deserts or Semi-Arid Zones, a Wide Zone of Transition

Regions affected by aridity may stay so during the entire year, or only seasonally, temporarily or also in a rather haphazard way. Dresch (1982) enumerated these areas:

> Semi-deserts or semi-arid areas (21 million km^2) are bordering the deserts and as a consequence, with their elevated precipitation, possess comparable characteristics: the Saharan Sahel in the true sense, the tropical and southeastern fringes of the Australian desert, the high steppes of East Africa and the Kalahari with summer rains and mild winters because of their latitude, the hot steppes of northwestern India and the Dekkan as well as the semi-deserts of South America (the caatinga of the Brazilian Nordeste and the Chaco), the plains of the central western United States with cool winters becoming colder to the north and with continental summer rains whereas in the west, the lee-side basins of the Rockies still experience winter rains just as the high steppes of North Africa and of Spain, and those of Anatolia, western Iran or the northern borderlands of the middle or central Asian deserts.

The seasonal aridity makes itself felt by:

- The dry season in the Mediterranean summer.
- The long dry season of the semi-arid and subhumid dry tropics.

The map of Alisson (Fig. 1.1) is again self-explaining. The Mediterranean summer drought is explained by the positions of the polar front in winter and in summer, respectively, in both hemispheres (lines P_w and P_s). The polar front in the Northern Hemisphere only traverses along the Mediterranean zone in winter and does not touch it in winter: there is thus drought. The line P_w cuts across the centre of Hokkaido. It marks the northern limit of intrusion of tropical humid air and at the same time the northernmost possibility for growing rice. In the same way, the dry season of the semi-arid, subhumid dry and subtropical zones is explained by the position of the ITF, the extreme positions of which are marked by the lines Ni_s and Ni_w which during the dry season do not reach the zones mentioned.

From the Atlantic to the China Sea the semi-arid and subhumid areas separate the drylands of the Sahara, Arabia, and India from the humid tropical zones. In China this transition zone is situated between the deserts of the northwest and the humid tropics of the southeast. The region corresponds to the final reaches of the monsoons: the African ones in the Sahel; the Indian ones in Ethiopia, the Arabian Peninsula and the northern part of India; and the Asian ones in eastern China. In Africa the monsoons extend to the 100 mm isohyet[2] and in Asia to the 300 mm isohyet, respectively. The regime of the monsoons entails a strong variability in the precipitation, a precipitation-evaporation ratio around negative values, a precarious hydrological and biological equilibrium, and pronounced crises or droughts.

From these degrees of aridity follows a simple but useful distinction for rural management programmes, namely those of:

- Tropical and subtropical dry climates without winters which permit several harvests per year with the aid of irrigation;
- Dry climates with severe winters and thus a shorter growth periods not permitting any harvests at all;
- Coastal climates without true seasons, falling into the first group.

1.2
The Concepts of Aridity and Drought in Drylands

With the spatial situation of the deserts having been defined and the causes and mechanisms of each of the five groups of deserts described, the climate shall not be described with all its facets but only by those of *aridity* and *drought*.

We have seen that continents including even Europe with its Mediterranean fringe possess drylands where aridity and drought are the two most apparent climatic facts. *Aridity* implies a *permanent* pluviometric deficit which is tied also to other specific climatic data like strong insolation, elevated temperatures, low air humidity and strong evapotranspiration. *Droughts* result from a *temporary* pluviometric deficit in relation to the normal precipitation. The Sahelian and the Sudanian ecosystems of Africa are those in which the equilibrium and the production systems are affected most severely by droughts.

The degrees of aridity and the different types of drought as well as the concepts of advanced and increasing desertification will be discussed for these African examples.

---

[2] *Isohyets:* theoretical lines joining points of equal mean annual precipitation.

1.2.1
Definition of Degrees of Aridity

Taking into account extent and variability of the arid, semi-arid and subhumid drylands, the FAO-Sida mission (October 1973–January 1974) proposed a subdivision for north-equatorial Africa according to its pluviometry P:

- $P < 100$ mm: Saharan zone
- $100 < P < 200$ mm: Saharan-Sahelian zone
- $200 < P < 600$ mm: Sahelian zone
- $600 < P < 800$ mm: Sahelian-Sudanian zone
- $800 < P < 1200$ mm: Sudanian zone
- $P > 1200$ mm: Guinean zone.

The first three zones, the only ones suffering from aridity, and the fourth one which may be added to them because of the droughts resulting from pluviometric deficits during respective fluctuations, make up the drylands.

Over this disposition of aridity across the latitudes is superimposed a gradation by altitude which was described by Cauvin (1981) for the Middle East: "a region in which water is scarce. ... We have to keep in mind, that there is a gradation of this scarcity from the well-watered mountain crescent through the intermediate steppe foot-hills still supporting dry cultures to the desert zone beyond the 200 mm isohyet." A different arrangement of these zones according to altitude is developed in Asia and China because the relief here is higher.

Under aridity we understand a state caused by mechanisms leading to a water deficit in air and soil by the feeble nature of the precipitation and by the intensity of evaporation, which represent the most important factors but by no means the only ones. The FAO/UNESCO soil map of the world is based on a quantitative criterion for differentiating the dry zones, viz. the *bioclimatic aridity index P/PET*, in which P = annual precipitation and PET = potential evapotranspiration.[3] The latter, expressed in millimetres per time unit, is not limited by the availability of water in the soil:

- *Hyperarid zone*: $P/PET < 0.03$, a zone corresponding to extreme deserts without vegetation except for some ephemerals and xerophytic bushes in the beds of wadis; it is inhabited only in oases.
- *Arid zones*: $0.03 < P/PET < 0.20$; this encompasses barren areas or those covered by a sparse vegetation of perennial and annual plants; pastoral nomadism is possible but not rain-fed agriculture.
- *Semi arid zone*: $0.20 < P/PET < 0.50$, covered by steppe (open vegetation cover) and tropical bush; perennial plants are most frequent here; extensive livestock breeding is possible.

---

[3] Potential evapotranspiration: quantity of water lost by evaporation directly from the soil and by transpiration of a plant cover.

- *Subhumid zone*: 0.50 < P/PET < 0.75, this encompasses the tropical savannah sometimes covered by bush, occasionally without trees, and the dry forest; permanent rain fed agriculture is practised here with cultures adapted to seasonal drought.

To subject aridity to such a scheme is over-simplistic as the various shades of aridity are rather subtle. This was shown with the aid of differences in positions, causes and mechanisms by Demangeot (1981) who stated that:

- about one third of the emerged lands is arid to different degrees;
- the proportion of hyperarid lands is rather low (4% according to planimetric estimates) ...;
- the Old World by itself harbours close to three quarters of the arid zones: This applies not only to its absolute surface area but also to its more massive shape. It is said that America as a whole is "fresher" than the Old World;
- the Northern Hemisphere, and especially in the Old World, contains the largest proportion of deserts. This is evidently caused by the respective continentality, a phenomenon that also explains the shifting of the thermal equator to the north;
- considering only the Saharan complex, aridity appears to follow a zonal arrangement and the line-up of the southern deserts south of the Tropic of Capricorn tends to confirm this;
- this zonal disposition is actually quite imperfect: the South American and southern African deserts are arranged at right angles to the thermal zones;
- there are deserts at latitudes where one would not expect them: close to the Urals and south of Patagonia ...;
- the presence of hyperarid zones along the oceans confirms that there are also non-zonal factors; in contrast to the Sahara, these deserts are more arid along coast than in the interior;
- with the exception of southeastern Arabia, the pronounced coastal deserts are found along the west coasts of the continents.

In conclusion it can be said that the "arid situation" results from a variety of factors: "... thermal zoning; continentality; position on the west coasts .... The feature common to all deserts, the fundamental fact, still is their scarcity of precipitation."

1.2.2
The Land Surface Affected by Aridity on the African Continent

In addition to the Sahara, there are two more deserts in Africa, i.e. the Kalahari-Namib south of the equator and the less known Chalbi in East Africa, around the equator from southern Ethiopia to northern Kenya. The lowlands extending from the north of Kenya, the south of Ethiopia and Somalia towards Sudan and the Sahara have probably been dry at least since the Tertiary.

For the drylands of the African continent Biswas (1987) estimated that 29% of the continent, with less than 100 mm precipitation annually, is desert or arid, 17% with 100–400 mm/year is semi-arid and constitutes the Saharan-Sahelian complex, 8% with 400–800 mm/year makes up the semi-arid ecosystems, the Sahel proper, 10% is

Sudanian-dry (subhumid dry) with 800 mm/year whereas the remaining 36% of Africa receiving >1200 mm/year is tropical humid. The 400 mm isohyet may be taken as the boundary of the zone with a vegetative period of more than 120 days and receiving sufficient precipitation to allow rain-fed cultivation of millet, sorghum and maize. Biswas (1990) estimated that 25% of the surface of the entire continent is suitable for rain-fed agriculture, to which may be added some 10% of marginal lands. The remaining 65% is not suitable for this type of exploitation. This estimate should restrain those for whom Africa should become more populated and for whom technology is able to overcome the harsh natural conditions.

Roche (1986) called aridity "a mean state in the ... climate of a region and of its consequences characterized by weak precipitation and poor vegetation. The concept of aridity depends on other factors, ... and notably on temperature. From the point of view of vegetation it is more correct to define aridity by a mean *hydrological deficit*. Aridity should not be confounded with drought." As we have seen, *aridity* as a long-term climatic phenomenon is different from the short-term phenomenon *drought* although both affect the drylands more in particular. Roche (1986) defined the hydrological deficit as: "the difference between real evapotranspiration and precipitation over a certain timespan" and real evapotranspiration (RET) as "the quantity of water ... taken by evaporation from a soil covered by vegetation taking into account ...the growth condition of the vegetation and the availability of water."

1.2.3
The Different Types of Drought

Droughts are harsh climatic events and natural disasters like floods, tropical cyclones and earthquakes (Chonchol 1989). Their impact on the environment, their socio-economic and political effects lead to disturbances of the equilibrium, to crises of the production systems, to a drop in foodstuff production and to social upheavals.

Drylands are affected in an irregular manner by droughts, the disturbing effects of which have become more severe since the third quarter of the 20th century. This aggravation is actually the result of an increasing demography and the growing inability of the states both to make provision especially for the food and socio-economic aspects of these disasters in the short term and to adopt anti-drought strategies in the long term. Since 1975 the Sahel has no longer been able to feed its population, which by 1980 had grown to 30 million.

The term drought is ambiguous like aridity as it refers to a non-zonal climatic situation and in particular to deserts, but may occur in any climatic zone. France experienced two severe droughts, in 1976 and 1985 and 4 years of rain deficit from 1988 to 1992. In 1993–1994 the autumn and winter rains were necessary for the aquifers to recover their levels of before 1976. As distinct from aridity, drought is always a relative concept for man and his needs, but both terms are close cousins.

For Roche, the term drought has two meanings: "... a climatologic one: period or year during which precipitation is considerably below the mean" and " ... a hydrological one: period or year during which the runoff is considerably below the mean." He notes that "drought manifests itself over time (dry period) whereas aridity is a spatial phenomenon (an arid region)."

It is not sufficient to define drought by quantitative criteria of the total annual pluviometric situation or by the volume of annual runoff, and thus other parameters have to be taken into account:

- The duration of the humid season.
- The starting point of the humid season. Toupet (1988) defined the start of the rainy season as the time when "the total precipitation amounts to at least 20 mm falling in a single day or in two consecutive days".
- The duration of the period of deficit.
- The severity of the deficit.

Beran and Rodin (in Sircoulon 1989) distinguished six types of drought:

1. *A deficit in runoff of several weeks or months* at the time of germination and growth of plants, with devastating consequences for irrigated cultures.
2. *Lower waters than normal and over an extended time*, a situation more frequent for small to medium-sized water courses with temporary flow. On allochthonous rivers the supply of water for cities is disturbed as in Niamey on the Niger in June 1985.
3. *A deficit in the annual amount of runoff* with effects on the production of hydroelectric power and irrigation from large dams.
4. *Lower than normal floods in rivers* and their devastating effects on management of hydroagriculture: submersion cultures of the *Falo* in the valley of the Senegal (Fig. 2.5).
5. *Shrinkage of groundwater resources*.
6. *Prolonged drought over several years*; the *Secas* of the Brazilian Nordeste and, in the Sahel, the years 1972–1973 and 1982–1984 when all types of water resources were affected.

Bernus (1989) introduced an additional grading according to the types of plants which can grow after a particular amount of rainfall: in the central Sahara rainfall that impregnates the soil at depth is sufficient to bring to life the annual ephemerals, the seeds of which had stayed dormant. In the Sahel zone perennial grasses come to maturity (formation of ears and seeds) when a series of rains close enough in sequence facilitates a complete cycle which is longer than that of the ephemerals. In both cases there is drought when the absence of rain does not allow the annual pastures to regrow, but in the Sahel in addition to the overall insufficiency there are also other parameters to be considered: the general distribution of rains, their succession, and their penetration into the ground. "It is the succession of dry years or, worse, the close succession of deficitary series which provoke the real crises".

Droughts cannot be explained, which caused Farmer and Wigley (1985) to admit with great honesty: "We neither know why the present droughts take place nor why droughts occurred in the past." It may appear naive to say that the pluviometric deficit is the first cause of a drought, but we shall see that according to its nature, be it a meteorological, hydrological, agricultural or edaphic drought, this concept may be seen independently from that of deficitary precipitation.

1.2.3.1
Meteorological Drought

Hydrologists call major droughts or dry spells those periods during which the relative distribution deviates by more than 20% from the mean (Durand 1988).

Drought is thus a meteorological phenomenon which occurs when the precipitation is below the average over one or several successive years. It is difficult to define a pluviometric deficit precisely. The means are misleading especially in the northern parts of the arid zones where total precipitation varies greatly from one year to the next. The quantity of rain alone controls the productivity of the vegetation only in part. Next to fertility and soil structure the distribution of the precipitation in time and space plays a crucial role and if this is satisfactory, even "below average rains" may permit entirely sufficient crops. However, "average" or "above average" precipitation is not synonymous with "average" or "above average" harvests when these rains are scattered and when dry periods alternate with periods of excessive precipitation (World Bank 1985).

Droughts are graded on three scales: local, regional, and subcontinental. Whereas they are usually only of local extent in Europe, they frequently reach the regional scale in Africa. The so-called Sahel drought of 1968-1985 has affected the countries of the Sahel from Mauretania to Ethiopia, especially between 1972-1973 and 1983-1984. Such a vast extent implies the action of mechanisms which are not yet understood. Another characteristic of the African droughts is their recurrent nature.

The central semi-arid part of the Brazilian Nordeste has a mean annual rainfall of 400-600 mm and an interannual variability of 0.35-0.40. "The most sensitive zones are situated below the 800 mm isohyetic contour (Fig. 1.6) and more in particular so

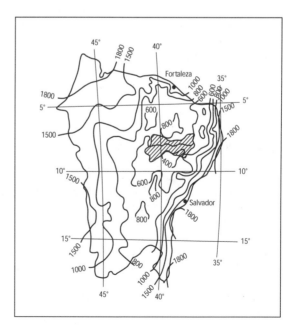

Fig. 1.6. Mean isohyets in the Brazilian Nordeste from 1912 to 1983 (after Molinier *et al.* 1989)

north of latitude 10° S and east of longitude 42° W where droughts are frequently very severe and the floods attain catastrophic dimensions" (Molinier *et al.* 1989). Between 1978-1983 the precipitation was deficitary with a true drought between 1981-1983, the devastating consequences of which were aggravated by the tremendous population increase. From 1984 onwards the rains were exceptionally strong and in 1985 this excess turned into catastrophic floods leading to crop losses and to the displacement of more than one million inhabitants in the state of Ceara alone (Molinier *et al.* 1989).

Petit-Maire *et al.* (1994) underlined "the synchronous nature of global warm phases with an intensification of the monsoon activity" in the semi-arid and sub-humid drylands. In the Sahel, the north of India and northern China the warming of the 1920s was accompanied by an episode of exceptional precipitation. In contrast to this, the cooling of the 1960s coincided with the persistent drought of the Sahel, the Arabian Peninsula, northern India, the Tibetan plateau and northern China.

1.2.3.2
Hydrological Drought

"The hydrological drought ... entails the diminution of the surface flow in the watercourses and a natural lowering of the groundwater level in comparison to a known normal value" (Boulanger 1990). To this is added the variation of the water level of lakes as a result of the couplet runoff/infiltration and evaporation. This is not a highly sophisticated indicator but shows any trends that might exist. Thus during the lower Holocene (10 000 B.P.) the levels of most African lakes, especially between 13° N and 2° S, were higher. A second general trend is the lowering of the lake levels since 2500 B.P., an indicator of a drop in precipitation.

The levels of East African lakes and amongst those in Kenya the Turkana, Beringo, Bogoria, Nakuru, Eleimentata, Naivasha, and Magadi, which had a higher level after the exceptional rains 1951-1952, have not ceased falling since. Lake Victoria, the level of which is recorded since the turn of the century, experienced a high level between 1960-1965, but has not ceased falling since. In the case of the Naivasha, a fresh water lake, overexploitation and deforestation are the main causes for the drop in the water level.

1.2.3.3
Edaphic Drought

This is defined as a decrease of the infiltrability of the soils and thus as an accentuation of the arid character of a certain landscape. Albergel *et al.* (1992) have drawn attention to this form of deterioration. In Burkina Faso they installed rain simulators in eight test lots of 1 m^2 each during 1981-1985. The rain simulators reproduced the natural conditions of intensity, duration, and frequency of rains of the region. By continuously recording the soil humidity they determined the total depth of infiltration, and the intensity of runoff and of infiltration as a function of time and of the water reserves in the soil. They could show that in the Sahelian and Sudanian-Sahelian zones the infiltrability depends almost exclusively on the condition of the soil surface.

Under the influence of droughts the density of the gramineous and bushy stratum decreases, faunal activity disappears, and the "glazed" surfaces and areas eroded by runoff and gullying increase in size. This increase in the physical mechanism leads to a severe deterioration of the environment which is sometimes irreversible, at least on a human time scale. Edaphic drought may be the consequence of a meteorological drought but depends also on the mode of exploitation of the soils.

During the 1950s, a mostly grassy vegetation cover fixed all sandy areas in Africa which received >150 mm precipitation annually. The plants colonized the dunes, trapping particles in saltation, holding back organic debris, thereby bringing nutritional components to the soil, ameliorating its structure, its water retaining potential and thus its fertility. Thereafter overgrazing exposed these sandy surfaces to aeolian processes during the dry period, resulting in a loss of the A-horizon. The vulnerability of the soils to erosion grew and nutrients were lost, leading to a lower growth potential for vegetation, to a degradation of the soil and to a reactivation of the dunes (Mainguet 1981; Mainguet and Chemin 1987; Bendali *et al.* 1991).

The eradication of the A-horizon exposed the B-horizon and subjected it to *scalding*, a polishing effect under the influence of saltating sand particles which renders it impermeable to the infiltration of rainwater. This makes it sensitive to the *splash* effect and thereby to water erosion. The B-horizon behaves from then on like a soft rock susceptible to sheet-runoff; this runoff favours the appearance of rill wash and aeolian erosion which leads to the appearance of striae of corrosion and aeolian depressions of the *blow-out* type.

All these degradational mechanisms leading to *edaphic* drought entail as consequences: a vegetation suffering under a deficit of moisture and oxygen and seed grains which, unable to penetrate the indurated upper soil layer, dry out and are carried away by the wind.

As we know that with an increase in the water content of a soil the cohesion of the soil particles grows and their potential of being lifted drops – according to Chepil, the erodability decreases with the square of the increase in moisture – edaphic drought will render the soils more vulnerable to aeolian erosion.

1.2.3.4
Agricultural Drought

An agricultural drought is defined in relation to water requirements. It occurs when the water resources necessary for agriculture become scarce. A short drought after several rain-rich years is softened in effect. It is not the same as a drought of several years duration. The most dramatic example known is that of the Sahel and notably of Ethiopia.

A meteorological drought leads to stress for the plant life whereas an agricultural drought is defined by needs resulting from human activities and consists of a deficit of water in the ground during the period of plant growth. It then leads to a disequilibrium in the agricultural economy. This may result from demographic pressures or from the introduction of unadapted plant varieties which are more demanding of water. In Kenya, for instance, the replacement of millet by maize has created such a disequilibrium in certain areas prone to droughts. Rains of normal quantity, but with a disadvantageous distribution, may also cause an agricultural drought. This was also the case in Kenya which was on the threshold of famine in 1984.

According to Le Houérou (1989), "the variability of the annual primary production in the arid zones on a worldwide scale is on average 50% higher than the variability of the annual precipitation." Thus the pluviometric variability of the Brazilian Nordeste, one of the highest in the world, increases the aridity factor and explains the xerophytic nature of the *caatinga*. The ecological and agricultural consequences of the drought here are more severe that in the Sahel.

Drought and climatic variability are parameters inherent to drylands to which are added the pressures resulting from demographic growth and modern economics, which in the framework of development are placing more emphasis on cash crops than on subsidence crops and thereby increasing the risks. The ecological disequilibrium caused by the droughts is felt all the more, as with the modernization of the ways of life, viz. sedentarism of nomads and the extension of agriculture into marginal areas, the demand for water increases.

1.2.4
Postglacial Aridification and Advance of the Deserts in Africa?

These two questions, amongst the most delicate ones, should be asked here, outlining also the controversies to which they gave rise.

Two hypothesis oppose each other. The one proposes that the droughts are episodic fluctuations characteristic of drylands. Since the start of the century four dry phases have occurred south of the Sahara: from 1895–1905, from 1910–1916, from 1938–1943 (Bernus and Savonnet 1973), and from 1968–1985. According to the other hypotheses, there is a longer-term trend to aridification: the postglacial aridification.

The climatic rhythms are explained by the theory of Milancovic cycles, cosmic cycles of 100 000, 40 000 and 20 000 years, onto which are superimposed shorter cycles of a dozen years, correlated with the sun spot cycles, which lead to the synchronous occurrence of glacial and interglacial periods in cold and temperate latitudes. This theory was disputed by Tricart, according to whom:

- The cold glacial periods are synchronous with increased aridity or hyperaridity in the dry tropical ecosystems;
- Whereas the interglacial warming of higher latitudes corresponds to more humid phases of progress of the steppe and of the savannah.

The Milancovic cycles are becoming increasingly rejected as an explanation for the climatic variations in the tropical drylands. From studies of hydrometeorological data and the discharge of water courses in West Africa (Niger and Senegal), Hubert *et al.* (1989) presented evidence that in the 20th century there were two humid phases, 1923–1935 and 1951–1970, and three dry phases, one prior to 1922, 1936–1950 and after 1970 (Fig. 1.7). As turning points are used the historic droughts of 1910, 1940 and 1970. The authors remarked that "the different climatic phases elaborated are part of a trend of aggravating aridity ... since about 4000 B.P." This trend was also observed from variation in the water level of Lake Chad since the start of the 18th century (Maley 1981), and is found also in the deserts of Asia.

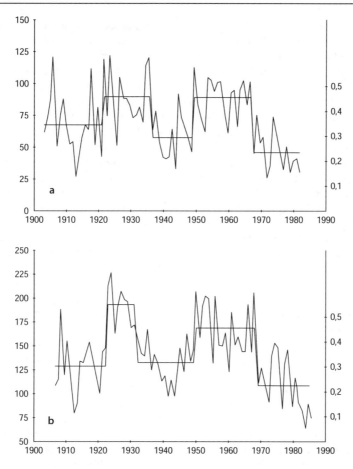

Fig. 1.7 a,b. Distribution and segmentation of mean annual discharges (after Carbonnel 1993). a Niger River at Koulikoro; b Senegal River at Bakel

The albedo, i.e. the degree of reflectivity of the soil, and its variation may be measured with the aid of remote sensing. There is a general consent on the relationship between erosion of soils, their degradation and an increase in the albedo. Charney (1975) advanced a theory of the aggravation of drought in the Sahel from an increase in the albedo of the soil surface as a result of degradation caused by rural activities which lead to a larger portion of the incident radiation becoming compensated by a greater subsidence of the air in contrast to the convective mechanisms leading to precipitation. Charney then concluded that human activities are the cause of drought in the Sahel, a situation contested by Courel (1989) for whom the droughts were more related to anomalies in the general atmospheric circulation than to point-like modifications in the state of the surface attributed to man.

Jackson and Itso (1975) rejected the theory of Charney by pointing out that the degradation of the vegetation cover has not led to the deterioration of the climate.

Friedman (1989) showed with the aid of Landsat images from either side of the border between the USA and Mexico, that the difference in the state of the vegetation cover caused by more intense overgrazing on the Mexican side has led to a difference in the albedo and to summer temperatures 4 °C higher south of the border without, however, having led to a change in the total precipitation.

Goudie (1972) proposed an increasing postglacial aridification around 20 000 to 18 000 B.P. after the more humid phases of the Pleistocene. Although younger, the planting sites and agricultural sites discovered along the edges of the Taklamakan desert in Chinese Sinkiang, in Turkmenistan around the Khorezm and in the Amu Daria delta, as well as in the Thar Desert made it possible to test their hypothesis This is especially valuable for the Asian deserts as these, in contrast to the African deserts, are frequently located on pediments. According to Lamb (1968), the melting of the glaciers and the supply of water to the aquifers rendered the margins of the deserts suitable for cultivation and thereby inhabitable in North Africa until the beginning of our era, when the point of exhaustion of the residual glacial reserves was reached. During the Quaternary, the Maghreb did not actually experience the arid phases of the Sahara and of its southern borderlands (Mainguet 1982). Since the beginning of the century, numerous scientists have subscribed to the hypothesis of an increase in aridity and an inexorable advance of the deserts in Africa. Wilson (in Goudie 1990) described how in 1865 the Kalahari "devoured" vast expanses of arable land, transforming them into sterile ground.

Some 125 years after this observation, even if there are aureoles of deterioration of the vegetation cover around agglomerations like Tsabong and Bokspit in Botswana, a strong reactivation of fixed dunes, and a deterioration within a circle of 10 km around waterholes, with the disappearance of perennial species, the Kalahari, nevertheless, remains a region of fixed sands. The prediction of Brown (1875, in Goudie 1990) that the Karoo would also become a desert like the Sahara is no longer valid. Based on rainfall data over several decades at St. Louis (Senegal) Hubert (1920) also showed a growing aridification but he did not ask himself whether the data for such a short time span would rather indicate only climatic fluctuations instead of real climatic crises. Chudeau has cast into doubt the data and conclusions of Hubert since 1921.

Renner (1926) and Bovill (1929), who observed the Senegal River and the northern part of Nigeria, and Stebbing (1935), who worked in Nigeria south of Niger and east of Mali, rallied under the same hypothesis of an aridification of the Sahelian border of the Sahara. It was Stebbing who conveyed formal status to the concept of the advancing desert with his article, *The Encroaching Sahara: the threat to the West African colonies*.

After Chudeau (1921), several studies during the 1930s, furnished arguments against the hypothesis of increasing aridification: Kanthack (1930) carried out studies in South Africa, Urvoy (1935) studied the terraces in the east of Niger, and Jones (1938) studied Lake Chad. Jones (1938) compared the high water level in Lake Chad to that found in the second decade of the 20th century, when the lake level was so low that in 1928 General Tilho expressed his fear that he would see the lake disappear.

This gave rise to the idea that instead of an advance of the desert, there is in reality an intense deterioration of soils as a result of human activities (Stamp 1940; Chevalier 1950), and Aubreville (1949) coined the phrase desertification, a concept that will be dealt with in the fourth part of this book.

1.2 · The Concepts of Aridity and Drought in Drylands

Hellden (1984, 1988), Olsson (1985) and Ahlcrona (1988) had the last word in this contest between supporters and opponents of this theory of the "advance of the deserts". Lamprey (1975) had stated for the Sudan that the southern limit of the Sahara had advanced at a rate of 5.5 km/year, based on a comparison of a vegetation map done in 1958 with aerial observations in 1975. The Swedish team of the University of Lund, however, was able to demonstrate with the aid of remote sensing and fieldwork that the reality was entirely different: in the region south of Atbara (south of 18° N) the albedo has changed only little. Together with the lack of quantitative changes in the vegetation, there were qualitative changes, i.e. a deterioration of the vegetation cover, but no formation of a desert landscape.

To support the advance of the desert into the Sahel by expressing its annual progress in figures is thus one of the scientific errors of our century. To the dynamic but unrealistic concept of the extension of the deserts we prefer the idea of propagating suitable mechanisms outside this zone in the Sahel, such as:

- intensified erosion by water;
- increased aeolian deflation with the loss of soil, more outcrops of fresh rock and more coarse alluvial deposits in the form of regs. The highest annual losses of soils were observed on the ferruginous soils of the Sahel with 200 t/km^2 (Berry 1974). Desertification has been frequently coupled with the development of active dunes, a concept that merits some consideration: if it can correspond to a reactivation of fixed dunes – replacing them by active dunes – then desertification leads to an exportation of sand, because of an increase in frequency and velocity of the winds, and not to its accumulation;
- disappearance of perennial plants from the vegetation cover;
- concentration of vegetation along the water axes.

Based on a study of rainfall data from the Sahel, Olivry and Chastenet (1989) stated that, through the fluctuations which probably were not of a cyclic nature, there has been a persistent diminution of water resources in the tropical regions since the second half of the 19th century (Fig. 1.8). The drought of 1968–1985 and in particular the two episodes of 1972–1974 and 1983–1984 have shown a shifting of the 400 and

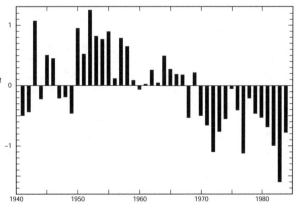

Fig. 1.8. Surplus or deficit of precipitation between 1941 and 1982 for 20 meteorological stations in sub-Saharan West Africa between 1° N and 19° N, and west of 10° E (Lamb 1985, in Glantz). *Ordinate* Deviation from the mean; *abscissa* dates

600 mm isohyets by an average of 2° and the presence between 11–17° N of a Sahelian-Sudanian zone similar to those of the Brazilian Nordeste and Rajastan, both of which are areas of high climate risk.

1.2.5
The Droughts of the 20th Century in North-Equatorial Africa

Systematic meteorological records did not start in the Sahara before about 1900.

The drought of 1910–1915 (Table 1.4), called *Grande Bere* or "long famine" in the Songhai calendar was called *Tasbane* or "calamitous" in the Peul calendar. At St. Louis in Senegal where it lasted from 1908–1915, the precipitation had fallen to 140 mm in 1914 instead of the mean of 330 mm, the mean of 1964–1974. Niamey received only 281 mm in 1915 instead of an average of 580 mm, representing a deficit of 48%. N'Djamena in 1913 received only 306 mm instead of the mean of 620 mm, representing a deficit of 50%.

The drought of 1940–1949 – interrupted locally during 1945–1946 – was called *wanda wassu* by the Songhai, i.e. "send your wife away" with the meaning of "free yourself of a mouth to feed."

The drought of 1968–1985 lasted initially from 1968 to 1973 and then, after two less dry years (1974–1976), continued with some interruptions until 1985 with the highest deficits in the years 1982–1984. This reveals the cumulative nature of the deficits but eliminates the spatial and temporal disparities. No dryland has functioned as a single entity. Pélisier (1984) talks about "pockets of drought next to rainy areas".

The maximum deficit of precipitation occurred:

- At Mainé Soroa (Niger) during 1969 with 56.2%, corresponding to a single rainfall of 229.9 mm
- At Tillabéry in 1971 with 56.5%, corresponding to a single rainfall of 250 mm
- At Zinder in 1973 with 59.9%, corresponding to a single rainfall of 291.5 mm (Durand 1988).

The Sahel droughts of the late 1960s, the 1970s and of the first half of the 1980s were all the more difficult to endure as the mean rainfalls of the preceding 1950s were 20% above the usual means. The exceptional precipitation of this decade caused the nomads to migrate north. The Peul people of northern Nigeria and southern Niger found themselves eventually in the Air. Also, the settled populations migrated into marginal areas frequented until then only by a dispersed and itinerary nomad population, i.e. into an area climatically unsuitable for the survival of a stable population above a certain density level. The catastrophe reached its climax in 1968 when the rainfall in the Sahara fell to 33% below the means of the 1950s and in 1972–1973 when it fell to 50% below. The pluviometric deficit continued, although in discontinuous fashion, until 1982 and another higher deficit occurred in 1983–1984. Seignobos (1984) found that in 1973 the 500 mm isohyet in the Chad came to lie in the position of the 800 mm contour during a normal year. The deficit resulted from the shortening of the rainy season and from a diminution of the number of rains with >40 mm precipitation.

1.2 · The Concepts of Aridity and Drought in Drylands

Table 1.4. Climatic fluctuations and drought periods of the 20th century in tropical Africa (from different sources)

1905 – 1908	Low level of Lake Chad
1912 – 1915	*Period of severe drought*
1916 – 1924	Humid period
1925 – 1928	Dry period
1929 – 1939	Humid period
1940 – 1944	*Period of severe drought*
1945 – 1946	Humid period
1947 – 1949	Dry period
1950 – 1961	Humid period
1962 – 1967	Intermediate period
1968 – 1973	*Period of severe drought*
1974 – 1976	Intermediate period
1977 – 1985	*Period of severe drought*
1987 – 1993	Period of favourable rainfall
1994	Severe drought in Ethiopia, indications of famine

In order to clarify the climatic fluctuations of the 20th century we should have a look at the movements of the 150 mm isohyet, i.e. the boundary between the Sahara and the Sahel on the southern rim of the Sahara. Its position at the end of the 1950s may be traced on aerial photographs by the change from active to fixed dunes and sand covers. Figure 1.9a represents the fluctuations of this isohyet, the southernmost position of which corresponds to the southern limit of presently overgrown dunes. The two positions are separated by up to 5°, illustrating that the period preceding the drought crises of 1968–1985 was comparatively favourable in comparison to the maximum period of droughts in the Sahel when the southern border of the Sahara around 20 000 B.P. was 5° S of its present position. This limit of 20 000 years ago was fortunately not reached again, not even after the last drought crisis.

Petit-Maire et al. (1994) have illustrated "the almost identical behaviour of the African, Indian and Asian monsoon systems". North of the semi-arid domain and of the area of influence of the monsoon the 100–150 mm isohyets of around 20 000 B.P. were located 5–7° N of their present position in Africa, the Arabian Peninsula, India and the Chinese part of Asia (Fig. 1.9b). A dry episode was experienced throughout the mid-Dryas (20 000–12 000 B.P.), the 100 mm isohyet was lying at around 13° N in Africa, the acacia savannah had retreated to 10° N, the discharge of the White Nile was reduced, evaporites were deposited in the central Sudan, and the Rub al-Khali and the Nefud formed in Arabia. Rajastan was also affected by a dry period and in Tibet the lakes became increasingly saline between 22 000–15 000 B.P., an indication of the same trend of increasing aridity.

During the Holocene, precipitation increased from 100 mm/year over the north of China and Tibet, from 100 to 400 mm/year over the Arabian Peninsula and from 200 to 300 mm/year over the southern Sahara and the Sahel, thereby pushing the 100 mm isohyet some 5° towards the north of its present position as shown by the limit of the vegetated dunes. This illustrates a fluctuation of the isohyets by 1000 km

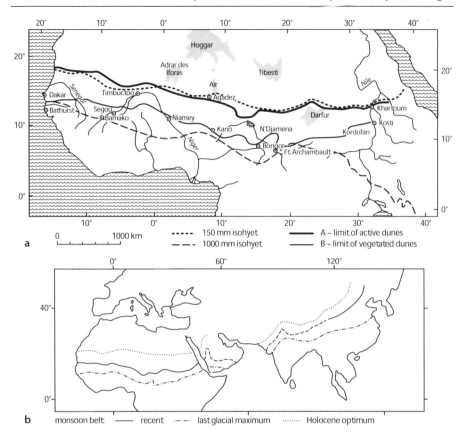

Fig. 1.9. a,b. a Shifting of the 1000 and 150 mm annual isohyets towards the north between the drought of 20 000 B.P. and the present. **b** Changes in the boundary of the monsoon domain since the last two climatic extremes (last glacial maximum at 20 000 B.P. and the Holocene optimum around 10 000 B.P.) to the present (after Petit-Maire et al. 1994)

in dry Africa and Asia and by 350 km in the north of India and thus indicates a latitudinal expansion of the monsoon region during the Holocene optimum. Between 5000–4000 B.P. a trend towards increasing aridification became established in the various areas of transition.

The positions of the isohyets of the borders of the drylands are illustrated in Fig. 1.10 on a smaller scale for western Mali from the Mauretania-Mali border (15° 40' N) to the Mali-Niger border (10° 50' N) for the various periods:

- A long period of about 60 years (1922–1980) which encompasses phases of excessive and deficitary rains.
- A 10-year period (1951–1960) of more humid conditions during which all isohyets were shifted north.
- A 10-year period (1971–1980) of drier conditions during which the isohyets migrated to the south.

1.2 · The Concepts of Aridity and Drought in Drylands

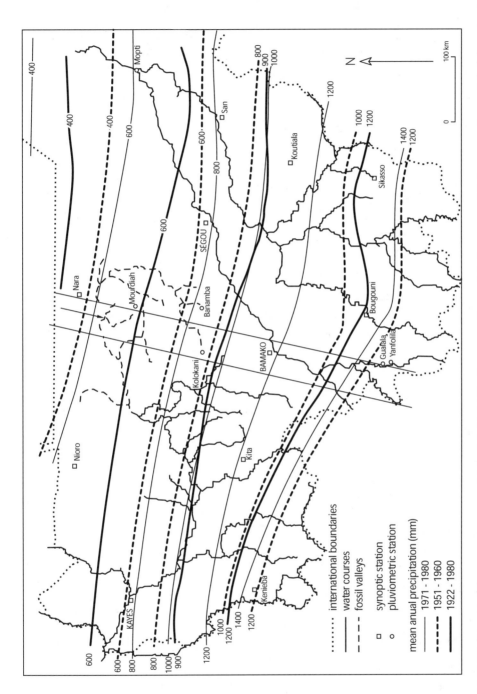

Fig. 1.10. Displacement of the 400–1200 mm annual isohyets in West Africa from 1922–1980, a period of average precipitation; 1951–1960, a period of above-average precipitation; and 1971–1980, a period of deficitary precipitation

Table 1.5. Amplitude of displacements of isohyets in Mali since 1922

	Amplitude of displacements of isohyets west of the longitude of Mopti (Mali)		
Isohyets (mm)	West (km)	Centre (km)	East (km)
400	–	–	125
600	135	120	100
800	40	40	60
1000	90	160	120
1200	75	150	140

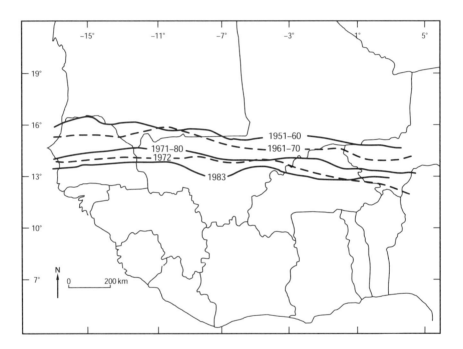

Fig. 1.11. Displacement of the 500 mm isohyet in West Africa prior to and during the droughts of 1968–1983 (after J.-P. Carbonnel)

The position of the 400 mm isohyet in Mali was thus subjected to a displacement of 400 km to the south during the discontinuous drought of 1968–1985 adding a previously semi-arid stretch of land to the desert.

Toupet (1972) has shown that between 1941–1942 and 1951–1952 the re-adjustment of the 100 mm isohyet added about 340 000 km^2 or roughly one third of the area of Mauretania to the Sahara (Table 1.5). The consequences were less damaging in the Sudanian-Guinean domain to the south where rainfall varies between 1200–1400 mm than in the Sahel to the north where rains ranged during this decade from 400 mm during a dry to 600 mm during a humid period. The different positions of the 500 mm isohyet in West Africa from 1951–1983 are shown in Fig. 1.11.

The 800 mm isohyet exhibited the lowest displacement which explains the resettlement of the population at this position when the climatic conditions in the north became unfavourable. It is also at these localities that we have observed on aerial photographs a near-saturation of the agricultural space since the 1950s. The pressure on the land here reached its maximum and was responsible for a shortening of the fallow periods, the disappearance of the tree cover in favour of a grassy cover and the extension of bare areas. In the pastoral areas of Hombori (Mali) an inverse evolution took place: The wooded steppe increased from 28 to 43%. A regrowth of the forest in northern Cameroon was observed by Hurault (pers. comm.) who ascribed this to the overgrazing which favours those young tree-shoots not consumed. Another explanation could be a transformation from extensive pastoralism to a less nomadic pastoralism around the centres of settlement.

The consequences of the period of dramatic pluviometric deficits of 1968–1985, combined with the southward displacement of the isohyets, with the pastoral pressure resulting from three to four times larger herds and a human population which had doubled in number and was in search of new land for cultivation are still difficult to estimate. Have they led to deterioration which could be reversible on the human time scale or are they irreversible? The question we should now concentrate upon is the real measure and the objective extent of the degradation.

A study of the 1984–1985 drought in Kenya furnishes a revealing clarification of the complexity of its causes, mechanisms, and consequences.

1.2.6
Study of the Drought of 1984–1985 in Kenya and Its Consequences

In 1984–1985 Kenya experienced a severe drought as a consequence of a pluviometric deficit: The vegetal production and especially that of fodder was so deficient that it led to insufficient production of food crops up to the threshold of famine. It was preceded by a year of normal precipitation over most of the country with, however, already a certain deficit in the regions of Kitui, Marsabit and Meru where the annual long rainy season was non-existent. For example, Kitui received only 25% of its normal rainfall. The herdsmen were thus ill prepared to overcome another drought in 1984. During this year, the main rainy season was poor over most of the country except for the west around Lake Victoria and a long strip along the Indian Ocean coast. The maize harvest was reduced by 34%, that of wheat by 39%, of beans by 79%, of sorghum and millet by 50% and the production of milk by 25%.

The herdsmen, already forced to a marginal existence by rigorous subdivision and demarcation of the land, had to abandon their traditional strategy of escaping a drought: the possibility of subdividing the herds and sending them to different areas had been lost. Points of permanent water, the pastures around which represented the last refuge in the face of a drought, became centres of concentration for cattle but had insufficient forage. Rain-fed agriculture no longer represented a possible additional source of income.

During the drought the pastoralists lost 80% of their cattle, sheep and goats, when help in the form of supplementary fodder for 2–3 months would have enabled them to be saved. The poorest families were the ones affected most. In regions farther away from the capital inefficient organization of commerce prevented the herdsmen from selling their livestock – sheep, goats and skins – to obtain money for purchasing the maize meal indispensable for their food.

The government responded to the drought by purchasing 393 000 t of maize from Thailand at a cost of US $65 million. The maize arrived in Mombasa in September 1984 when the national reserves had fallen to below 1 month's supply, and was immediately trucked inland. At the same time slaughtering at the two national abattoirs rose from 7000 to 23 000 head.

During the ensuing months, the sale of the food organized by the state went well, but it was difficult to secure a food supply for the poor and for herdsmen lacking cash.

Among the climatic constraints suffered by drylands the recurrence of droughts is the most traumatic one. In East Africa statistics have revealed a repetition factor for droughts of 0.25, i.e. one in every 4 years. An early warning system should be part of the necessities for development. Its components were described by Finkel (1992):

- Permanent access to drought indicators: meteorological, agrometereological, moisture equilibrium of the soil, areas affected (delineated e.g. by scouts), populations concerned, indices of malnutrition and famine, price increases for foodstuffs, indicators for the end of the drought.
- Rapid collection and analysis of data.
- System of rapid transfer of data to decision makers.

After the diagnostic phase comes the plan of action. In Kenya in 1984–1985 the information was obtained rapidly but its transfer to the decision-making authorities took rather a long time. It was only when they received the information that the sale of cereals had risen drastically to above the normal 90 000 t/month that the danger was recognized and decisions rapidly taken.

Finkel detailed that the plan of action should incorporate:

- Identification of the institution to be addressed in case of a food shortage.
- Continuing the necessary food aid to the most needy.
- Continuing the commercial exchange between the sale of livestock and the purchase of grain.
- Determination of the threshold of available foodstuffs below which aid would become necessary.
- Identification of the means by which the population and in particular the most needy may furnish themselves with daily needs.
- Identification of the critical levels beyond which the normal ways of distribution of foodstuffs are no longer sufficient.

Finkel recommended the establishment of a manual for officials confronted with a drought, a manual that should:

- Describe all stages of the strategy to be adopted by the administration.
- Include a list of equipment (e.g. transport, pumps) and supplies (e.g. grain, medicines, etc.) required in urgent cases.

- Contain precise instructions for human and animal food rations.
- Foresee modes of transfer of funds and personnel.
- Outline the different responses in relation to the severity of the drought.
- Foresee strategies for the time after the drought and the return to normal.

The example of Kenya may serve as an illustration for the complexity of the problems caused by a drought at a point in time and space, of the complexity of its analysis and of the necessary responses.

1.2.7
What Lessons Do We Learn from the Last Drought in the Sahel?

Poncet (1974) estimated that at the start of the drought in the Dallols of Niger the population north of 13° N was affected most and the pastures deteriorated as natural reseeding did not take place due to the lack of rains. The years of drought then saw the cattle disappear, depriving the herdsmen of their direct supply of milk, meat and other income. South of this parallel the pastures remained, but rain-fed agriculture, which requires a regular distribution of precipitation between germination and maturation, was highly deficient. The farmers experienced a considerable drop in the harvest of subsidence and of cash crops. The drought led especially to a dearth of foodstuffs (Photo 1.3).

During this drought, harvests shrank by 40–50% and the herds lost more than one-third and even one-half of their numbers. The Sahel would not have survived, were it not for help from the outside. We are then faced with the question: is the deficit in foodstuffs a corollary of the drought and its consequences or is it rather the expression of another more severe evil, the improper management of resources, an erroneous concept of development? It appears that the economic disequilibrium is but the visible part of a fourfold ecological disequilibrium.

The first, resulting from the side of livestock breeding was aptly described by the World Bank in 1985 immediately after the drought crisis:

> The traditional social structures in the pastoral zones underwent a profound evolution since the start of the colonial area. Since then they ceased to maintain the privileged relationship which they enjoyed with the farmers which subjected themselves to their authority and the pastoral societies lost a large part of their authority and at the same time of the control which they had exercised over the access to land. The period of relatively abundant rains between 1950–1965 had encouraged the farmers to settle also in the marginally arable lands of the Sahel zone and it had been barely possible to ban them from doing so even there, where like in Niger, a zone exclusively for pastoral activities had been officially instituted. The herdsmen were thus forced to return for larger periods every year to more marginal pastoral lands north of the Sahel.
>
> From the 1920s onwards, veterinarians performed vaccinations which led to the herds growing in size and to the menace of overgrazing in these pastoral terrains. Breeding projects of the last three decades placed emphasis on the health of the animals and on the supply of water for breeding purposes, mostly in the form of open wells of large diameter or on tubular wells and deep bore-holes using mechanical means of pumping. These new wells were installed in public places and therefore accessible to everybody which reduced the efficiency of the traditional rules regulating access to grazing on which the owners of the water wells had a right of use. Even today the public wells are still constructed without consultation of the local population and frequently installed in areas in which one erroneously assumes a rather abundant supply of grazing. The results here were an increase in size of the herds, an expansion of the cultivated ground at the expense of pastures and a retreat of the traditional management practices of grazing (World Bank 1985).

Photo 1.3. Very young ecological refugee from Chad in the Sudan. In March 1985, just after the drought of 1983–1984, she sorts wild berries to feed her family. Many natives of the Chad had by then settled down in the Bindisi region, south of Zelingei in the Sudan. The berries are usually not consumed as they are slightly toxic, but are eaten during periods of famine but they have to be cooked for a long time, requiring a large amount of wood (Photo by M. Mainguet)

The *second* cause of the disequilibrium is found in the agricultural activities based traditionally on long fallow periods for the fields, enabling the soil to recondition itself. At present, the population growth restricts the available arable land and the fallow periods are reduced without a change to a better productivity of the cultures and without any real remedy for the impoverishment of the soils.

The disequilibrium caused by the disappearance of the trees as a consequence of their exploitation as fodder and as a source of energy is also the result of the population explosion and of the greater consumption of wood, leading to pronounced deforestation, the consequences of which cannot be measured yet. This question shall be taken up again in the study of degradation of the vegetation cover by human activities (Photo 1.4).

And finally, with the increase in urbanization and the growth of the urban populations, the demand for foodstuffs has risen, but the unattractively low prices for cereals do not entice the farmers of the drylands to increase their production, and furthermore fertilizer and pesticides are too expensive. In the farming world there is thus a trend to self-retreat, the farmers feeling themselves ecxluded from the processes of development.

Photo 1.4. The tree steppe after the drought in Mali. In the semi-arid zone of Mali near Bengo on the longitudinal red dunes system the overgrazed and deforested steppe is moribund after the severe drought of 1983–1984, as indicated by the fallen trees and the dry branches. Shortly after this photo was taken, the first rains of June 1985 revitalized the trees and let the grass cover come back within only 2 weeks, confirming the strong resilience of this vegetation cover (Photo M. Mainguet)

In conclusion, we should keep in mind that after every drought and, assuming a pluviometrically less deficient regime, there will be several questions:

- Does the drought represent a mere climatic fluctuation or is it a manifestation of a progressive aridification? We have seen that in the past there had been other, comparable droughts and that they existed already during biblical times as illus-trated by the allegory of the seven lean and the seven fat cows. These repeated droughts should cause the states to prepare themselves for survival. Such a task, however, is difficult as the equilibrium between production and demand, aggravated already by demographic growth, is interrupted and not re-established: food aid continues until the present time.
- To what degree can the environment regenerate itself? Has it crossed the point of no return?
- What will be the "memory effect" of this drought for the soil?
- Is there, after each drought crisis or series thereof, a return of the vegetation cover to the previous state, even though deterioration of the cover is always more rapid than its rehabilitation (Bernus 1989)?
- Will the drought exacerbate the poverty of the rural population or will it teach the populations to live together more efficiently by putting cultural habits to the test?

1.3
The Thin Vulnerable Soils of Low Agricultural Potential

Soil is the result of interactions between the source rock, topography, meteorological (rain, temperature, wind) and biological factors. This transformation of the surface layer of a rock takes a long time in drylands. Many soils have been inherited from humid phases preceding the dry episodes and are thus relict soils, vulnerable because easily exhausted and of low agricultural potential. The aim of this study of soils is not to present a pedological analysis but, after the definition of aridity and droughts in the preceding sections, to illustrate how these two climatic parameters control their formation.

In the drylands, where available water is missing and where pedogenesis progresses slowly, the soils are thin, sandy, pebbly and salt-rich. Organic matter and humus are rare which explains their specific physical and chemical properties.

1.3.1
The Dryland Soils of North-Equatorial Africa

One should not discuss soils of the drylands without mentioning the *FAO/UNESCO Soil map of the World* (1971–1979) on a scale of 1 : 5 000 000 in which the soils are subdivided into 28 classes of which 9 occur in drylands (Table 1.6).

The climatic zones control the distribution of soils of the face of the Earth. For the drylands of Africa south of the Sahara, Casenave and Valentin (1989) proposed the following succession from drier to more humid zones:

- *Little-evolved sub-desert soils:*
  - Raw mineral soils: little differentiated A-horizon overlies C-horizon (altered rock) or R (unaltered rock).
  - Lithosols, unaltered hard rocks
  - Regosols, penetrable to roots
- *Soils developed over a sand cover*: brown subarid and red-brown subarid soils which are characterized by the penetration of well-humidified organic matter in the profile over a depth of 30–40 cm. These are the soils developed on aeolian sands.

Table 1.6. Families of soils in Africa (FAO/Unesco 1978, Soils map of the World, 1 : 5 000 000)

	10^6 ha	Total (%)
Desert soils	620	21.8
Sandy soils	577	20.3
Salty soils	67	2.3
Acid soils of tropical lowlands	509	17.9
Tropical mountain soils	39	1.4
Dark clayey soils	99	3.5
Tropical ferruginous soils	194	6.8
Mediterranean soils	87	3.1
Poorly drained soils	276	9.7
Thin soils	376	13.2
Total	*2844*	*100.0*

- *Ferruginous tropical soils,* leached to varying degrees, depending on precipitation. Casenave and Valentin distinguished three types:
  - Slightly leached soils over aeolian sands
  - Leached or depleted soils with a contrast in colour and texture between the sandy grey or bleached horizons A and E (formerly AZ) and accumulations of iron and clay of red-ochre colour in their B-horizons and frequently with some ferruginous concretions
  - Reworked tropical ferruginous soils with a high content of ferruginous pebbles

To this succession have to be added:

- *Hydromorphous soils* which are not restricted to certain zones; their characteristics and evolution are controlled by an excess of water; they may be recognized by their rusty, bluish or greenish colours (gley).
- *Salsodic soils* are defined, still after Casenave and Valentin, by the presence of soluble salts (chlorides, sulphates and carbonates of sodium and magnesium) and correspond to the solonez of the Russian classification.

The soils described above correspond to the zonal arrangement of the climates in tropical Africa north of the equator. This description does not, however, cover all the soils of the temperate latitude drylands in Asia. Because of this we shall return to the Turan but, wherever possible, we shall make reference to other dryland soils, notably those of Africa.

1.3.2
The Soils of Turan

In the hyperarid and subhumid dry parts of Turkestan there are a number of soils which are also known from other drylands but which have a nomenclature specific to this desert. Most of the soils of the Turan – within the natural oases – have been formed under a paleoclimatic regime different from the present one. The organic matter they contain has frequently been inherited from earlier more humid periods. Under recent conditions this organic matter finds itself in a rather precarious equilibrium. Another negative aspect of the steppe and desert soils of the Turan is their high salinity which leads to special vegetational types: halophilic, hyperhalophilic (supported by salinities above 300 g/l), euryhaline (indifferent to moderate or low salinities), and alkalinophilic. The following compilation of the main soil types of the Turan uses the Russian nomenclature when there is no English equivalent.

Chernozem (Black Soils). Soil of continental semi-arid climate with precipitation of 400–600 mm/year. The most typical ones are found in the Ukraine, Russia, Manitoba and Dakota. This type of soil is missing in the Sahara and its periphery. In northern Turan where precipitation is actually barely above 200 mm/year there are the palaeochernozems formed during the Lavlyakian (12 000–6000 B.P.), an epoch that was warmer by 2–3 °C and more humid with a mean precipitation of 250 mm/year.

The chernozem is an isohumic soil with a deep incorporation of the humic matter, a little differentiated profile, rich in organic matter with a high proportion of humic acids, a saturated adsorption complex, a crumbly structure and accumulations of powdery limestone in the uppermost 125 cm. At depth indurated gley horizons or limestone crusts may be developed. The "black soils" of the arid regions of the former USSR contain 0.8–2% (in weight) organic carbonate (Kononova 1979), and the total of organic matter ranges from 4 kg C/m^2 in the chernozems to 10 kg C/m^2 in the dry grasslands. These values are three to four times lower than the values in soils of grasslands from temperate climates. In the less watered and thus less leached areas the humiferous horizon is thinner and lower in organic matter, and the carbonated horizon lies closer to surface.

These black soils are the most fertile soils on earth, excellent for the cultivation of maize, but mediocre for forests as mycorrhizae (fungi forming nodules rich in nitrogen on the roots of plants) are totally absent from them. As deep soils with a porosity of 70% they possess an excellent capacity for aeration and for the retention of moisture. They are rich in exchangeable cations, assuring a good mineral potential for plants. The supply of nitrogen to the plants is facilitated by a rapid mineralization of the organic matter and by active nitrification (Duchaufour 1965).

Chestnut Soils (the Ustolls of the American Classification). They are found farther south in the Turan in drier steppe ecosystems where precipitation is not above 240–400 mm/year, and in the dry climates of North Africa. The surface horizon is thinner and poorer in organic matter than in the chernozems and frequently overlies a brown Ca horizon with a prismatic structure and an accumulation of limestone. The soil profile is frequently carbonated to a variable thickness. When the environment becomes drier, the thickness of the humic layer decreases to a dozen centimetres, whereas the carbonated layer becomes thicker and richer in gypsum. With a higher iron content the soils become brown, light-brown and red-brown as in the Mediterranean steppes. Bearing flax and mugwort in North Africa, they are frequently covered by a limestone crust. These soils possess a low fertility.

Grey Desert Soils or Slerozems. The most widely distributed soils in the Turan; they may be subdivided into the series of isohumic sub-desert soils (desert margins) in which the surface horizon contains only 1–3% of organic matter (Duchaufour 1991) as the short rains only permit a low and scattered vegetation. Slightly decarbonated on the surface, the sierozems possess a crumbly, lamellar or sometimes compact structure on the surface which becomes polyhedral at depth.

Saline Soils. "… they cover vast semi-arid and arid areas in the African countries. In the less saline areas traditional techniques of irrigation by furrows partly leach out the salts. The bottom and the sides of the furrows are leached (the salts accumulating by evaporation on the intervening dams) it is possible to cultivate salt-resistant plants on the slopes. … The management of saline soils is all-the-more difficult as they are found in closed basins where drainage poses a problem. Their utilization thus depends on the possibility of irrigating them copiously in order to leach them and to drain off on surface the excess water to eliminate the salts" (FAO 1986). The saline soils of the drylands are the *solonchaks, solonets* and *takyrs* of the Soviet soil classification (Photo 1.5).

1.3 · The Thin Vulnerable Soils of Low Agricultural Potential

Photo 1.5. Salinization of an Algerian sebkha. Algeria is rich in saline depressions of the sebkha type in its steppic zone. The one shown here, at Merouane ben Djeloud between El Oued and Touggourt, dried out seasonally in March 1984 and its sediments swollen by saline efflorescences and sheets of gypsum became concave under the influence of the crystallizing salts (Photo M. Mainguet)

Solonchaks or White Saline Soils. Solonchaks is the Russian term for salt swamps). Solonchaks are derived from the floors of ancient lagoons or *sor*, the equivalents of the Saharan sabkhas, where the soils are enriched in salts by infiltration of underground waters. When the piezometric level of the aquifer is low, the soluble salts will rise only a little or not at all. When waterlogging (filling of a soil with water to the point of saturation) takes place close to the surface, evaporation of the water rising under the influence of capillary action will lead to the deposition of salts on or below the surface; a secondary solonchak then forms the salt efflorescences or crusts assume the shape of cauliflowers. The surface of a solonchak is characterized by white saline efflorescences. The sodic soil with a calcic-adsorbing complex of crumbly structure possesses a little differentiated soil profile, flocculated clays and a humiferous A-horizon referred to as *mull*. Its pH does not exceed 8–8.5 (Duchaufour 1965). Research has not so far been able to find ways of ameliorating the mediocre qualities of these soils.

Solonets. These sodic leached soils with a well-developed profile belong to the group of saline to alkaline soils. Below their greyish surface with a muddy texture the B-horizon (or natric horizon of the American nomenclature) possesses a compact structure of small columns covered by sodic humates and amorphous silicate gels. The pH is around 7 on the surface and increases to 9–10 in the B-horizon.

Seasonal movements of the salts transform the solonchaks and solonets. Lateral or vertical drainage carries away the superficial salts, leaving on the surface a mud consisting of colloidal clays with little humus which on drying becomes hard and impermeable. The reaction between sodium and the dissolved carbonates leads to a highly alkaline pH. The mineral and organic particles held at depth form a compact impermeable layer rich in iron, silica and illuvial humus. Under the effect of the changing seasons this indurated layer becomes increasingly thick and deep, because in each dry season a similar layer forms on the first layer, eventually producing a puff-pastry effect as it were, breaking up in polygons with a columnar structure. The alkaline solutions percolate between the columns, leading to the collapse of the structure.

Takyr. This is a soil of the drylands, little evolved and frequently salty, derived from the sedimentation of fine particles (clay, mud, fine sand) carried away by runoff or deposited at the end of intermittent floods on the final fans of water courses. During drying out of temporary water bodies, the takyr, a type of raw xeric soil forms slabs which during the dry season break up into polygonal flakes. Being hard, uniform, and compact with less than 18% porosity it is not easily infiltrated by water and until recently made up the tracks favoured by the caravans. It sometimes contains a layer of gypsum at depth. The takyrs are slightly richer in humus (1%) than other desert soils. When they are irrigated in oases, their humus content may reach a value of 2% down to a depth of 80 cm, whereas in other soils the humus layer is not thicker than 25–40 cm. Receiving water during the annual floods or from intermittent sources on the downstream side of alluvial cones, they are the preferential sites for development, despite their only modest potential, when the more fertile lands are already being utilized.

Loess. Loess is a soft, porous rock consisting of particles of quartz, mica, feldspar and 35–40% calcium carbonate with a modal size below 30 µm. On alteration these particles liberate exchangeable elements (Na, K etc.) explaining their fertility for agriculture.

Although a locss is a rock – a formation of the margins of hot or cold deserts – it was included in this chapter as it behaves like a soil. On older loess, during more humid paleoclimatic phases, chernozems formed, which in the Turan degraded to chestnut or red-brown soils. In the Turan, along the northern foothills of the mountains to the south (Fig. 1.12), there is a loess band some 100 km wide and up to 200 m thick. Aeolian in origin it owes its formation to the accumulation over thousands of years of dust winnowed by winds from the alluvial deposits of Aralian and Siberian water courses, the sandy blankets of the deserts and the calcareous plateaus from Ust-Urt in the west to Lake Baikal in the east. The loess, which formed in Turkestan in great thickness during the Quaternary, is periglacial in origin. Brought by (catabatic) winds originating over the ice fields to the north and northwest, they were deposited in the region to the southeast where air pressure was lower. The periglacial zones surrounding the glaciers covering the Pamir massif probably also represented a source of loess.

The high water retention capacity of the loess and its position on the foothills allow them to benefit from the diffuse runoff from the neighbouring slopes and allow their conversion into good agricultural land. When deposited in less favourable sites (depressions) they develop gypsum crusts or evolve into saline soils.

1.3 · The Thin Vulnerable Soils of Low Agricultural Potential

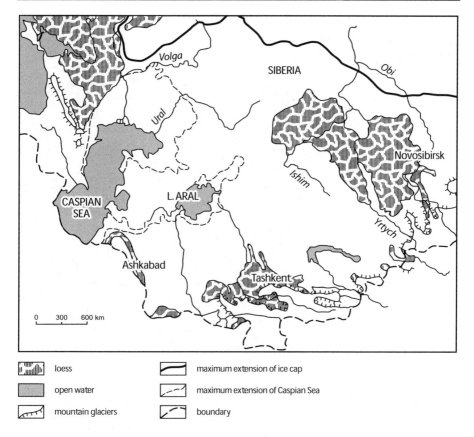

Fig. 1.12. Loess deposits of glacial periods with the position of maximum extent of the polar ice cap in the Caspian Aral Basin (after Ganelin *et al.* 1961)

The most extensive loess areas with a mean thickness of 100–200 m and even up to 400 m (Tamura 1992) are found in the Chinese deserts, in the semi-arid centre of the Yellow River Basin, which at present receives 300–450 mm/year precipitation (Photo 1.6). The loess plateau covers some 580000 km^2 and some 60 million people live on it in utter poverty. The material of this loess had been winnowed from the deserts to the southwest over more than 1 million years. The plateau is known as the cradle of Chinese culture and agriculture. A desertification caused by overexploitation has trans-formed the areas of the plateau into a succession of bare hills separated by deep gullies.

The porosity of the loess may reach 50%. Fertile when there is sufficient water, loess also represents the most vulnerable rock (Barrow 1991) as its granulometry is optimal for aeolian transportation. The losses from erosion by wind and water may reach their highest values here. Loess is sensitive to compaction and to waterlogging when cultivated and presents a higher risk of gully formation and erosive tunnelling on steep slopes.

Photo 1.6. Secondary salinization in China. In the background hills of loess dominate the irrigated plains of the Huangho (Yellow River) in the Lanzhou region. Irrigation here is responsible for a strong secondary salinization and a desertification rendering the soils sterile (Photo M. Mainguet)

Stony Pavement Soils and Regs. These infertile, repulsive soils cover some 40% of the Turan and are also well represented in the Sahara. The term stony soils is preferable for soils made up of angular fragments (Ust-Urt) (Photo 1.7), the *hamada* on the limestone plateau, the *tassili* on sandstone plateaus where the substrate is mechanically fragmented, as well as for the soils of the *reg*, also made up of pebbles (some parts of the Karakum and the Kyzylkum), and the *serir*, where the pavement consists of finer-grained material. The *regs* are frequently underlain by vesicular structured formations which further lowers their water retention capacity and thereby their potential fertility.

The pavements represent coarse material concentrated by deflation, by leaching and by vertical migration through the mass of debris during alternating moistening and desiccation. Their genesis is related to fracturing under the influence of ice and of salts. Aeolian deposition of clayey salty aerosols in large quantities is one of the prime mechanisms favouring expansion and contraction leading to polygonal soils, to the fragmentation of the rocks and to the formation of desert pavements.

1.3 · The Thin Vulnerable Soils of Low Agricultural Potential

Photo 1.7. Hamada near Atakor in the Algerian Hoggar in 1984. Deflation carried away the sandy matrix of the soil leaving only a pavement of the hamada-type made up of slabs of a volcanic rock covered by desert varnish (Photo M. Mainguet)

Sandy Soils. These are the sandy covers of sand sheets, dune fields and sand seas. Nearly 25% of the Turan is covered by sand, which was winnowed from vast accumulations of alluvial material reworked by the wind.

In the Sahara, sand covers only about 20% of the surface. In contrast, the Sahel south of the Sahara possesses a sandy cover, the thickness of which may reach 60 m north of Zinder in Niger. This seemingly paradoxical situation is easily explained if we accept that the Sahara and the Sahel are inseparable in their external dynamics. The Sahel receives its water courses and fluvial material from the small mountain massifs like the Air and the Adrar des Ifoghas bordering the Sahara. Furthermore, the Sahara and Sahel are part of the same global system of aeolian action (GAAS) dominated by the harmattan winds, which possessed a greater efficiency during drier phases and transported large quantities of sandy aeolian material from the Sahara towards the Sahel, and dust even further.

Fixed and active dunes can retain great amounts of water because of their high porosity (40%), a fact that has attracted the attention of agronomists in e.g. the Taklamakan, the *Grand Erg Occidental* and southern Egypt for the development of irrigated agriculture. "These soils possess a little differentiated profile. Their low natural productivity is caused by a general lack of nutrients, ... their mediocre cation exchange capacity and their lack of oligo-elements" (FAO 1986).

Alluvial Soils. These are the soils of mainly fluvial alluvial deposits (terraces and alluvial cones) and of lacustrine ones, both of which are inherited from more humid paleoclimatic phases. Where present, the coarse components are rounded. In drylands these soils only rarely possess a differentiated profile. Their fertility depends on their texture, i.e. on the relative abundance of the fine-grained loamy-sandy matrix. According to the FAO (1986), Africa possesses 101.4 million ha of alluvial soils, especially in the valleys of the Chari, Niger, Senegal and the Nile. These alluvial soils are the main supports of agriculture in the drylands of the Middle East.

Hydromorphous Soils. These soils are frequently inundated and always waterlogged. On drying they are transformed into takyrs or solonchaks, depending on the composition of the subground, the nature of the underground waters and the percolating waters as observed e.g. in the Turan.

Strongly Cracked Argillaceous Soils. In Africa

> "almost half of these soils occur in the Sudan (42.2 million ha). They are abundant in the Chad (7.7 million ha), Ethiopia (11.2 million ha), Morocco (1 million ha), Somalia (1.8 million ha) and Burkina Faso (1.3 million ha). These heavy soils are difficult to work as on moistening, the clays which they contain start to expand. On drying they shrink and crack leading during the dry season to hard soils which become plastic after the rains making the use of ploughing utensils difficult. On the clayey soils millet is grown in the Chad, in Mauretania and in Burkina Faso. In the intensive irrigation systems in Gezira in the Sudan 1.8 t of cotton, 2.6 t of sorghum or 1.4 t of wheat are produced per hectare. In Morocco wheat, barley, clover and cotton are produced on these soils" (FAO 1986).

1.3.3
The Specific Characteristics of Dryland Soils

The soils of the drylands possess specific geomorphological features:

- *Polygonal soils* and *gilgui* (an Aboriginal term) are the most widely distributed soils in Australia, but occur also on the other continents. Consisting of mounds 1–3 m in height and separated from each other by flat ground sometimes outlined by pebbles, they depend on the occurrence of swelling muds and clays. Their origin is still under discussion.
- Warm deserts are the sites of the so-called *duricrusts*. These duricrusts overlie residual reliefs or make up the higher parts of landscapes. There are three different types, viz. calcareous, gypsiferous or siliceous ones:
  - Calcareous crusts or *calcretes*, according to Yaalon (1981), cover 13% of the emerged land. According to Goudie (1983), they are characteristic of drylands receiving 400–600 m in annual precipitation and are found in Mediterranean Europe, the Middle East, semi-arid India, the dry states of the southern USA and the south and central-west of Australia and its Northern Territory (Dixon 1994). The calcretes are rather diverse in habit: laminated crusts, incrustations, petrified roots, pulverous limestones, efflorescences and nodules (Durand 1949). The carbonates most frequently are of aeolian origin, having been brought in as dust.

1.3 · The Thin Vulnerable Soils of Low Agricultural Potential

Photo 1.8. Reg and desert varnish in the Wadi Feija (Morocco). Rounded block of sandstone from a reg in the Wadi Feija (Morocco), a tributary of the Draa; the block bears a thin dark-brown desert coating where previously above ground. The light-coloured lower part had been embedded in the sandy matrix of the reg (Photo by M. Mainguet)

- Gypsiferous crusts or *gypsicretes* are found in areas receiving 200–250 mm of annual precipitation (Watson 1983), and occur in all subtropical areas. They are very frequent in the Kyzylkum desert. Watson described three different modes of origin: evaporites, layered crusts resulting from the deeper incorporation of aeolian gypsum (Israel, Tunisia, Namib) and surface crusts resulting from evaporation.
- Siliceous crusts or *silcretes*, consisting essentially of silica (up to 99%), from either in situ by dissolution of the enclosing material and reprecipitation or in an *allochthonous* way from aeolian supply or from runoff and infiltration or by transport in underground waters.
- Desert varnish is a shiny or dull dark-brown coating on the surface of rocks, made up of ferrous and manganiferous materials, the origin of which is still under discussion. These coatings possess different characteristic features depending on the site of occurrence. However, an aeolian supply of the constituents of desert varnish is most frequently advocated (Photo 1.8).

In *conclusion* let us recall that the stability of a soil depends on a high content of organic matter and humus. When the humus disappears, either by too frequent and too deep ploughing or by overgrazing, the soil particles lose their cohesion and get carried away by the wind, forming dust devils or dust storms. In Algeria soil erosion progressed in proportion to the number of multi-disc ploughs.

In the soils of the Turan worked without green fertilization (mostly stalks, twigs and leaves of cotton) the organic layer disappears more rapidly especially during deep ploughing, the curse of all modern agriculture; these problems also occur in Europe.

The biomass of living plants in the former USSR (Reiners 1973) is some 1400 g/m^2 in the grass steppe of the temperate zones and some 350 g/m^2 in the solonets of the arid steppes. Thousands of years of extensive overgrazing in the Turan have depleted the soils in humus and where natural rejuvenation is critical, these soils have become fragile. The premature utilization for grazing by transhumance during the short rainy season endangers the stability of the soils, their replenishment with humus and flowering before re-seeding (Uncod 1977).

Whereas with proper irrigation loess and alluvials are fertile, the other characteristic soils of desert regions are low in humus, require fertilization and support only specific crops. Another of their difficulties from an agricultural point of view lies in the low content of mineral salts indispensable for plants, and in particular of nitrogen and to a secondary extent of phosphorus. In the drylands of central Asia and China the farmers of ancient times improved their soils by mixing in loam, mud from rivers and canals and by sand for the takyrs and in all cases by organic residues. They succeeded in establishing an equilibrium, albeit on a moderate scale.

Another difficulty lies in the intensity of the evaporation. With the groundwater, continuously drawn to the surface and carried along dissolved substances are ions of sulphate, carbonate and chloride, of sodium, calcium and potassium, as well as sometimes rare elements like lithium and minerals like borates and nitrates, forming mineral resources exploited because of economic importance. These substances, gypsum, rock salt, sulphates and carbonates of sodium and magnesium, are deposited in the upper layers of the soils, where they modify its structure (calcium carbonates forming "caliches" or calcareous crusts) and prove to be devastating for the vegetation. This phenomenon of salinization, studied in the third part of this book, is the main threat to poorly drained dryland soils and one of the greatest obstacles which the United Nations try to resolve through their organizations: Food and Agriculture Organisation (FAO), United Nations Environment Programme (UNEP) and United Nations Development Programme (UNDP).

1.4
Vegetation Covers: Economic Potentials and Droughts

Coming from the hyperarid desert – like the Erdis in Chad – the traveller in 1968 had the delight to observe the first tufts of open vegetation cover in areas corresponding to the 150 mm isohyet in the dry Wadi N'Gaoulé to the north of Ennedi, the rainy season being the warm season of the Sahara. It is here that one enters the semi-arid domain, the Sahel. In contrast, north of the Sahara and the Negev, the boundary between the arid and semi-arid regions is marked by the mean position of the 100 mm isohyet, the rains falling in the cooler season. According to Bernus (1994), who confirms Chevalier (1938), the Saharan oases and the margins of the desert are the native areas of millet, sorghum and of certain spontaneous varieties of rice.

In this chapter only those features of the vegetation cover which are of importance for human activities will be dealt with and not their botanical characteristics. Let us recall that the potential of drylands up to the 350–400 mm isohyets – depending on

1.4 · Vegetation Covers: Economic Potentials and Droughts

the nature of the soils – rests in pastoral activities. "Outside the irrigable areas one may just migrate and not cultivate" (Collective 1974, TCTF Technical Centre of Tropical Forestry). It is only with increasing precipitation that these areas become suitable for agriculture. The main characteristics of the heterogeneity and adaptability to the conditions of aridity and drought will be recalled before we briefly return to some key features of the vegetation along the dryland diagonal from the Sahara to China. Because of the importance of land utilization, the analysis will touch upon the pastoral activities of the semi-arid steppes, then the value of its foraging potential and its carrying capacity will be covered and eventually the behaviour of these covers in the face of a drought.

1.4.1
Heterogeneity of the Vegetaion Cover in Drylands

In contrast to the general belief, deserts are by no means barren. According to Abrahams and Parsons (1994), the most common cover of grasses and bushes occupies 59% of their surface. Whatever their latitude, tropical or temperate, the drylands possess open vegetation cover.[4] The causes of this discontinuous cover are:

- *Insufficient supply of water in the soil* because of the low total precipitation, its seasonal distribution and its irregularity, which go hand in hand with an unequal infiltration of water into the soil. This infiltration is itself controlled by the structure of the soil surface and the distribution of the precipitation over the slopes and the lowlands (Cornet 1992).
- *The low temperatures* which adversely affect the raising of water by the plants from the soil, leading to a combination of drought and cold.

The irregular distribution of plants changes the physical and chemical properties of the soils and exacerbates their spatial heterogeneity. The soils are richer under plants than in the barren ground. Francis (1994) mentions Romney *et al.* (1977) and Rostago *et al.* (1991) who observed higher concentrations of exchangeable cations (sodium and potassium), organic carbon and nitrogen under plants. The decomposition of organic matter is favoured around the plants by shading, higher humidity and biological activity.

The variability of the dynamics of the water on and in the soil, the diversity of which can be seen from spatial differences and from differences in the availability of water to the plants, leads to a heterogeneity of the vegetation cover and of the functioning of the various components of a phytocoenosis.[5]

The mosaic aspect of the vegetation of drylands depends on these data, which result in spots of different surfaces, with variable stages of evolution and in an unstable equilibrium (Cornet 1992).

---

[4] *Open formation*: vegetation cover, in which the canopy of trees and bushes and the foliage of the herbs do not meet, leaving areas of bare soil between them.
[5] *Phytocoenosis*: "Types of spatial organization of a vegetal community", resulting from a group of variables leading to a number of characteristic responses which permit the identification of the structures" (Poissonet *et al.* 1992).

1.4.2
Adaptability of Plants to Aridity and Drought

Evenari *et al.* (1975, in Floret and Pontanier 1982) subdivided plants into two groups depending on their adaptation to aridity and drought:

- *Arido-passive species*, the photosynthetic tissues of which are inactive during the dry season.
  - Annuals with their vegetative cycle restricted to the rainy season, the seeds being resistant to desiccation during the dry season.
  - Perennials which lose their leaves during the dry season and rapidly remobilize their reserves at the onset of the next rainy season.
- *Arido-active species* which have at their disposal reserves of water allowing them to function at a reduced rate during the dry season and to preserve their entire biomass throughout the year. In particular, they encompass persistent species characterized by a small number of stomata and small leaves, generally in the form of spines; they may be trees, bushes or annuals.

With regard to the germination and exploitation of water the flora of the drylands is well adapted to drought:

- *Annuals* produce seeds in large quantities with a long period of dormancy as they frequently only fructify just before the end of the rainy season.
- *Perennials* are adapted through their root systems which differ between the various species and depend on the thickness of the mobile layer. They possess two types of roots: superficial ones utilizing mostly the surface waters and deeper ones exploiting the aquifer. One adaptation of the plants to drought is the length of this root system so that a certain area may harbour only a limited number of individuals. "The vegetation cover itself, with regularly distributed tufts is highly reminiscent of extensive grazing" (Gillet 1974). The reproduction of the perennial species takes place via seeds or shoots from stumps or in both ways.

1.4.3
The Vegetation Along the Dryland Diagonal Sahara-China

In hyperarid and arid regions with less than 150 mm in annual precipitation, the vegetation is referred to as *contracted* (Photo 1.9) as it grows preferably in wadis with inferoflux (underground runoff) and in depressions with shallow water levels. This vegetation, characterized by *Chamaephytes*, covers some 10% of the soil, reaching up to 50% thanks to the annual precipitation during years of favourable rainfall. Wehn not limited to wadis or to the sites of rain-wash concentration, the vegetation is referred to as *diffuse*[6] (Photo 1.10).

Huetz de Lemps (1970) wrote: "Despite its immense area, the Sahara proper has a poor flora consisting of only 1200 species. The number of species per unit of

---
[6] Diffuse, open.

Photo 1.9. Steppe vegetation on a reg in semi-arid Morocco. This reg to the north of Tamgroute on the plain of the Draa (June 1993) is cut by small barely incised sanded-up wadis harbouring the only trails of vegetation, which are contracted here and arranged in tufts, forming aureoles each of which is creating a microclimate economizing in water drawn from the inferoflux. The northern escarpment of Djebel Bani is in the background (Photo M. Mainguet)

10 000 km² surface is about 150, compared to 3000–4000 in humid tropical regions and 2800 in the Maritime Alps. In contrast to this the number of genera is rather high, as a genus frequently is represented by one species only."

Trees are rare in the Sahara except in its mountain massifs – and man is certainly responsible for this scarcity. As one moves from the fringes of the Sahara to the Sahel and the summer rains surpass 150 mm/year, trees make their appearance in the plains. They are relict trees of the Mediterranean highlands but mostly Sudanian. These trees are frequently the suppliers of fodder for the Sahelian margins, but also serve a variety of other purposes. As the drylands have been occupied by man for a long time already, the vegetation cover observable now is by no means a "natural" one. But how should we know what is a natural cover in an ecosystem in which climatic fluctuations are the rule and occupation by man that old?

The Karakum, a temperate latitude sand sea, belongs to the Irano-Turan botanical province with a flora of more than 1200 species, the most characteristic of which are members of the Chenopodiaceae, Leguminosae, Cruciferae and grasses. Trees make

Photo 1.10. Discontinuous cover of the flowering Negev steppe. On Mount Ramon in the Negev, close to the ancient cisterns of *Hemet Well* the desert is in bloom (*Pistachis atlantica*) several days after a downpour in December 1993 (Photo M. Mainguet)

up 2% of the total cover, bushes 11%, shrubs 14%, perennial herbs 22% and annual herbs 51% (Kalenov and Muklanmedov 1992).

The relict plants of this semi-arid area bear witness to older arid conditions going back to the Oligocene after the retreat of the Sarmatian sea leaving behind the hamadas on which a vegetation cover was developed, some plants of which subsist in an endemic state on the Miocene Ust-Urt plateau which is a hamada. During this dry phase, the vegetation acquired a great resistance to conditions of restricted humidity, to high temperatures and to the salinity of the surface and ground waters. Many of these plants exhibit a vast ecological span and grow on a variety of substrates. *Haloxylon aphyllum* (*locally named saxaul*), the most widely distributed plant, grows on clayey, loamy, sandy soils, on sandstones and frequently in areas of salty soils from Asia Minor to central Asia, China and Mongolia. In the Karakum vast areas are covered by *Calligonum* (for example *salsa*) and varieties of *Ephedra*. Salty deposits have favoured the growth of a halophilic vegetation. On the Quaternary piedmont loess south of the Aral basin other groups of plants make their appearance.

Three simple observations should be pointed out:

1. Drought will not lead to the same vegetation cover in tropical and temperate drylands.
2. Dryland is not characterized by the absence of any vegetation but by a vegetation even genetically adapted to the drought, i.e. a xerophilic one.

3. The relatively low grazing potential of the Sahara offers limited opportunities, enjoying contracted vegetation only in parts, such as: the floors of wadis possessing underflow and/or receiving floods at irregular intervals; and on some slightly moister slopes or levels of mountains like the Eglab, Hoggar-Tassili, the Tibesti and its southern semi-arid environs, the Adrar des Ifoghas, the Aïr massive and Djebel Mara. In the deserts of central Asia the situation is different as e.g. in the Karakum and Kyzylkum around the position of the 100 mm isohyet the transition from arid to semi-arid takes place, and grazing potential is higher than in the Sahara and high frequentation leads to an advanced stage of desertification.

1.4.4
The Open Vegetation Cover of the Sahel

In Africa the steppe of the Sahel covers wide areas from West Africa (Mauretania, Senegal) to Ethiopia and Somalia. Its northern limit on the Nile lies at the latitude of Khartoum. In the west the steppe zone runs out some 300–400 km to the north and towards the east, in East Africa, it encircles virtually the entire Ethiopian massif. Although lying in equatorial latitudes, Somalia is Sahelian in nature. Its equinox rains facilitate the appearance of open dry forests, with spiny plants like *Acacia senegal*, *Acacia tortilis* and *Balanites aegyptica*, or of thickets on better soils or soils with better water-holding capacities.

The short duration of the rainy season and the interannual irregularity lead to a seasonal rhythm for the vegetation: After the first showers the plants develop rapidly and store water allowing them to survive the dry season. The period of active life of the pastures ranges from 1–3 months between the north and the south of the Sahel starting in August in the north and ending in June in the south. The average forage value of the pastures also varies along this north-south gradient.

According to Boudet (1974),

> The Sahel proper, characterized by a continuous thorn steppe, may be delineated by two isohyets: 100–200 mm/year and 500 mm/year (normal annual rainfall): in the north from the bend of the Senegal and of the Niger to north of Lake Chad, in the south from Dakar, to Mopti, Niamey, N'Djamena. ... To the south of this boundary we find the domain of the perennial grass savannah and deciduous bush, truly a Sudanian domain. The ideal area of non-irrigated cultures should barely exceed the latitude of this line. This predominantly agriculturally used Sudanian zone should nevertheless contributes essentially to the evolution of the pastoral world of the Sahel.

To define the Sahel zone by fixing its limits as a function of the isohyets is risky in view of the known fluctuations resulting from the alternation of humid and dry phases (Fig. 1.10). These fluctuations apply to the Sahel as a whole but for the Sahara they are only a temporary feature at the expense of pastures utilized for extensive livestock breeding. As the Sahel zone is spatially not fixed, its capacity for harbouring humans and livestock is also variable.

The vegetation of the Sahel is made up of open formations of grasses (in particular *Panicum*), bushes and trees, the latter as single individuals or in groups. This vegetation is called thorn steppe as the trees have developed thorns: Combretaceae or *Mimosa*, in particular acacias or gum trees (*Acacia seyal, Acacia rediana*). The fringes of the Sudanian savannah are marked by the baobabs.

Like the plants of the deserts, those of the Sahel steppe, in particular in regions receiving 100–250 mm/year in precipitation, concurrently developed morphological, physiological and genetical adaptations for limiting the loss of water by evapotranspiration, for improving the search for water in the ground and acquiring vegetative and reproductive cycles that are as short as possible. There are three categories:

1. Those having short life cycles, escaping the drought by entering into dormancy during the dry season.
2. Those growing little or not at all when pluviometric conditions are adverse, but remaining alive and ready to resume growth after the arrival of the rains.
3. Those escaping drought by accumulating water reserves.

Gillet (1974) proposed a convincing distinction:

- *The North Sahel*, where low precipitation permits a herbaceous vegetation of the steppe type; each tuft is separated from its neighbours by barren spaces and the vegetation cover represents less than 30% of the edaphic surface.
- *The South Sahel*, when trees start to leave to their usual habitats along the rivers and spread into the countryside. The South Sahel is tree-covered whereas the North Sahel is a gramineous or shrub steppe. At the same time the density of the annual gramineous stratum increases and becomes waterlogged during the rainy season. Here is the heart of the livestock breeding areas.

1.4.5
The Vocation of Grazing the Sahel Steppe

The Steppe of the North Sahel. Here, the vegetation is contracted at sites favourable for the collection of water from runoff. The gramineous layer remains thin with a predominance of xerophilic perennial grasses (*Aristida pallica, Passicum turgidum*). The gramineous cover is absent on skeletal or argillaceous soils and diffuse on sandy soils where the herbaceous biomass reaches 40 g/m^2 with yields of 400 kg dry matter per hectare. It becomes denser over loamy soils with a biomass of 50 g/m^2 and yields of 500 kg/ha.

The Typical Sahel Steppe. Here, the woody coverage[7] is below 5% with the exception of thickets on clayey-loamy soils. The gramineous layer is dominated by annual xerophilic grasses. The biomass ranges from 200 g/m^2 on the slopes of sand dunes to 300 g/m^2 in the interdunal spaces. With 30% of the soil barren, the mean biomass is 100 g/m^2, equivalent to 1000 kg/ha. On clayey or loamy soils it varies from 100 to 300 g/m^2, dropping to 180 g/m^2 on skeletal soils, but as 75% of these soils are barren the mean production is only 80 g/m^2.

In the oases of Tunisia where the precipitation (mean of 180 mm/year) falls in winter from September to March and the soils are argillaceous-sandy-gypsiferous with

---

[7] *Coverage:* relationship between the projection of the plants onto the horizontal plane and the total area used for the analysis of the vegetal formation.

0.4–0.7% organic matter, the *Rhanterium suaveolens* steppe produces 1000 kg/ha annually, or 5.5 kg ha^{-1} mm^{-1} of precipitation. During years of favourable precipitation, the production grows to 1550 kg/ha, whereas during years of deficit it falls to 1060 kg/ha (Thornes 1994). From 1940–1950 the coverage of the Algerian steppe was 40–50% and the biomass reached 1000–1200 kg/ha. Today, the coverage is only 2–5% and the biomass barely 200–300 kg/ha, an indication of the pronounced degradation of the vegetation cover (Mourad 1993).

The South Sahel Steppe. Here, woody and herbaceous north-Sudanian species join the Sahelian species. The woody stratum reaches 60% on loamy soils and 15% on sandy soils. The gramineous biomass is 150 g/m^2 dry matter on sand dunes and 300 g/m^2 on the loamy-argillaceous depressions. Both the trans-Sahelian rivers Senegal and Niger and Lake Chad are bordered by aquatic grassland which becomes inundated at the end of the rains. The *bougoutières* of *Echinochloa stagnina* have a biomass of 600–1.700 g/m^2 of which 1300 g/m^2 is due to the submerged stems. Their strong regrowth makes them a valuable fodder during the dry season.

Along the climatic north-south gradient characterizing the Sahelian domain species of limited extent blend with others of a wider distribution which take advantage of the rainy seasons to enlarge their implantation but are more affected by the dry period: the perennial grass *Andropogon gayanus* and woody species like *Balanites aegyptica*, *Commiphora africana* and *Guiera senegalensis* are examples.

The woody cover does not compete with the gramineous layer as its light shade creates a microclimate which is favourable for shade-loving grasses, which produce 160 g/m^2 dry matter in the shade against 65 g/m^2 on an insolated sand dune in the north-Sahelian sector of the Senegal (Boudet 1974). Light shading by *Acacia* actually leads to a microclimate favourable to the more productive mesophilic grasses. In contrast, the rapid growth of herbs after the first rains is damaging to the growth of young trees as the osmotic pressure of the former is high and their potential for drawing water high. The grasses grow well on fine- to medium-grained sandy soils, the optimum porosity of which imparts to them an excellent water retention capacity. The grasses exploit a large volume of soil by forming a dense and long root network, and they draw water from the superficial layers. The trees need a continuous supply of water even after shedding their leaves, but they require less per unit of time. Their root systems allow them to search for water at greater depths.

With regard to the South Sahel Gillet (1974) remarked: "The term *extreme seasonality* could not be applied better than to this south-Sahelian zone where the contrast between the rainy season and the dry season is dramatic," and leads to differences in the pastures according to the pluviometric regime.

The Pastures of the Rainy Season. The young tender grasses spread in areas with stagnating water and on loamy depressions during the humid season. The populations of *Echinochloa colona* (which may reach 1000 blades/m^2) or of *Panicum laetum* are eaten avidly by animals. These annual species, even where intensely grazed, produce sufficient seeds to assure their reproduction as long as the substrate stays moist. Most annual species of the South Sahel are hygrophytes, requiring permanent humidity and, save for their seeds, are not drought-resistant. As the dry season is early, drought or only isolated rains lead to an impoverishment of the pastures of the Sahel.

In the fixed sandy covers the sand remains moist down to a depth of more than 50 to 60 cm during the dry season, permitting the gramineous vegetation to survive. Gillet (1974) noted that the psammophilic[8] plants, more than those on hydromorphic soils, possess a "neotenic" faculty of reproducing in the young stage like certain species of *Brachiaria* (*B. hagerupii* or *B. deflexa*) and *Cenchrus biflorus*, the famous *cram-cram*.

The Pastures of the Dry Season. These essentially assure nourishment for livestock during the dry season. Gillet (1974) distinguished the pastures of falling floods, of dry matter or straw, and aerial pastures, these last being represented by forage trees.

The River-Fall Pasture. This pasture type is fed by the waters of allochthonous rivers. Such rivers are not subject to severe seasonal climatic variations, which occur along temporary rivers or wadi tributary of regional rains. Along them good pastures briefly support the livestock during its migration to the south. During dry years, these stations prove to be disappointing for the herds, leading again to its retreat. Therefore the species sought after, like certain species of *Panicum* and *Cyperus* (*Pycreus tremulus*), cannot be grazed, much to the disadvantage of the consumers.

The river-fall pastures are able to grow as long as there is water on the surface and the soil stays moist. At the start and sometimes well before the dry season when the water reserves in the soil are high, these pastures furnish a considerable phytomass, which nearly always consists of highly "alibile" species[9]: during their active period the heliophytes produce leaf-bearing shoots from rhizomes in the mud. Because of this the inner delta of the Niger is a place for grazing, the livestock using the leaves and leaf-bearing shoots of the *bourgou* (*Echinochloa stagnina*) and the related *E. pyramidalis*.

Straw Pastures. The Sahelian cattle consume dry straw during more than 6 months of the year. Not all grasses furnish straw of the same quality. The straw should be fine, and is much appreciated by the cattle when the desiccated foliar parts and the shreds of the flower stands still contain seeds rich in protein.

Aerial fodder. Such fodder are supplied by forage trees. The leaves of certain evergreen trees represent valuable fodder during the dry season. Bovids consume them and wild animals, in particular gazelles, eat them avidly. During the height of the dry season, they are frequently the only source of green fodder and proteins. Weight for weight the leaves of trees are much more nutritional than dry hay. Both evergreen and deciduous trees are present:

- Truly evergreen trees. The main examples are the thorny *Balanites aegyptiaca* and the thornless *Maerua crassifolia*. According to the Veterinary Institute of Maisons-Alfort, the protein content of their leaves is high, 30% weight in dry matter for *Balanites* and 21% for *Maerua*. Therefore these trees are also browsed in pastoral areas, the accessible young leaves being eaten as soon as they appear. Although

---

[8] *Psammophilic:* preferably growing on sand.
[9] *Alibile:* a substance suitable for nutrition, digestible and also nutritional.

Balanites has not developed a sharp canopy as a defence against livestock, it is able to resist livestock because of its bulbous growth with strong spines. *Maerua* resists with the aid of interweaving branches. These trees regenerate poorly as the young specimens incessantly consumed, do not get a chance to develop. The situation is aggravated during periods of drought. A useful measure would be protection by wire mesh fencing until they have reached a sufficient height.

- Deciduous trees. Most of the Sahelian forage trees lose their leaves at the onset of the dry season. Bille (1972, 1973) has shown that the woody stratum furnishes more than 100 kg of dry matter (leaves and fruits) per hectare in northern Senegal. This applies to the most sought after trees: *Commiphora africana*, to all species of *Grewia* and *Ziziphus*, *Cordia rothii* and *Combretum aculeatum*. Others, like certain species of *Acacia* (*A. raddiana*) and the tamarind, keep their foliage during part of the dry season, at least in normal years. The leaves of *Acacia* contain at least 20% proteins in dry matter and are thus much in demand (Photo 1.11).

In a normal year the gap in supply is bridged by trees with dense foliage, especially in trees growing in depressions. In dry years, however, these trees shed their leaves early and aerial fodder becomes less abundant. During periods of famine when the leaves have disappeared, the pastoral nomads cut back the green trees to

Photo 1.11. Soil denudation in a village in semi-arid Mali. Huts and shelters in the sedentary village Erebongou in the heart of the Mali Sahel in February 1985, illustrating the double effect of cultivation of millet and raising of livestock. Under the *Acacia raddiana* with its tropical flat canopy which is close to 50 years in age, there is a circle of goat droppings and a drying platform. The dusty atmosphere is caused by aerosols during the height of the dry season, when dust storms are frequent (Photo M. Mainguet)

supply fresh foliage to their cattle and goats, cutting them back sometimes completely without any chance of regeneration. The disappearance of the trees deprives the nomads of shelter for their animals against the sun. In addition, young specimens, which would otherwise develop under the trees where they are sheltered from excessive insolation, do not grow. Trees in such favoured positions are more active in reproduction than those growing in the open sun.

The trees suffered considerably during the persistent drought of the 1970s and 1980s. In certain parts of Niger and Chad, more than half the *Acacia raddiana* died. Numerous specimens of *Capparis decidua* perished in eastern Chad, mostly as victims of the sinking groundwater level.

1.4.6
Potential Foraging Value and Carrying Capacity

The nutritional value of pastures depends essentially on the advantage the livestock derives from it and is characterized by the following pasture characteristics:

- *Its productivity*, expressed in kilogram dry matter or energy per hectare. The herbaceous biomass produced during the period of active grazing is not conserved in the dry season (Tables 1.7 and 1.8). Between the end of the rains and March, even when not utilized, the amount of forage drops by 20–50% and even by up to 65%. Bush fires are responsible for this, extending over large areas as soon as the herbaceous biomass produced during the active period increases to more than 1 t dry matter per hectare. Because of this a rain-rich year favourable for the production of forage may still be insufficient for the maintenance of the herds (Boudet 1974).
- *Palatability*[10] of the species; the productivity may refer to the entire vegetal production or only to the edible species as whole.
- *Consumable quantities* of fodder and the quantities actually consumed; the daily consumption is estimated on average as 2.5 kg dry matter per 100 kg live weight. Agrostologists (specialist for pastures) estimate the carrying capacity from the aerial production of edible herbaceous cover. According to Boudet (1974) only 50% of the production is consumable under optimal conditions because of the drop in productivity resulting from grazing during the growth of the annual species, because of trampling and because of the necessity to preserve a protective cover against erosion by wind and water.
- *Concentration of indispensable nutritional elements* (proteins, vitamins, minerals, macro- and oligo-elements) in the ingested parts.
- *Digestibility of species consumed* which determines their content in usable energy.
- *Content of nitrogen compounds* which frequently is a limiting factor. It varies with the phenological state of a plant and may be e.g., in the grass *Cenchrus biflorus* in percent of dry matter:
    - 16% during growth under grazing in the rainy season
    - 4% in the straw in November
    - 2.6% in the straw in April

---

[10] *Palatability:* exciting appetite.

1.4 · Vegetation Covers: Economic Potentials and Droughts

Table 1.7. Decrease in herbaceous biomass in kg/ha after its phase of active production between August and October in Mali (Bille 1971)

Type	October	March
Dunes	1000	350
Sandy plateaus	1500	750

Table 1.8. Evolution of herbaceous biomass in Senegal in kg/ha (Valenza and Dialla 1972)

Type	Aug.	Sept.	Oct.	Nov.	Dec.	Feb.
Dunes	150	400	600	650	550	380
Depressions	300	3000	3400	2800	2400	2400
Shaded surface	400	1600	1800	1600	1000	800

In *Acacia albida* in March the content of nitrogen compounds is 11% in the pods and 18% in the leaves. The requirements of cattle are satisfied when the nitrogen content is 5%. According to Boudet (1974) in the Sahelian domain the dominating annual grasses possess a sufficiently high energy value until the dry season, but the nitrogen content is highest at the end of the rainy season during fructification. The supplementing supply of indispensable nitrogen during this period is assured by trees and especially by leaves and fruits.

As plants the perennials are more interesting than the annuals. According to Granier (1975):

- The biomass of a perennial association is 2.5 times higher than that of a therophytic one[11] corresponding to 2.5 t/ha dry matter against 1 t/ha.
- Their productivity is also spread out in time: aftergrowth at the start of the dry season and reactivation when the hygrometry rises again in spring.
- Their regrowing parts are full of carotene and nitrogen which one does not find in the straw of the annuals.
- The persistence of their stumps enriches the soil in organic matter and protects it against aeolian erosion by tying down the mobile surface layers.
- Perennials are furthermore utilized for maintaining pits and wells and for the construction of palisades around huts and shelters.

The therophytes are sensitive to early grazing: because of their shallow roots they are easily trampled down or pulled out and are thereby reduced in density. The consumption of the underground parts entails a shorter extension of the roots and thereby leads to reduced access to water. Grazing during early growth reduces the chance of seed production. During fructification, immature seeds are consumed and the potential for seeding is reduced. When grazing takes place during the whole of the rainy season, the annual grasses disappear within 2 to 3 years. *Zornia glochidiata* flowers and fructifies within 3 weeks and is able to multiply despite grazing during the first years. Thus within 2 or 3 years only *Zornia* will persist. In the Senegalese

---

[11] *Therophytic:* annual herbaceous species which stays in grain during dry season.

Ferlo this excellent fodder legume was overgrazed and died off rapidly, and grazing became impossible.

The first rains permit the uninterrupted development of grasses up to maturity when the water retention capacity of the soil is good. A rainfall of 3 mm, followed by a similar one not more than 1 week later, will suffice to allow the plants to germinate (Gallais 1975). Towards the end of June the Sahel turns green again within a few days after the intertropical front (ITF) passes abruptly to the north of 20° N. The great downpours impregnate the soil and lead to germination of the seeds which, however, proves to be devastating when the ITF stays away, as one period of drought is enough to let the Sahel turn yellow and the plants die. When the rains return at the end of July only the perennial herbs which are more resistant but less nutritional will turn green again. This explains the spells of famine during years of perfectly normal or even above-average rainfall when, however, such rainfall is poorly distributed in time and furnishes pastures of low value only. Such a localized advance of the ITF is without consequences for the following years and anemochorous[12] seeds will reseed the region. However, when the advance takes place across the entire Sahel, it takes several years of good rains to overcome the effects of this interruption of the rains.

In the part of the Sahel utilized for grazing, the rainy season is rather short, lasting from 15 July to 31 August (Bernus and Savonnet 1973). The considerable variations in the quality of the grazing throughout the year are controlled by the short duration of the period favourable to the vegetation, viz. July to mid-September, during which the herbs are green, abundant in supply, rich in water and only little lignified. At the start of the dry season the grass wilts rapidly, become hard as the season advances and at the same time they become scarcer. At the end of the dry season, the vegetation is dormant except where growing close to points of permanent water or along the rivers.

The *carrying capacity* is a theoretical concept difficult to quantify, especially for such a heterogeneous ground like the drylands in which, as we have seen, the ecological dynamics are controlled by the variability of the precipitation already on the scale of the slopes and by climatic accidents such as droughts. The rainfall – an unforeseeable factor – in addition to civil wars and changes in the organization of land tenure make the access to grazing and water rather haphazard.

The mean carrying capacity, in the absence of fires, ranges from 25 kg/ha annually in live weight for the dunes and sandy-loamy depressions or plains of the northern Sahel via 50 kg/ha per year in the typical Sahel to 60 kg/ha per year in the southern Sahel. In the *bourgoutières* it will reach 2500 kg/ha per year. Grouzis (1987) estimated that the livestock charge in the Sahel proper varies in time and space between 3–20 TLU/ha per year (TLU = tropical livestock unit), but is generally low with an average of 6.75 TLU/ha per year. In comparison, in the Marne valley of the Champagne (in France), 1 ha will be sufficient for 1–1.5 animals of 600 kg/year, whereas in Normandy where an ox of 700 kg will grow within 2 years, 1 ha will carry four animals (Table 1.9).

According to Boudet (1974), maintaining 100 kg/year of live weight of herbivores (in particular, cattle) requires by the end of the active season a production of the vegetation cover equal to 1825 kg/ha in dry matter ($2.5 \times 2 \times 365$).

---

[12] *Anemochorous:* borne by wind.

Table 1.9. Vegetal biomass (in kg) required for maintaining live weight (in kg) per hectare per year in Gourma, Mali (Boudet 1974)

Sector	Fixed sand sea		Fairly thin heterogeneous soils		Muddy depressions	
	b[a]	lw[b]	b	lw	b	lw
Desert	300	15	250	20	900	50
Sahel with 100–200 to 400 mm precipitation	1200 – 2900	65 – 160	400	20	2700	145

[a]b, Vegetative biomass.
[b]lw, Live weight.

Valenza and Fayolle (1965) measured a biomass of 1300 kg at Dara-Djoloff (Senegal), where the precipitation was 430 mm/year. Where the precipitation was 200 to 300 mm/year at Fété-Olé, Bille (1971) estimated a biomass of 650 kg/ha for dunes, 2800 kg/ha for depressions and 1600 kg/ha for the shaded parts of dunes. The last value confirms that shading increases herbaceous production in the Sahel zone by nearly 250%.

1.4.7
Weakening of the Vegetation Cover by Human Activities: Different Estimates

Until 1994, nearly 10 years after the accepted end of the last drought crisis in 1985, the literature produced a number of different estimates of the weakening of the vegetation cover, which culminated in the optimistic paradigm of Warren and Khogali (1992). We shall concentrate here on the apparently reasonable aspects, taking into account place and time of the respective observations in relation to the drought crisis. Among the causes of the destruction of the vegetal associations, the most damaging ones are:

Cutting of Trees for Domestic Firewood. The deficit is such that one may talk of a firewood crisis heightened by reduction of the fallow periods. Forests as defined by the FAO (1993) are areas of more than 10 ha in which trees (woody plants of >7 m height) cover more than 10% of the ground and in 1990 forests covered 1.715 million ha. Of these, 245.9 million ha occurs in drylands comprising deciduous forests, of which 155.3 million ha are found in Africa, 40.8 million ha in Asia and 38.8 million ha in America. These are lowland trees at altitudes below 800 m. Included are also the mountainous regions of northern Mexico and parts of the Andes, and parts of Ethiopia and Pakistan, but excluding the Mediterranean parts of the Near and Middle East. The rate of deforestation is 1% in Asia and 0.8% in the semi-deciduous dry forests of Africa. However, the population density is 247 inhabitants/km^2 in Asia compared to 24 inhabitants/km^2 in Africa. The subhumid dry forests of Africa are thus being cleared at a 21 times higher rate than in Asia or, expressed per inhabitant, at a rate of 150 m^2/year in Africa, 7 m^2/year in Asia and 100 m^2/year in America which, in combination with the population explosion represents a real danger for the dryland forests of Africa.

Photo 1.12. Sedentary village of millet culture in Mali. The village of Diébégou in January 1985 was typical of the Sahel in Mali. The young baobab tree on the right signifies a south-Sahelian latitude. The huts and shelters, covered with the stalks of *Andropogon gayanus* and wooden framework are evidence for a severe demand for wood for construction. The group of wooden mortars for stamping millet illustrates that this task was still carried out manually and collectively in 1985. The rather long pestels are leaning against the wall of the hut to the left (Photo M. Mainguet)

Cutting of wood around villages and in rural environments is a problem which all drylands have in common. The diameter of the sphere of wood removal is proportional to the size of the agglomeration: around Abéché (Chad) there has no longer been any dead wood within a 10 km radius, and 100 km around Kharthoum since the mid-1980s, a dimension which will be even larger today. The rural communities, poorly supplied with electricity, have to take recourse to wood and agricultural debris much to the detriment of the environment. The use of solar and aeolian energy devices and of biogas is not progressing at a sufficient rate and it has to be pointed out that the utilization of saxaul wood for firing the steam locomotives of the trains in the Turan was the reason for the degradation observed there. The governments of China and India have promoted the use of solar stoves which possess the following advantages:

- Independence from conventional energy.
- Decrease in deforestation.
- Reduction of the labour of women and children for gathering and carrying firewood.
- Less pollution by smoke which is so dangerous for the lungs.

1.4 · Vegetation Covers: Economic Potentials and Droughts

In the search for construction materials the best trunks of the species most resistant against termites are selected like, from the banks of the wadis, *Anogeissus beiocarpus, Hyphaene thebaica* and *Acacia scorpioides,* with A*cacia raddiana* and *A. seyal* for the underconstruction of the roofs of grass huts. These species have thus become rare throughout the Sahel (Photo 1.12).

In drylands, solar energy should represent the preferable source of energy instead of wood, which should be protected because of their nutritional value, their leaves as fodder for livestock and their fruits for man, while they are still considered as a source of energy by the managers or a mean of combating the ill effects of the wind. In the Negev the Israelis have perfected the domestication of fruit trees in drylands.

The tree should also be utilized as a source of nutrients for the soils which unfortunately is not the case in the short term, in places replanted with eucalyptus. This tree sucks up water and minerals and is not suitable for drylands in the long term. In contrast to *Acacia albida*, it has no nitrogen-fixing nodules and neither its leaves for animals nor its fruits for man are edible. After two fellings, the regrowth of new wood becomes slower. Being Australian in origin, the eucalyptus does not represent a worthwhile substitute for the various autochthonous trees of the African drylands. Its gluttony should ban it to derelict fringes of ground and one should favour the use of autochthonous trees (Pélisier 1989).

Planned and Accidental Bush Fires. These fires destroy the humus which favours the retention of water, the wealth of mineral salts as well as the microbial life in the upper soil horizon, They also temporarily destroy any chance of regeneration, since they leave a barren soil vulnerable to aeolian erosion. The bush fires in the steppe do not destroy the trees, but momentarily change their productive capacity and their physiology. In the second part, and especially at the end of the drought period, they are particularly harmful. However, a few restrictions have to be made here: in areas with few edible perennial grasses the bush fires are useful in helping the livestock to gain access to fodder of a better quality.

In the subhumid dry zones the effects of overgrazing combine with those of bush fires. The density of the savannah is reduced by the former, leading to a reduced spreading of the latter and to a regrowth of the bush and tree vegetation, leaving a false impression of luxuriant conditions (Hurault 1975). The disappearance of the basal layer under the bushes and shrubs favours rill wash and water erosion. When they no longer can find good fodder, the livestock abandon these sections, dry matter accumulates and when the bush fires come back, they are more violent and thus more damaging than before (Bell 1987).

Debranching. At the end of the dry season when the herbs have disappeared from the pastures, the herdsmen turn to the foliage and husks of the trees. To make fruits and foliage more accessible they debranch the trees, cut off the canopy or even cut the trees down completely. All mimosas are treated in this way, in particular *Faidherbia albita*, the foliage of which forms during the dry season, and *Acacia scorpioides.*

Land Clearing. The virtual absence of fertilization requires the rejuvenation of the cultivated lands. The growth of the population and the demand for an increased production entail a shortening of the fallow periods leading to an exhaustion of the soils. Around villages the fallow grounds with rotational cycles of 5–10 years have nearly disappeared in regions of intensive cultivation. In the Hausa country of Niger fallow periods of 20–30 years have recently been replaced by shorter ones. After several years of utilization of the same sections, an entire village had moved and settled down in the vicinity, not returning to the previously utilized site before several other moves. This rotation within one generation no longer exists. Around major centres where fallow lands have disappeared, the impoverished soils become invaded by characteristic species like *Calotropis procera*. Along the wadis the production of vegetables like tomatoes and onions is responsible for the lowering of the water table, endangering the regeneration of the species on the banks.

During the half-century preceding the great drought in the drylands of Africa north and south of the Sahara at the end of the 1960s, the improvement of the sanitary conditions led to a considerable increase in the number of herdsmen and their livestock. At the same time programmes for water supply to herds and the use of mechanical means of pumping have facilitated the exploitation of new pastures and led the nomads to abandon their migrations in favour of settling down close to the wells and the banks of the larger water courses. This aspect will be dealt with in the third part of this book. All this resulted in areas of overexploitation, areas of vegetal disequilibrium and changes in animal health.

The overall increase in livestock numbers in these areas is also considered as one of the major causes for degradation of the environment. Barrow (1991) estimated that between 1955–1976 the number of cattle increased by 34% in the developing countries and that of sheep and goats by 32% each. Overgrazing and increased trampling were not without consequences for the environment:

- Overgrazing has exerted negative effects on the density of nourishing and palatable plants. Degradation makes itself felt by the disappearance of these plants in favour of plants more difficult to consume or thorny. The effect is less pronounced in areas of lower plant densities. Overgrazing in subhumid dry zones leads to the substitution of the usable grass cover by a vegetation of trees and bushes without interest to the livestock.
- Trampling, via a ploughing effect, may be beneficial to the regeneration of certain plants. But depending on type of livestock, texture and structure of the soil and degree of overgrazing, trampling may entail a compaction of the soil, which results in a loss of the infiltration capacity for water and an increase in runoff, thereby resulting in an intensification of the erosion. In contrast to this light soils are loosened by trampling and can be mobilized more easily by water and wind erosion, which will lead to dust storms from the barren soils and to an increase in airborne dust. This is considered to be one of the reasons for subsidence of the air and thus for reduced precipitation.

Man increases his sphere of activity by means of his livestock which is responsible through trampling for the destruction of certain areas causing gregariousness in animals. Because of grazing the natural climatic associations become replaced by impoverished floral associations. The proliferation of nitrophilic plants (*Boerhavia repens, Gymnaudropsis gynandra, Amaranthus graecizans*) is encouraged as is the multiplication of *Acacia*, because their thick-skinned seeds germinate more readily after having passed through the digestive tract of the ruminants.

Around the drinking points where the livestock charge surpasses the limited grazing capacity the ill effects of overgrazing are sometimes combined rather drastically with those of over-trampling, and degradation will be at its maximum here. During the long dry season the herds which need water daily, will move up to 9 km only from the water supply in their search for exploitable grazing and up to 15 km and more when the herdsmen take them to drinking only every 2–3 days (Chevallier and Claude 1990). The term *pyosphere* was introduced by a Russian agrostologist in ex-Soviet Asia to describe the sequence of concentric rings with decreasing degrees of degradation around watering holes. The results of Perkins and Thomas (1993) from the Kalahari (Fig. 1.13) show that the average diameter of the affected area is about 4 km. In this system the limiting factor is the absence of surface water. The results obtained there cannot be transferred directly to the Sahel where the limiting factor is rather the amount of available fodder. Furthermore, the breeding livestock, in contrast to the numerous wild ungulates, cannot search for its fodder at too great a distance from the watering holes outside the area of overgrazing.

In the northern Ferlo of Senegal where the mean precipitation is 300 mm/year, Poissonet *et al.* (1992) studied the changes in composition and productivity of the pastures in relation to the distance from a borehole for two areas around the water-

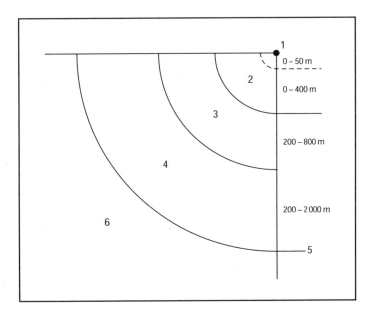

Fig. 1.13. Pyosphere in the Kalahari (after Perkins and Thomas 1993). The area affected by degradation of the vegetal cover and of the soil depends on the distance the livestock can move away from the bore holes. *1* Bore holes and surroundings where the excrement of the animals leads to a higher supply of nutrients to the soil; *2* derelict area; in *1* and *2* the vegetation is destroyed and the soil structure broken up, subjecting the ground to aeolian erosion; *3* area in which nourishing edible grasses dominate; *4* sphere of invasion of a bush vegetation; *5* limit of major effects of grazing; *6* grazing reserve

ing points of Kabgar and Namarel, where permanent drinking facilities have been installed. They observed that the monthly variability of the rainfall, and especially that in July, led to modifications of the environmental conditions controlling the frequency of appearance of the different vegetal species. The other primary factor was the distance from the well.

> The trampling by the animals which is intense within a radius of 2.5 km from the watering point leads to modifications of the environmental conditions ... which may be recognized from variations in the frequencies of the species. ... To explain this evolution the following hypotheses are ... proposed:
>
> - Excess of organic matter around the permanent watering places favours monospecific communities with *Boerhavia diffusa*.
> - Overgrazing and trampling favour bispecific communities with *Zornia glochidiata* and *Dactyloctenium aegyptium*.
>
> One has to move 4 km from the watering hole to ... observe the start of an equilibrium in communities which is eventually achieved after 7 km and corresponds to an equilibrium between the number of animals per unit area and the transformable production (into milk and meat) of the corresponding vegetal communities without under- or overgrazing (approximately 5 ha per tropical livestock unit).
>
> This proximity to the watering point entails a disequilibrium in the grazed communities by favouring ... the dominance of 1 or 2 species, but it does not correspond to a lowering of the productivity in the agronomic or pastoral sense of the term. Numerous investigations ... have proven that in the sense of pastoral productivity related to the amount of organic matter the best results are obtained within a range of 0.4–1 km around a well from a bispecific vegetation ... consisting of the leguminous *Zornia glochidiata* and the grass *Dactyloctenium aegyptium*.

This description of the state of rangeland around a watering point carries an optimistic note for the evolution of densely frequented sites. The lowering of the foraging productivity so frequently emphasized in terms of pastoral productivity is contradicted by the observation that the best results are actually obtained within a range of 0.4–1 km around a well. However, the precipitation had improved since 1985. The damage to the vegetal communities by livestock grazing nevertheless has resulted in a reduction in density of the perennial edible plants in comparison to the less edible ones, to an elimination of the consumable shrubs and to the complete eradication of the vegetation cover around certain wells. The zoochora species and among these the diaspora[13] like *Cenchrus biflorus* are carried on the skin of the animals and thereby favoured to the degree that they become dominant. The nomads in particular exert an influence on appearance, composition and dynamics of the vegetal formations by harvesting trees, bark and fibres, fruits and seeds for themselves and their livestock and by spreading with their clothing seeds of plants with which they have come into contact, like *Cenchrus biflorus, Pupalia lappacea, Setaria viridis* and *Aristida fumiculata*.

Boudet (1974) remarked furthermore that the irregular distribution of faeces leads to the appearance of areas with nitrophilic species in association with spots of trees, the seeds of which possess a higher level of germination after having been digested by the animals. Boudet ascribed the different reactions of the pastures to the period

---

[13] *Diaspora:* part of a plant carried away from the mother plant which resumes its development, thereby assuring the dispersion of the species.

of utilization. When exploited during the rainy season they were sensitive to overstocking and above a critical level the herbaceous cover will deteriorate with:

- The disappearance of the ears of the annual grasses which no longer are able to fructify.
- An increase in the species consumed with a short life cycle, which supports their multiplication even under grazing (*Zornia glochidiata*). The same phenomenon is observed in the mowed grasslands of the Sahel (Valenza 1970). However, these species disappear after the end of the rains and the evolution favourable for grazing during the active season becomes damaging in the case of exploitation during the dry season or throughout the entire year.
- The appearance of herb-covered micro-dunes on sandy soils separated by areas which have been denuded and sterilized by aeolian ablation.
- The appearance of denuded, damaged and asphyxiated areas, the death of the trees and the removal of fine-grained soil particles by sheet erosion. After aeolian remobilization of the sandy particles, the denuded areas may lead to the extension of the "brousse tigrée" or striped bush facies so frequent in the Sahel.

The evolution of the vegetation depends on the pressure exerted during the rainy season. Abundance incites the herdsman to no longer control the exploitation of grasses and Granier (1975) noted that

the influence of the livestock manifests itself at different levels:

- Trampling of the young seedlings;
- Outrooting of the young seedlings;
- Trampling of ears and liberation of the seeds leading to the loss of protection by the shells;
- Dispersion of certain diaspores;
- Defoliation.

Exploited only during the dry season, the Sahelian pastures preserve all their potential productivity after the dissemination of the diasporous grasses, save for the areas subjected to excessive trampling. The zoochorous species with strong productivity multiply and raise the production of fodder. Sectors invaded by young trees, the seeding of which is favoured by faeces, can even serve for local regeneration of the tree cover provided that the young shrubs and trees are excluded from grazing for a number of years. They could then be exploited as afforestation for the production of firewood.

1.4.8
Behaviour of the Vegetal Cover During the Drought of 1969–1973

In conclusion it may be said that the low coverage of the soils of 10–20% and sometimes less represents the main factor for the fragility of the Sahelian lands during a drought. Runoff acts without constraint and its concentration by tufts contributes to the uprooting of plants and to the removal of fine-grained particles. The coarser ones remain in situ and form a pavement. The action of deflation also makes itself felt. The environment thus reacts in a fragile manner against the two preponderant mechanisms of external dynamics, viz. water and wind erosion which will be discussed in the second part of the book.

The increased vulnerability of the vegetal communities of the Sahelian cover usually observed in years of drought finds its expression in:

- Retarded start of the formation of the foliage.
- A later appearance of the fruits.
- An early shedding of the leaves and thereby for certain species a shorter period of foliage production.
- A reduction in the percentage and duration of flowering.
- A reduction in leaf length and in the number of pairs of pinnules.[14]
- A reduction in weight of dry matter and of foliar production, especially for the youngest trees.
- A reduction in or absence of fructification which becomes more pronounced in younger trees

In certain areas, the drought leads to the almost total disappearance of perennial grasses like *Andropogon gayanus* and *Cymbopogon proximus*, and of the sedge *Cyperus conglomeratus*. The annuals will replace the perennials when there is overgrazing and a moisture deficit. In contrast, the extension of the perennials in the Sahelian-Sudanian domain indicates a more humid phase. Furthermore, one of the most astonishing consequences of the drought is the disappearance of three tree species. *Acacia senegal, Commiphora africana* and especially *Guiera senegalensis*. The oldest specimens are most resistant in the first two and least resistant in the third. The fruits of *Adansonia digitata* (bread fruit tree), *Sclerorya birrea* and *Ziziphus mauritania* serve as food for man and *Acacia senegal* as fodder for animals. Thus their absence will lead to a pronounced scarcity of food.

The lowering of the density of the vegetation cover, as described above, increases the physical transformations of the soils:

- Encrustation of the soils through splash, leading to a layered surface skin, compaction and reduction in the infiltration of the precipitation, increased erodability and especially strong inhibition of germination
- Concentration of organic matter and nutritional substances in the uppermost part of the soil, aggravating the effects of water and wind erosion which carry away the rich upper layer

In a section from the north of Mali to the south, from the Mauretanian border to that with Guinea, the comparison of aerial photographs of the 1950s with those of the 1980s, supported by fieldwork, confirmed that:

- The vegetative cover diminishes during dry years.
- The incrustation of the surfaces grows in bare areas.
- Rill wash increases.

---

[14] *Pinnules:* leaflets of a composite leaf arranged on either side of a common stalk.

- Numerous trees, especially in higher lying areas, die off.
- The density of the gramineous vegetation cover recedes.
- Paradoxically, the effects of the drought, the increase in rill wash, and the increase in colluvial deposits into lower lying areas, resulting from deforestation from the higher areas to the slopes and the interfluves, correlate with regrowth of forests in the lower parts of the landscapes, as observed by De Wispelare (1980) in the Senegalese Ferlo.

1.5
Conclusions

The dry lands of tropical latitudes comprise the arid, semi-arid and subhumid dry environments where precipitation is not registered for more than 6 months/year. It extends from equatorial latitudes in eastern Brazil and East Africa (Chalbi desert) to temperate latitudes in Asia. Its open vegetation is of the steppe type or xerophytic bush which is nearly absent or only contracted in arid and hyperarid regions. The drylands are subdivided according to the degree of the water deficit which controls the presence or absence of a vegetation cover. In addition, they are subdivided according to the temperatures which distinguish between tropical deserts and the deserts of temperate latitudes. The dry cold areas were not dealt with because frost here is the factor limiting plant growth and pedogenesis.

The sensitive nature – or vulnerability – of the vegetation covers of drylands is the result of the weak persistence of the covers and their poor stability. This reflects both their dependence on precipitation, especially on the variability of the latter, and their resilience, i.e. their capacity for rehabilitation when favourable conditions are re-established.

Today, many scientists are of the opinion that the state of the rangeland in semi-arid environments is more likely to be the result of repeated droughts inherent in such ecosystems, rather than the result of human activity. With the present state of our knowledge any decision between these possibilities is difficult. Whatever the decision may be, the anthropogenic effects are difficult to evaluate. One – possibly optimistic – vision suggests that overgrazing in areas of extensive grazing has been overestimated as a reason for degradation. Warren and Khogali (1992) gave seven reasons for this:

1. The herds rarely reach the carrying capacity of an area.
2. When too numerous, they cannot be maintained during the dry season and because of this the pastures of the rainy season will not be damaged.
3. The carrying capacity is difficult to estimate where the herdsmen move constantly between heterogeneous areas of highly variable quality.
4. The pastures are rehabilitated rapidly after droughts and thus not damaged in the long term.
5. Despite the numerous assertions about a degradation of the vegetation the herds have continued to grow over the last few decades.
6. Where agrostologists have in the past observed severe damage around watering points, recent studies have noted that areas have improved by the supply of dung from the livestock coming to the surrounding pastures.

7. Many pastoral communities have improved their strategies of utilizing the pastures. The knowledge of the pastoral nomads makes them experienced users of the qualities of their pastures. It is in those areas where pastoral and agricultural populations are in competition with each other that the damage is at its maximum.

The spatial position of the drylands, their characteristic features aridity and drought, their poor soils and their open vegetation are sufficient to consider them as high-risk areas. We shall see in the second part of the book that they possess specific constraints limiting the utilization of their resources.

CHAPTER 2
Resources vs. Hydrological and Aeolian Constraints

The dry ecosystems being defined by their localization, their characteristic features aridity and drought, and by their soils and vegetation, we shall examine in which way these environments are victims of a fundamental handicap by presenting an inventory of their limited resources, especially with regard to the search for *sustainable* development (as proposed by the Brundtland Report 1987), and by presenting an analysis of the specific constraints in which the hydrological and aeolian conditions oppose an optimum management of these resources.

In drylands as well as in other environments these constraints are physical (or environmental) and human (financial, economic, social, legal, religious, cultural, political) in nature. The reader should understand that the study of the former will be limited only to the hydrological and aeolian constraints. With water being of poor availability as well as unequally distributed in time and space, and wind being omnipresent – are these not the two decisive physical parameters with which man is confronted but which he has the technical potential to deal with? To deal now only with these two physical constraints has also two other reasons: their specific nature in drylands and the possibility of fighting them on the short term. The human constraints, and especially the social, religious and cultural ones, entail certain taboos which must be overcome on the medium to long term. They shall not be dealt with here among the other resources, but will be discussed more extensively in the third part of the book under the aspect of human ingenuity.

2.1
Inventory of Resources

The notion of the environment or the framework of life is supplemented for man as for any living being by the concept of resources, i.e. of anything within this environment that is able to fulfil his needs and demands. In relation to the rest of the living world, man is certainly the being who knows how to look for his resources near and far or, in other words, how to extend his sphere of life to the limits of the planets. Has he not already started to extend beyond these limits several decades ago?

The spatial distribution of man and his activities is actually being transformed: areas of concentration (towns and megalopolises) are established, separated from each other by vast areas deserted because of a possible temporary loss of fertility or even an apparent economic sterility. A spatially homogeneous geography is replaced by a network-type geography of flux enclosing deserts (in the sense of areas without man).

However, the concept of resources depends on the ecosystem, i.e. on climate, geographical situation (relief, hydrography, continentality, insularity, etc.), on historical development, on cultural wealth, etc. The list could be continued, the analysis prolonged. The natural and human resources of the drylands have to be classified based on this complexity:

- *Permanent resources*, not degradable by erroneous management: solar energy, wind energy, gravity, wave energy and geothermal energy. Note that these are all energies. The exceptional insolation of the deserts is a major trump for the vegetation from which other ecosystems do not profit to the same degree.
- *Renewable resources* which, on the human time scale, are able to regenerate when the intensity of human activities or of natural degrading mechanisms does not exceed their regenerative capacities: air, fresh and salt water (the mass of water is a planetary constant), soils, flora and fauna. These resources are renewable on a historic time scale but not unlimited on a planetary scale.
- *Non-renewable resources* which are present only in defined quantities or the replenishment of which is too slow compared to their rate of exploitation. These are most of the mineral resources and the so-called fossil underground waters, i.e. waters which infiltrated into their aquifers thousands or tens of thousands of year ago, like the *Continental Intercalaire* horizon of the Sahara.
- *Human resources*, intrinsically tied to the very nature of man: his know-how, his institutions, his management which utilizes the first three categories of resources. In drylands the population density is low, e.g. only 1–10 inhabitants/km^2 in the semi-arid regions south of the Sahara (Chevallier and Claude 1989), and is accompanied by the mobility inherent in nomadism and transhumance.

Teilhard de Chardin (1956) stressed "the four properties" of human resources:

- "an extraordinary urge for expansion
- an extremely rapid differentiation
- an unexpected persistence in the potential to multiply
- and finally a capacity, unknown before in the history of life, to interconnect 'between the branches of the same bundle'"

These properties – resources of the (possibly privileged?) temperate world turned out to be constraints in the drylands, standing at the start of the demographic explosion of the 20th century and of its consequences, the degradation of the environment and the famines. This situation differs according to the resources:

- The drylands are well situated globally with regard to the permanent resources; wind and solar energy are more readily available here than elsewhere.
- Among the renewable resources water, soils, fauna and flora are less available, more fragile or, in the ultimate deserts, not available at all.
- With regard to non-renewable resources the drylands are also disadvantaged: even when they contain minerals of great value, their extraction and transport to centres of transformation are sources of particular difficulties and high costs. The fossil waters, however, possess here a value much higher than anywhere else; they

should be here still more respected and their exploitation to exhaustion represents a virtually irreversible step.
- Finally, human resources are smaller in drylands. It has been shown by the investigations of the UN during the 1950s that raising the temperature by even a few degrees will lower human capacities. Because of his naked skin as well as his low resistance against dehydration man is just as poorly protected against excessive cold as against excessive heat, and he will not survive a loss of water above the equivalent of 20% of his body weight. How is it then possible that there are signs of his presence during earlier prehistoric times in arid and semi-arid regions?

After this rapid analysis we shall arrive at the first conclusion that the drylands offered man a whole series of resources, albeit overshadowed by certain constraints which for him were challenges to be overcome, just as much as occasions for putting to the test the particular characteristics which distinguish him for the rest of the living world.

2.2
Hydrological Constraints in Drylands

Water, together with light, air, and temperature is one of the bases of life. This truism takes up unforeseen dimensions in our modern societies for which the abundant supply of water is a social and economic necessity. The situation becomes more complicated in drylands where there is a whole array of degrees of water deficit between hyperaridity and the subhumid situation. This leads us to the statement that the scarcity of surface waters here is the decisive factor limiting any development. We shall examine the mechanisms: the haphazard nature of precipitation (variable in time and space); the intensity of the other climatic factors (insolation, temperature, evaporation); the state of the surface waters; and finally that of the underground waters.

2.2.1
Specific Hydrological Features

Thornes (1994) summed these up in seven points:

1. The precipitation is very intensive, its annual total is low and it falls at irregular intervals, with irregular starting points of the rainy seasons and high interannual variability.
2. When the rains fall on partly barren soils, they lead to compaction which affects infiltration; evaporation is then increased.
3. Infiltration is highly dependent on the nature of the substrate: barren rocks; rocky, sandy or salt-rich soils; complex surface roughness ... It is counteracted by algal crusts and encrustation compaction of the soils, but is supported by vegetation and the presence of organic litter in the soil.
4. Evaporation losses depend on the water pressure in the soils, on the pedogenic profiles and on the atmospheric data. Thornes (1994) estimated that the loss of water from the soil in vapour form can amount to 1500–2000 mm/year close to the surface, dropping to 100 mm/year at a depth of 2 m and becoming negligible at 3–4 m.

5. Inundations after downpours occur rapidly over a very rough soil topography.
6. Runoff is ephemeral and depends on the losses within the alluvium.
7. The underground waters obey the same rules as those in temperate ecosystems, but the role of the underflow within the alluvial beds is more pronounced.

2.2.1.1
Scarce Precipitation

This is one of the main causes for the difficulties of survival in the drylands. Man can locate drinking water only with difficulty. He will not immediately find directly available surface water, without which he cannot cultivate the food crops he requires. The scarcity of precipitation entails long time intervals between two successive rains, increased interannual variability, increased aridity and more aggressive rains.

In addition to the effects of the poor precipitation, caused essentially by the absence of frontolysis and by the effects of obstacles to the penetration of humid air, which have already been mentioned, the following mechanisms also influence the scarcity of rains:

- *Dynamic subsidence*: the *jet stream*, an air current from west to east at a height of 12 km, depresses the cold air of higher latitudes to the ground on its right, thereby causing dynamic tropical anticyclones which are stationary above the large deserts.
- *Thermal subsidence*: above a cool surface the air becomes denser and sinks to the ground, explaining e.g. the thermal winter anticyclone over Siberia.
- *Orographic subsidence*: on the lee side of orographic obstacles the air descends, becoming drier because of a foehn effect, leading to the formation of piedmont and intra-montane deserts. This has already been described in the first part of the book, when dealing with the great basins of the western US, followed to the south by the Mohave, the Sonora and the Chihuahua Deserts and the Punas de Atacama of Peru.

In the southern part of Israel in an hyperarid climate with less than 50 mm/year precipitation, Sharon (1972) observed at Nahel-Yeal over an area of several hundred km^2 that 50–60% of the precipitation is rather local in nature, as it is tied to small convective cells of 5 km diameter, which are clearly separated in time and space. These cells are distributed without a recognizable pattern and affect only 20% of the area every day. This localized arrangement is more pronounced at the end of spring and in autumn. In winter the distribution of the rains is more uniform in space. There are two types of rains which are also representative of the dry Saharan areas:

1. Rains of low to medium intensity lasting several hours are spatially uniform.
2. High intensity rains lasting only a few minutes occur in localities which correspond to small convective cells.

The tropical deserts are subject to a precipitational regime which is degraded when compared to that of the margins. Thus the northern Sahara receives its rains during autumn like its Mediterranean margin whereas the southern Sahara receives them during the boreal *"hivernage"* in August, depending on the circulation of the intertropical convergence front.

2.2.1.2
The Interannual Variability of the Precipitation

This variability caused Toupet (1992) to talk about the tyranny of the rains. Climatologists base their studies on the data of the WWR (*World Weather Records*) of the National Center of Atmospheric Research in Boulder (Colorado, USA) which utilizes the data of the NCC (*National Climatic Center*) in Ashville (North Carolina, USA) which, in turn, receives its data from the synoptic stations forming part of the WMO (*World Meteorological Organisation*, Geneva, Switzerland).

On a map by Pettersen (1941) of the interannual variability of rains around the globe, the areas of greatest irregularities correspond to the tropical deserts. The pronounced deserts actually exhibit a variability of above 40%. These are the Sahara, Arabia, the Thar as well as the deserts of central Asia and China in the Northern Hemisphere, and the Namib and the desert of Chile and Peru in the Southern Hemisphere (Fig. 2.1). Demangeot (1981) has observed values of 80–100% for the central Sahara and more than 100% for the Libyan Sahara, reaching 150% at Dakhla between Koufra and Kharga.

The variability of precipitation in the Sahel zone may be illustrated with the aid of standard deviations and coefficients of variation:

$$\text{standard deviation } s = \frac{(Sx^2 - T^2)/N}{N-1},$$

where the standard deviation is an indication of the dispersion of measurements around the mean and is obtained by calculating the square root of the variance; Sx^2 = the sum of the squares of the total rains for x years; T = the arithmetic sum of the measurements; N = number of samples.

The coefficient of variation $V = (s/x)\,100$ is expressed as the ratio of the standard deviation to the mean (x) in percent.

Throughout the Sahel the variability increases from the south towards the north. The coefficient of variation increases northwards with decreasing rains, varying from $V < 20$ in the south of the Sahel around Kayes to 71.7 at Fderik and 97 at Nouadhibou.

2.2.2
The Fugacity of the Surface Waters

After a note on the causes and mechanisms controlling the most determining factor in drylands, i.e. the acute deficit in water, we shall look at the forms of the surface water and the characteristics of the underground water. Margat (1992) defined three classes of water resources:

1. *Class 1*: the surfaces resources – rill wash and runoff – and the underflow accompanying the water-courses. In tropical dry environments, in regions of the crystalline basement and in basins of temporary runoff the latter are the only sources of permanent water. We prefer here to include the underflow along rivers with class 2 as their exploitation requires pumping.

84 2 · **Resources vs. Hydrological and Aeolian Constraints**

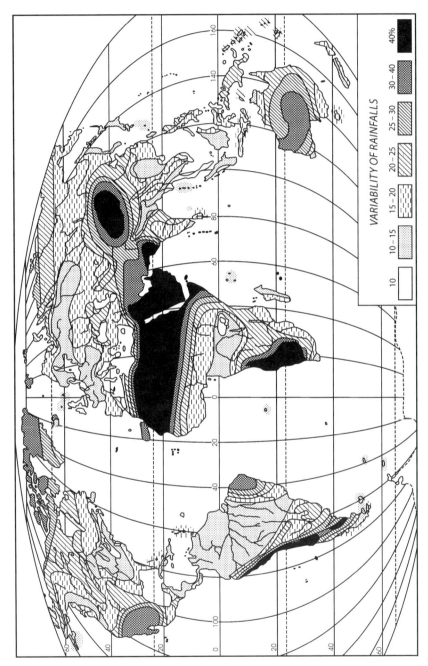

Fig. 2.1. Interannual variability of rains on the surface of the Earth (after Pettersen 1941, Introduction to Meteorology, p. 281). The rainfall variability is highest in the driest tropical and subtropical zones

2. *Class 2*: predominantly renewable water resources with potential for exploitation to satisfy permanent needs, within certain limits.
3. *Class 3*: predominantly non-renewable underground water resources.

For his survival man utilizes surface waters in four forms: allochthonous courses, wadis with temporary runoff, rill-flow waters and endorheic bodies of waters. The arid and semi-arid zones possess only resources of classes 2 and 3 for their development. The low rate of precipitation in fact combines with the severity of the potential evaporation (Table 2.1) in limiting the resources of surface waters. Real evapotranspiration is around 2500–3800 mm/year at a precipitation of only 400 mm/year and the evapotranspirational deficit thus is in the range of 2000 mm/year (Chevallier and Claude 1989).

The surface waters are considered here only to the extend to which they serve as supplies of water for man. We shall only briefly refer to the mechanisms of drainage organization and the dynamics of running water as morphogenetic phenomena even if we know that despite their present scarcity the surface waters are efficient agents of land forming and of erosion in drylands and that these environments have experienced more humid climatic phases during which runoff has created a network of valleys reactivated in our days after accidental rains.

The runoff is characterized by short duration, irregularity, high velocities, the absence of clear limits of the drainage axes – explaining the usual flooding beyond the poorly marked banks – and especially by the importance of the underflow in the whole water system developing when there are alluvial deposits. The distinguishing mark of the drylands is the absence of permanent runoff except for the allochthonous water-courses traversing them.

2.2.3
The Allochthonous Water-Courses

These are water-courses the source and upper reaches of which are located in an ecosystem which is moister than the one they are later traversing. These waters receive little or no tributaries at all during their traverse through drylands and are greatly reduced by evaporation.

2.2.3.1
The River Nile

The Nile, as the principal if not only source of surface water in Egypt, is at present still referred to as an example of an allochthonous river. It allowed the birth of the

Table 2.1. Potential evaporation (PET) in semi-arid environments

Locality	Authors	PET (mm)
Bol (Lake Chad)	Ponyard (1985)	2170
Mare d'Oursi (Burkina Faso)	Chevalier et al. (1985)	2860
Lake Bam	Ponyard (1985)	2343

Egyptian civilization, the prosperity of which, until modern times, was based on an agrarian economy. It is the longest river on Earth with a length of 6694 km; it originates in equatorial latitudes and enters the desert at the latitude of 16° N near Kharthoum, in Sudan (Fig. 2.2).

The Nile feeds a spectacular irrigation system, the advantages and disadvantages of which we shall be seeing later. It flows through a vast alluvial plain bordered by

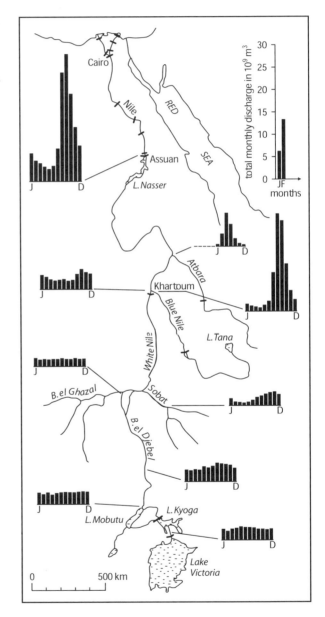

Fig. 2.2. Discharge of the Nile at Kharthoum (after Bakre et al. 1980). The confluence of the White Nile – conserving from its equatorial origin and its traverse through the lakes a balanced water regime – with the Blue Nile – marked by a pronounced tropical regime and a maximum in August to September

terraces, and its delta extends for some 200 km into the Mediterranean. Prior to the construction of the first dam below Cairo in 1861 the river here split into the seven branches forming the delta. The active sedimentary load at Cairo was 120 000 t/year of which about half spread onto the cultivated lands, whereas the rest was transported to the sea, explaining the size of the delta.

2.2.3.2
Allochthony in the Middle East

In the Middle East, the Euphrates and Tigris, the *Perath* and *Hadakat* of the Bible, are the main rivers of Asia Minor and Mesopotamia (Greek for "between the rivers") forming the cradle of Western civilization. Their sources lie in Turkey, they traverse Syria and Iraq and form the Chott-el-Arab and the Basra estuary at the northern end of the Persian Gulf. The Euphrates carries 31×10^9 m^3/year towards Syria and the Tigris 17×10^9 m^3/year to Iraq. They are examples of allochthony, a purely geographical concept that has taken on a major geographical note, taking into account the antagonisms between the rapidly developing countries of the area with their high population growth.

The Atatürk Dam constructed on the Euphrates in Turkey some 110 km north of the Syrian border was inaugurated in 1992 and forms the last step in the most important project in southeastern Anatolia, the *Güneydogu Anadolu Project* (GAP). It extends over seven provinces of the southeast with a total area of 75 000 km^2 or nearly one tenth of the land surface of Turkey, the most advanced country of the area with a constant undersupply of electrical energy. Co-ordinating 13 supplementary projects for irrigation and production of electrical energy, the GAP entails the construction of 22 dams and 19 hydroelectric power stations on the Tigris, Euphrates and their tributaries.

As the ninth largest dam in the world with a height of 169 m the Atatürk Dam alone is able to store four times the mean annual discharge of the Euphrates. The project was conceived as a source of alternative energy during the oil crisis of 1973. The region which it will transform is one of the most arid and most disadvantaged parts of Turkey with a mean gross product amounting to only 47% of the national average. It is entirely financed from the national state budget and has already cost $9 billion. A further $23 billion cost is foreseen by the end of the century together with $10 billion for associated projects. In total the project absorbs 10% of the annual state budget, not counting the credits offered to incite private investors to come to the region (Sablier 1992).

In this way Turkey is able to control the upstream parts of the two largest rivers of western Asia and their tributaries, which obtain most of their water from snow and rain in the Taurus and Zagros mountain ranges. Turkey estimates that with the GAP it would reduce the discharge of the two rivers by about 20% and in 1981 undertook to supply Syria with 500 m^3/s, at a time when Syria would have required 700 m^3/s. This drop in water supplies has devastating consequences for Lake Assad on the Euphrates in Syria which supplies that country with most of its electrical energy. Because of a water shortage throughout the summer of 1991, the Syrians experienced cuts in water and electricity supplies which led to reduction in the quantities of oil destined for export, the proceeds of which would have been used to buy cereals. The

Turkish thirst for water revives the fear of the authorities in Damascus and Baghdad of falling back into their historic dependency upon their northern neighbour.

After fighting for land, the peoples become aware of water which might develop into a new motive for wars. Was the *Six-Days-War* not perhaps started by Israel after it learned that its Arab neighbours had convened a conference in which they studied the deviation of the waters of the Jordan River? And is Egypt not watching with great attention the hydraulic projects carried out on the Nile by the Sudan? And one of the difficulties about the present situation in the Aral basin – the home of five Republics – lies in the allochthony of the Amu and Syr Daria rivers.

2.2.3.3
Natural Allochthony in Central Asia and One Example of Anthropogenic Areism (i.e. the stopping of runoff by human activities)

Meltwaters of mountain glaciers form the rivers Amu and Syr Daria, the courses of which end naturally in Lake Aral after flowing through the sandy Karakum and Kyzylkum Deserts (Figs. 2.3 and 2.4). Other allochthonous water-courses in Asia also obtain their water from bordering mountains covered by glaciers: those supplying the town of Urumqi (Sinkiang, China) descend from the Tienshan massif to the south. The Tarim coming from the more than 7000-m-high Kuenlun massif, ends in the Lop Nor depression at 780 m, the lowest point in the basin, after a wide circle through the northern, hyperarid part of the Taklamakan desert.

The Amu and Syr Daria were the only natural permanent water-courses in the hydrological system of the Aral and the only ones which until the start of the large hydro-projects of the 1960s still reached the lake, but which since the runoff are annually stopped becomes of excess water withdrawal. The valleys of temporary runoffs like the Kasha-Daria, Zerafzan, Mourgab, Tedjen, Tchu and Talas on the territory of the CIS and the Hulm, Balkh, Sary Poul and Shirintagar of Afghanistan also are part of this surface runoff system and together with the Syr and Amu Daria contribute about 120 km^3 annually (Fig. 2.4). Few examples exist for such large water-courses feeding an endorheic lake like the Aral, save for the Tarim and Lop Nor in the Sinkiang depression. In Africa the system of the Chari-Logene and Lake Chad is an example albeit on an smaller scale.

2.2.3.4
The Allochthonous Water-Courses of the Sahel

The fluctuating state of the regime of the allochthonous water-courses of the Sahel and the Sudanian-Sahelian zone (Chari, Logone, Niger, Senegal) affects in part the precarious nature of the settlement capacity of the region. We have to remember that according to the United Nations the population in the Saharan-Sahelian, Sahelian and Sudanian-Sahelian zones south of the Sahara had risen to 31 million by 1980, corresponding to a density of six inhabitants per km^2 if we exclude the Saharan and Saharan-Sahelian parts. This might appear low but we shall see how risky it is to take only such mean densities into account in development programmes.

The allochthonous water-courses of the Sahel possess permanent flow, their rainy season is concentrated in the month of August, with a peak of the floods in Septem-

2.2 · Hydrological Constraints in Drylands

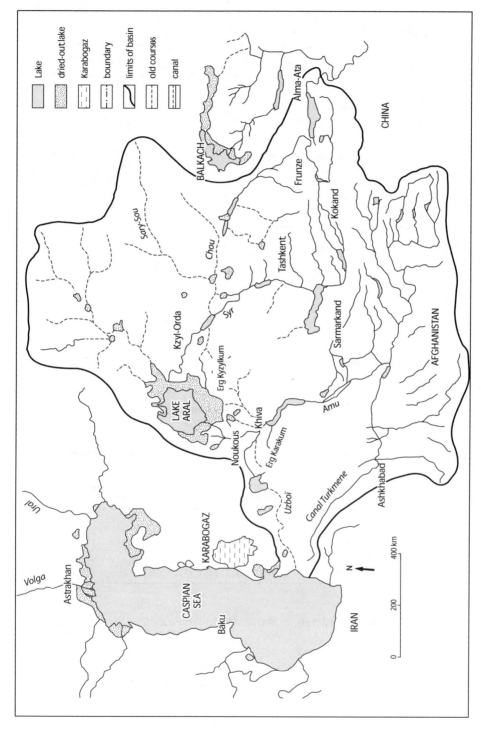

Fig. 2.3. The international nature of the Aral basin

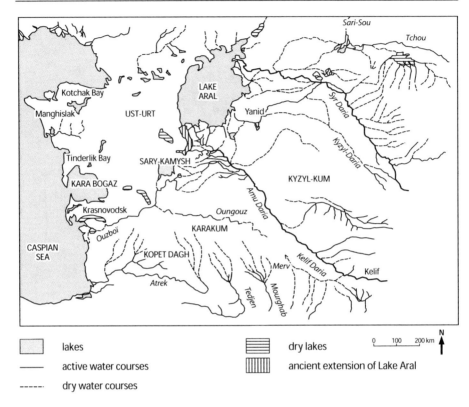

Fig. 2.4. Recent and older water-courses in the Turan (Letolle and Mainguet 1993)

Table 2.2. Discharge (in 10^9 m^3) of allochthonous water-courses in the Sahel (after Sircoulon 1992)

	Senegal at Bakel	Niger at Koulikoro	Bani at Douna	Chari at N'Djamena	Total
Start of observations 1985	22.3	46.2	17.3	35.1	120.9
Start of observations 1967	24.7	48.7	22.1	40.4	135.9
Period between 1968–1985	13.7	37.7	8.3	24.6	84.3
Year 1984	6.9	20.1	2.2	6.3	35.5

ber/October and the lowest levels in March/April (Table 2.2). The mean annual discharges are:

- The Senegal from Bakel to St. Louis 24×10^9 m^3 (Fig. 2.5).
- The Niger at Mopti 70×10^9 m^3, at Diré 36×10^9 m^3 and at Niamey 30×10^9 m^3.
- The Chari 40×10^9 m^3.
- The Logone 15×10^9 m^3.

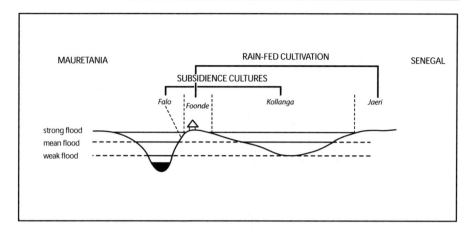

Fig. 2.5. Schematic section through the Senegal valley (after Olivry and Chastenet 1989)

According to Sircoulon (1987) the evolution of the allochthonous water-courses of the Sahel from the start of the great drought in 1968 until 1985 exhibited:

- A lowering of the total annual discharges.
- A drop in the maximum height of the floods.
- A drop in the low-water levels: the Senegal at Bakel, the Niger at Niamey and the Chari above its confluence with the Lagone ceased to flow completely.

All water resources of the Sahel must be considered as endangered by the climatic variations. Sircoulon (1992) stated furthermore that the improvement of the precipitation since 1985 has not led to a return of the discharges of the large rivers to normal levels in the following years.

2.2.4
Rill Wash and Runoff in Tropical Drylands

There is only one Anglo-Saxon term for the mechanisms by which water moves on a surface, viz. *runoff*. French geographers use the word "*ruissellement*" for the movement of water on slopes of whatever gradient for which we propose using the English rill wash or rain wash, and "*écoulement*" for the movement of water in a valley or on a plain for which the English word runoff can be used.

2.2.4.1
The Dynamics of Runoff on the Slopes

Under runoff we understand two actions:

1. The rill wash leading to water erosion especially on cultivated lands
2. The rill wash forming the initial supply of runoff.

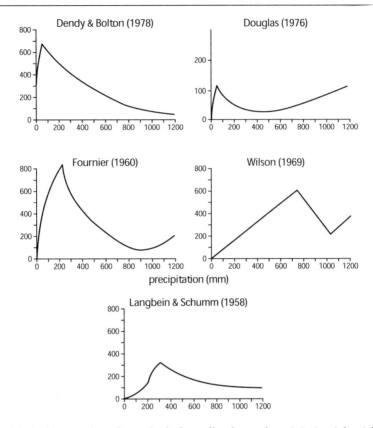

Fig. 2.6. Relation between the sediment load of runoff and annual precipitation (after Schainberg and Morin 1987). For the *four top diagrams* the *abscissa* gives the mean annual flow in mm, in the *bottom figure* the real precipitation in mm. The *ordinate* shows the sediment load in t/km² per year. N.B. There is an increase in the sediment load with increasing precipitation in the arid zone; from the semi-arid and the Sudanian zone the sediment load drops together with water erosion; in the Guinean zone the sediment load and the water erosion increase again

The main features of rill wash and runoff should be noted, as the absence of springs initiates sheet flow, and in the drylands controls the recharging of the underground aquifers which is in particular assured by the downward percolation of water from the beds of the water-courses. Paradoxically, rill flow and runoff are strong erosive agents. Schainberg and Morin (1987) presented five graphs (Fig. 2.6) illustrating the relationship between annual precipitation and the mass carried away by water erosion. With the exception of the graph by Wilson (1969) there is a common trend in the graphs:

- The maximum water erosion correlates with low precipitation in arid to semi-arid regions ranging from 70 mm/year (Dandy and Bolton 1976; Douglas 1976) to 20 mm/year (Langbein-Schumm 1958; Fournier 1960).
- After this maximum is reached, water erosion paradoxically decreases with increased rainfall: the curves of Langbein-Schumm, Fournier and Dendy and Bolton

agree and exhibit an erosive minimum in the Sudanian climate with 800 to 1200 mm annual precipitation. But, because of the high temperatures, water density and viscosity are lower, long-distance transport increases at an equivalent flood velocity.

In a large number of semi-arid and subhumid dry areas rill wash in the cultivated areas in assisted by the existence of a less than 1-mm-thick surface crust resulting from the combined action of the kinetic energy of rain drops falling onto the ground and of the dispersive effect of these drops. The permeability of the surface is reduced by thin compacted layers or crusts, sometimes in combination with algal biocrusts (Casenave and Valentin 1989). The former block the pores of the soil and reduce infiltration whereas the influence of the latter on infiltration is difficult to interpret: limitation of the splash effect and an increase in infiltration or, on the contrary a decrease by water-repellent action. Yair (1990) suggested that in the Negev the crusts might protect the dunes against erosion. Investigations carried out in the *Tengger Desert* of China revealed that the fixation of dunes starts via an initial stage of organic encrustation.

Shainberg and Morin (1987) observed that the formation of a crust and its impermeability depend on the percentage of exchangeable sodium ions (PES) at the soil surface and on the concentration of electrolytes in the percolating waters. The permeability decreases as PES grows and the concentration of electrolytes decreases. The depth of the humidification front and the distribution of the humidity have little effect on the rules of infiltration of encrusted soils.

In the 1930s Davies proposed a classification for the various forms of runoff which is still valid:

- Laminar or film-like runoff, which equals *sheet-wash*;
- Diffuse runoff, which is called *rill wash* when organised in rills;
- Runoff in sheets, which becomes *sheet-wash when the rills overflow and join to form layers.*
- Concentrated runoff in ravines or gullies.

Dubief (1953) observed in the Sahara that a downpour of 5–8 mm with minimum intensity of 0.5 mm/min suffices to start rill wash. The rill wash coefficient is higher over dryland soils and is furthermore increased by the vast areas of barren rock. Dresch (1982) reports coefficients of 55% on sandstones at Ennedi (Chad), 45% in the American West and even values reaching 50–80%. The coefficient of infiltration is high on porous rock and flat surfaces, but where there is even a slight slope or a cover of desert varnish, the rill wash with its strong erosive potential will predominate.

Tables 2.3 and 2.4 illustrate the variability of rill wash as a function of the respective bedrock and of the state of the soil surface as shown by Casenave and Valentin (1986).

Tixeront (1965) distinguished between "saturation rill wash" when the soil is water-saturated after prolonged rain and "intensity rill wash" when the intensity of the rain exceeds infiltration velocity. In the region of Wadi El Kebir in Tunisia where the precipitation is about 500 mm/year the soil will become saturated by a rainfall of 100 mm.

Table 2.3. Mean runoff in Sahelian basins (after Chevallier and Claude 1989)

Bedrock	Mean precipitation (mm)	Mean runoff (mm)	Runoff coeff. (%)
Granite, gneiss			
Abu Goulem (Chad)	400	24	6
	500	35	7
	600	46	7.5
Gagara-West (Burkina)	300	48	16
	400	68	17
	750	172	23
Schists			
Po (Mauritania)	400	72	24
	500	104	26
	600	172	28.5
	300	113	37.5
Ader Doutchi (Niger)	500	200	40
	750	300	40
	350		12
Mare d'Oursi on gabbros, granite and migmatite	350		37

Table 2.4. Drainage vs. soil texture for a mean precipitation of 400 mm (after Chevallier and Claude 1989)

Texture	Clay	Clay/mud	Mud	Sand
Drainage (mm)	6.8	10.2	13.5	26
Drainage coeff. (%)	1.7	2.5	3.4	6.5

The depth to which the humidification front descends in drylands is small and the recharge of the aquifers is negligible, except under water surfaces and along valleys with runoff where infiltration through alluvial deposits may be higher, leading to a notable recharging of the local aquifers.

2.2.4.2
The Wadis – A Runoff Discontinuous in Time and Space

The terminology of Martonne and Aufrére (1928) for characterizing the different types of runoff in drylands still serves its purpose:

- *Arheic*: areas without runoff like the basement pediments, regs, sandy regions and sanded-up depressions.
- *Exorheic*: areas with runoff leading to the seas.
- *Endorheic*: areas in which the runoff ends in a continental depression like the Aral basin.

In the subhumid dry Sudanian-Sahelian regions runoff decreases when precipitation drops below 700–800 mm/year. The 750 mm isohyet for hydrologists represents

Photo 2.1. Wadi Ferran in the Sinai. A small palm grove benefits from the underflow of the temporary regime of Wadi Ferran, surrounded by densely fractured crystalline rocks

the boundary of the drylands. This degradation is caused by a number of complex factors:

- The length of the dry season during which the gramineous layer disappears; runoff is no longer slowed down and infiltration no longer supported.
- The short duration of runoff, which is insufficient to remove all material eroded by water erosion. The bed of the rivers becomes overloaded, the main course is divided into several channels, becomes choked and may terminate its course in an endorheic basin.
- The shallow nature of the slopes and the vast surfaces of sandy covers piled up into hillocks.
- Competition between fluvial and aeolian mechanisms, the latter forming dunes blocking the valleys. An example for this competition is furnished by the inner delta of the Niger where the supply of sand in the form of E-W oriented longitudinal sand ridges blocks the general SW-NE flow.

In the arid and semi-arid zone the runoff, neither a regular nor a permanent feature was characterized by Sircoulon (1992):

- "by its intermittent nature (because of the concentration of the rains over several weeks or months);
- by its pronounced heterogeneity in time and space resulting equally from the rather heterogeneous nature of the precipitation and the highly variable permeability of the soils ... [Photo 2.1]

In such regimes certain useful parameters lose their significance and interest: what is the significance of e.g. a mean annual discharge? Or low water in a water-course that is dry for eleven months of a year? In contrast to this the concept of runoff volume, duration of runoff and flood period (maximum, shape, duration) take on a particular importance."

The fluvial networks do not exhibit any long-term trends (over several decades) but are subjected to variations over years or a decade. This illustrates that they are not heading towards an equilibrium. Furthermore, the concepts "long-term" and "equilibrium" are not well understood. The short-term variations are responsible for the frequent changes in bed configuration, characterized by meanders and anastomosing laterally braided channels, with catastrophic effects for the environment and for human activities (Baker *et al*. 1988). They make such waters difficult to utilize for development.

The violent nature of the floods and their heterogeneity in time and space may be explained by climatic, geographical and biological factors:

- *Climatic*:
  - Short duration of convective downpours lasting only a few minutes to a few hours.
  - Intensity of downpours; Sircoulon (1982) reports values of 150 mm/h over 5 min at a total annual precipitation of 100–200 mm/year.
  - Scarcity of downpours, one dozen annually in the Saharan-Sahelian zone to 50 at the margin of the dry world around 750 mm/year.
- *Geomorphological*: variations in the slope values, nature of the soil, roughness and especially different degrees of permeability. Because of the lack of moderating factors, the runoff reflects the irregularities of the pluviometric regime. The rains, whether downpours or just a few drops, are localized in space and short in time, and these two features are little conducive to the organization of runoff into wadis. However, runoff does not have any other source than precipitation, springs are virtually absent, the thin and skeletal soils possess only a low capacity for water retention and evaporation always exceeds the pluvial supply.
- *Biological*: low density of the vegetation cover.

After the rapid formation of concentrated intense rill wash on the slopes the water does not encounter the intercepting effects or the roughness caused by a vegetation cover. The rocky barren reliefs of the inselberg type favour the rapid gathering of the water and appear to be engulfed by curtains of water during the violent rains as we have seen in the Ennedi of the Chad.

In the Saharan regions receiving less than 100–150 mm/year the few floods are seasonal, occurring throughout the two months of the rainy season between July and September, the time of the strongest rains. At Tamanrasset 39.5 mm were registered during the single day of July 1st, 1979 when the Wadi Abalessa came down in flood (Gribi *et al*. 1992). The piezometric level of the aquifer of the wadi rose by more than 1.5 m after this flood which, however, did not counteract the general drop of the aquifer by 4 m between 1972 and 1983.

In the desert zone two parameters support the formation of floods: the steep slopes (mountainous relief) and the impermeable nature of the soils. Under such conditions floods will take place every year.

2.2 · Hydrological Constraints in Drylands

Photo 2.2. Wadi Archei (Ennedi, Chad). The western edge of the Ennedi massif made up of pink palaeozoic sandstones at Archei. This wadi with its contracted vegetation of acacias on the sandy bed, lies at the foot of the sandstone massif which is weathered to pinnacle shapes. This wadi has an "accidental" regime, it flows only very rarely. In its upper reaches it contains the "gueltas" famous for their dwarf crocodiles. Precipitation is in the range of 100 mm/year. The gueltas are watering points for livestock. Several thousands of camels could be seen here at the same time in 1965 before the Sahelian drought (Photo M. Mainguet)

The Bardagué *enneri*[1] at Bardai in the Tibesti (Chad) with a catchment area of some 4000 km² may develop floods annually. Nevertheless, geographers from Berlin University, who studied the area prior to 1968 (the start of the Toubous war), have shown that even during the more humid phases from 1950–1960 floods did not occur every year.

The start of runoff was studied by Braquaval (1957) on the pediment below the sandstone massif of Ennedi (Chad) at the boundary between the arid and semi-arid zones. All floods here take place between mid-July and mid-September. They were launched in the wadis by rainfalls of intensivities above 20 mm over 15 min and 80 mm/h during the first few minutes and which on average lasted ten hours. At the Fada station in the same area evaporation rates of 10 mm/day have been measured.

In the Saharan-Sahelian zone (subdesertic) in the Air mountain massif and on its piedmont with a precipitation rate of 150 mm/year, several annual floods may occur during the two months of July–August (Sircoulon 1992). He mentions as an example "the little degraded basin of Kori Teloua in the Air (1300 km², 150 mm/year rains). Ob-

---

[1] *Enneri*: local term for wadi.

servations carried out since 1956 and quantitative determinations taken in 1959, 1960, 1964 and since 1975 have shown that the annual number of days with runoff fluctuated around a mean of 25 days/year. It reached 50 days/year in rain-rich years (1989) and dropped to 5–6 days/year during deficit years like 1972. However, in 1984, the driest year with an estimated mean rainfall of 15 mm over the basin against 169 mm in 1980, the total duration of the runoff was not more than 35 hours. This appears to indicate that the area has not experienced any runoff at all only once in a century."

Sircoulon (1992) observed furthermore that (Photo 2.2) "in the Sahelian zones a series of floods, frequently with continued runoff during the intervals is observed every year over a period which may surpass three to four months. In the small basins drought conditions do not appear to influence much the total duration of the runoff. In Burkina Faso, e.g. the Gorouol at Koriziena (2500 km^2) experiences 80–100 days of runoff per year even during years with rather meagre precipitation. In contrast to this in larger basins the duration is much more dependent on climatic variations. In the Komadougou basin (120 000 km^2) at Gueskérou in 1984 runoff was recorded only for four months from August 4 to November 30, whereas it usually lasted for nine months from July to the end of March because of a supply from underground aquifers".

Runoff, irregular in arid zones, becomes really seasonal in the Sudan-Sahel zones, from 400 to 600 mm/year.

2.2.5
Ponds and Water Bodies: The Third Type of Water Reserve in Drylands

Because they are less frequent in the drylands than in more humid ecosystems, such bodies of water assume a greater value here and the functions they serve are rather diverse:

- Sources for recharging the underground aquifers;
- Areas of attenuation and storage of floods;
- Areas for spreading out the runoff, dissipating the forces of water erosion;
- Catchment areas for sediments and nutrients eroded on the slopes;
- Areas for filtration of nutrients;
- Areas of more permanent production of natural or cultivated forage plants which are especially valuable in dry years, and frequently points of competition between livestock breeders and agriculturalists;
- Areas with fishing potential;
- Conservation areas for floristic heritage and wild fauna.

2.2.5.1
A Considerable Variety

Among the bodies of water in drylands, aeolian depressions, low points in sloping basins and valley floors, those which collect rain or rill wash water are the most frequent, although there are also some which are fed by underground waters derived from moister areas upstream. And, as outlined above, the supply of water may also come through floods from higher lying catchment basins. Their degree of permanence depends on precipitation, underflow, underground waters, evaporation and the impermeability of their floors. The main adverse factors are their shallow depth and the intensity of evaporation which in most cases prevent them from surviving during the dry season. Chevallier and Claude (1989) stated that the 15×10^6 m^3/year of the upper 2 m of the Oursi pool in Burkina Faso are taken up entirely by evaporation.

On the dry prairies of western Canada there are aeolian depressions dating from the last retreat of the glaciers. Since the arrival of the white man some 1.2 million ha have been dried out and taken under the plough, leaving in 1982 only 3.2% of such depressions compared to 1928. Conversion to agriculture entailed cutting of the tree belt and as a consequence a lower capture of snow, a drop in the concentration of water, a diminution of the recharge of the water-bearing layers and, in certain depressions with a shallow depth of the phreatic layer, a salinization which affected the habitat of water fowl (Hollis 1989).

Lake Kineret, the main water body for Israel, is supplied by the River Jordan flowing through Lake Hula, another example of a depression. Lake Hula, a peat-swamp, was drained in 1956 in connection with an irrigation project. As a conse-

Photo 2.3. Irrigated reafforestation in the Jordan Valley. Various tree species in irrigated cultivation 15 km north of Jericho, in a part of the Jordan Valley with a rainfall of 200 mm/year

quence the natural filtration by trapping of sediments and nutrients by the lake and its swamps ceased, leading to an accelerated growth of the Jordan delta in Lake Kineret, the eutrophization of which led the Israelis to devise a project to bring part of the water back to Lake Hula so that it may again play its role as a trap for sediments and nutrients and keep down the excessive deposition of sediments in Lake Kineret (Photo 2.3).

In Africa such bodies of water, depending on their degree of salinity, are referred to as *dayas* or *sebkhas*. As temporary bodies of water the *dayas* in the Sahara form on low-lying argillaceous areas after rains. The *sebkhas*, an Arab term describing saline floodable depressions, are known from central America under the name *salinas*, *salar* or *playa*, in Iran as *kewir* and as solontchaks in the deserts of central Asia. Lake Eyre in Australia with a surface area of some 9000 km^2 is one of the largest depressions of this type. Downwind of sebkhas on the *crescent-shaped* aeolian edifice (with a concave opening into the windward direction) and the *chotts*, saline pastures are suitable for livestock grazing.

The *pans* of the Greater Kalahari belong to the most astonishing examples of closed depressions and deserve special mention as sites of ephemeral bodies of water. They are present throughout those parts of southern Africa receiving less than 500 mm/year in precipitation, except Lesotho and Malawi, and especially in areas underlain by shales and unconsolidated sands. They are a prominent feature of the northern Kalahari up to Zambia where precipitation increase to 1000 mm/year (Thomas and Shaw 1991). Their diameters range from a few metres to 200 km, or e.g. 170 km in the case of the Makgandikgadi structural depression extending over some 37 000 km^2. Their modal size is between 1–16 km and their depth around 20 m. Because they are round, elliptical or kidney-shaped they are characterized by crescent-shaped dunes at their southern or south-eastern ends, depending on the predominant wind direction, and are located on a sandy substrate (Lancaster 1978).

That such depressions frequently result from hydro-aeolian action is proven by the existence of dunes made up of gypsum sand (parabolic dunes opening into the windward direction) like around the Linga Lakes in the western part of the state of Victoria in the south and south-west of Australia (Hammer 1986) and by certain depressions in the Algerian and Moroccan steppes. In many case the deflation is only a simple accentuating feature emptying depressions of an originally tectonic origin: The Fayum depression of Egypt was hollowed out during the upper Tertiary but its origin is rather complex, involving epeirogenic and fluvial processes (Lotze 1957 in Hammer 1986). The interest in such hydro aeolian depressions results from their water resources for livestock breeders and from the fact that in their vicinity grazing is available for a longer time into the dry period. Lake Chad, probably the most curious example of these hydro-aeolian depressions, possesses phragmites, areas of phanerogam type representing excellent grazing and reserves of forage for the dry season.

2.2.5.2
Lake Chad, a Shallow Tropical Endorheic Lake

As just a relic of a vast palaeoclimatic lake, this lake extended during the Holocene over nearly 350 000 km^2. According to Schneider (1992), the Greater Lake Chad was of mid-Holocene age around 9200 B.P. It formed part of a group of large shallow tropical endorheic lakes, clearly deserving special attention in this book. It is situated between 12–14° N and 13–15° E on a submerged sand sea with little differentiated topography. The northern half of the basin is a monotonous plain at an altitude of 400 m.[2]

Save for the Benue which runs through a gap, the Chad basin is entirely surrounded by mountains: in the far north-west the 2450 m high Hoggar, in the north between Libya and the Chad the volcanic Tibesti attaining a high of 3400 m, in the north-east the small 1400 m high sandstone plateau of Ennedi, in the Sudan to the east the 3020 m high Djebel Marra and to the south the Adamana and Mandara (>2000 m) and the Jos plateau ranging in altitude from 1000–1500 m.

The lake is divided into two parts occupied by open water and separated from each other by a line of shallows, the "great barrier". Both parts are bordered by islands, representing the tops of dunes of a sand seawhich is increasingly submerged from east to west. In the south these submerged sand bodies carry aquatic phanerogams (phragmites and papyrus) forming islands of vegetation or "islet banks". The plant barriers and the islet banks developed during a low stage in the water level.

The altitude of the lake floor varies between 275.8–280.1 m with a mean of 278.21 m. The mean depth is 3.6 m when 18 000 km^2 are covered by water,. The northern part of the lake is deeper than the southern one where the floor is rarely below 278 m. The open waters of the southeast appear to represent a perideltaic area depressed with respect to the surrounding areas. Lake Chad occupies the centre but is by no means the deepest point of the Chad basin. At an altitude of 160 m this is located some 500 km to the NE at the end of the Bahr-el-Ghazal in the lowlands of the Chad around Largeau, an area that is presently dry. This curious situation explains why the flow of the Bahr-el-Ghazal was able to take place into both directions during geological times.

The Climatic Framework. Between the limit of the arid zone in the north and that of the Saharan-Sahelian zone to the south, the lake experiences a dry season from November to March alternating with a short rainy season (depending on the position of the intertropical front); it receives a mean precipitation of 100 mm/year. The average annual air humidity is 36%. The evaporating power of the air as measured with the aid of a Piche evaporimeter (Mao station) and a dug-in Colorado basin at Bol gave a potential evaporation rate of 2200 mm/year (Maglione 1975).

The Hydrological Framework (after Sircoulon and Schneider 1994). As a basin without an exit Lake Chad receives the waters of the Chari augmented by those of the Logone downstream of N'Djamena. This water-course represents 95% of the fluvial supply, the remainder being contributed by the El Beid (4%), the Komadougou and the Yedseram (1%). Some 10–20% of the water supply comes from rains.

In 1909 General Tilho drew attention to the low mineral content of the lake, despite its endorheic nature and its subjection to strong evaporation. The size of the

---

[2] *Holocene:* The youngest period of the Quaternary, between the Pleistocene and the historic era. According to Schneider (1992), it started around 12 000 B.P.

lake surface fluctuates considerably, especially because of the changing supply from the Chari. With the aid of satellite imagery Mohler *et al.* (1989, in Schneider 1994) measured the areas given in Tables 2.5 and 2.6.

After Lake Aral, Lake Chad is one of the largest endorheic lakes. The absence of any exit in both cases represents a problem for handling sewage water. The lakes have experienced pronounced changes in their hydrological equilibrium during the last three decades. For Lake Aral these resulted from human activities; for Lake Chad these changes resulted not only from human activities, but also from the drought resulting in a reduced discharge of the water-courses. This climatic situation cannot, however, hide the negative effects of the irrigation projects for rice and cotton started at the beginning of the 1970s interrupting the seasonal runoff, reducing the grazing areas and, as a common consequence, resulting in overgrazing and soil erosion which themselves were aggravated by the movements of the population during the civil war.

2.2.6
The Underground Waters

"We know ... that the underground waters represent virtually the entire stock of liquid waters present ... in the emerged lands [98–99%] and a considerable part of the total fresh water runoff of our planet. Nearly one third of the total water flux circulating in the terrestrial environment is subterranean over a shorter or longer time span and over variable distances of the water-courses. The underground waters furthermore, as the most stable and most extensive phase of the terrestrial fresh water, represent a worldwide intensely utilized resource. Some $600-700 \times 10^9$ m^3 are drawn annually from underground, representing more than one fifth of the waters used for all purposes together. This proportion naturally is highest in drylands where surface waters are rare and highly irregular" (Guillemin and Roux 1992).

We have seen that in certain parts of arid and semi-arid regions surface waters are completely absent, whereas in others they are rare or only episodically present. This dearth of water is aggravated by coarse alluvial deposits, by accumulations of sand forming barrages and by irrigation (Loup 1974) which cause man to turn to the underground waters as frequently the only source of water.

The underground waters of deserts and semi-deserts were subdivided into two major groups (Guirski, in Schoeller 1959) depending on their origin:

1. *Autochthonous*, fed from the deserts themselves and originally derived from infiltration below the beds of water-courses and below bodies of water, permanent lakes or temporary swamps, and to a lesser extent from precipitation. The supply of water directly from infiltration of precipitation plays a significant role in those deserts in which the rainy season falls in the cold winter season, as it is not then removed entirely by evaporation. In truly dry areas the supply from rain is less than that from water-courses because of the floods and inundations. The supply is a function of the lithological nature of the bed, the state of the banks and the extent of the floods:
2. *Allochthonous*, i.e. originating in the neighbouring regions and coming into the deserts via permeable beds.

Table 2.5. Fluctuations of Lake Chad from 1800–1984 (various sources)

ca. 1800	The lake level was very high (286 m?)
ca. 1823	The lake level was around 282 m
1823 – 1850	Period of drying out
1850 – 1900	Flood period: level between 283–284 m
1900 – 1950	Lake level recedes and oscillates between 281–282 m, with a minimum of about 280 m between 1908–1914
1953 – 1964	Short flood exceeding 283 m at the beginning of 1963. The surface of the lake then was 23 500 km containing 105×10^9 m^3 of water
1965 – 1971	Low stage of lake with a surface area of 19 000 km^2 and a water volume of 49×10^9 m^3
1972 – 1973	Very low floods of the Chari
1973	The lowermost levels of Lake Chad since 1965 are reached
1984	Very low floods of the Chari; Lake Chad reaches its lowest level with a surface of not more than 3000 km^2

Table 2.6. Variations in surface area of Lake Chad between 1966 and 1986

June 1966	22 772 km^2
September 1973	1752 km^2
November 1982	2276 km^2
August 1984	1655 km^2
November 1984	1653 km^2
April 1985	1653 km^2
January 1986	4452 km^2

Loup (1974) distinguished three types of aquifers:

1. *Aquifers in alluvial layers or underflow*, resulting from the infiltration from floods or any other form of flow in the alluvial deposits. "Although the surface flows are only of short duration, they support … the recharging of the alluvial underflow." In the arid and semi-arid zones they are "… generally tied to precipitations which are more or less regularly spread over one or two rainy seasons per year or, as in the Sahara, to violent … precipitations exceptional on a time scale of several years" (Guiraud 1989). Certain perennial flows of this type supply population centres like Kano in northern Nigeria or Alice Springs in Australia, and the town of Massawa in Eritrea on the shores of the Red Sea exploits such an aquifer with an underground dam.
2. *Lateral underflow supplied by perennial rivers below the water-course.* The lateral aquifer of the Nile at Wadi Natrun, WSW of Cairo, some 30 km from the closest diffluence of the Nile, lies about 8 m below sea level. When in flood the river recharges the horizon over a width of about a dozen kilometres on both sides of its bed. The infiltrated discharge during strong floods amounts to 300 m^3/s in upper

Egypt and to 100 m³/s between Assiut and Cairo, where the total annual recharge is some 5×10^9 m³ (Loup 1974).
3. *Deep aquifers*; among these we find fossil aquifers, trapped in geological layers, and artesian underground water, contained in storage rocks between impervious layers.

The underground water resources are controlled by the fluvial balance, which itself depends on the precipitation, by the runoff and infiltration coefficients, by surface flow and underflow, by evaporation, evapotranspiration and variations in water reserves (snow and glaciers), by heightening and lowering of the levels of lakes and by free water ponds. The balance is complicated by deep artesian waters belonging to a hydrological cycle that is longer than that of the shallower aquifers. Schoeller (1959) calculated that in the Albian aquifer of the Sahara water flow over 3000 km required 0.5–1 million years. This appears right as the water velocity is in the order of magnitude of the diffusion of water, namely 1.5–3 m/year, which would give a duration of 1 million years. The balance thus may not be established on the basis of preceding years, but rather as a function of the much longer geological times.

Artesian waters are utilized in increasing quantities in tropical dry Africa, which possesses one of the largest aquifers in the world (Balek 1983) in the form of the Nubian Sandstone, the *"Continental Intercalaire"* (CI) and *"Continental Terminal"* (CT) which we will deal with in the context of oases in the fourth part of the book. The former, the CI, is a terrigenous porous formation upon which the sea deposited an impermeable clayey layer some 100 Ma ago. The second porous terrigenous horizon, the CT, also a good aquifer, was deposited after the start of the Tertiary some 70 Ma ago. In the Nubian Sandstone and the arenaceous CI some 600×10^3 km³ of water are stored over an area of 6.5×10^6 km². Recharge is negligible at present. Having been emplaced at the end of the Pleistocene, the waters are 30 000–40 000 years old. Isotopes give an age above 20 000 years, with the recharge interrupted between 20 000 and 16 000 years ago (a dry phase on a regional scale), and then resumed between 10 000–2000 B.P.

Waters fresher than the surface waters may be encountered under pressure at depth. This is the case in particular in the deserts of the Aral basin. In the Sahara, the waters of the Albian horizon are fresher than those of the saline horizons of the Miocene–Pliocene and of the Quaternary (Schoeller 1959). To understand these situations we have to analyse paleoclimates and tectonics: Rain-rich phases in deserts favour the supply of fresh water whereas warmer dry seasons are responsible for more saline waters. At the same time orogenic periods are accompanied by leaching processes and by removal of salts from rocks together with a higher salt content in the rivers, in closed basins and in the water-bearing aquifers during phases of uplift.

2.2.6.1
Formation of Underground Aquifers

Hydrologists are of the opinion that the water supply of phreatic aquifers in desert regions depends on local geographical conditions. This opinion, however, no longer holds true when the aquifers are at greater depths and especially when we are dealing with captured underground waters derived from outside the deserts like the Albian

2.2 · Hydrological Constraints in Drylands

aquifer of the Sahara which is an example of such deeper aquifers fed from the Atlas and thereby from outside the desert under which it is located.

The conditions for feeding the underground waters in deserts are of an intermittent nature, concentrated over reduced surface areas and thus difficult to quantify (Chevallier and Claude 1990). That part of the pluvial supply stored in the *unsaturated zone* of the soil or found in the aquifers after the removal by evapotranspiration determines the deep drainage, for which only few estimates are available. At the latitudes of the Sahel proper only a small proportion of the precipitation of 400 mm per year reaches this deep drainage level. Chevallier and Claude (1990) proposed the following schematic hydrological balance for the Sahel:

- "5–6 times more than the annual supply are susceptible to evaporation
- 1/20 to 1/30 of the annual supply run off on surface
- less than 1/50 of the annual supply may be stored in the underground reservoirs"

They concluded that the largest portion of the available waters is surface water and that this occurs almost exclusively in the form of rill wash and less as the rapidly waning flow in wadis. When there is no blocking effect by a vegetation cover or by a thick soil layer, infiltration will be controlled more than anything by lithology. It will be highest in rocky deserts and on the regs, which are virtually non-cultivable as water is not retained there. In contrast, in clayey deserts where water will stagnate, infiltration is almost non-existent.

The accumulation of water from rill wash in topographic depressions is beneficial for recharging the aquifers. Dubief (1963) remarked that "only the portion of rain water gathered in rill wash will, because it is constricted at great depth and over a rather long period will facilitate a deeper infiltration susceptible to recharging the underground aquifers of the deserts." The ephemeral regime of the wadis of the Sahara and the Saharan Sahel will not recharge these aquifers. An intermittent or seasonal regime like that of wadis of the Sahelian or Sahelian-Sudanian climatic regime, however, benefits such aquifers.

Bouwer (1989) estimated that a hydrographic network covers 1% of the drylands, that runoff lasts about ten days per year and that infiltration amounts to 1 mm/day on average. The recharge of a basin of 100 km^2 will thus be: $1 \times 10 \times 0.01 \times 100 \times 10^6$ = 10^7 m^3/year or 0.1 m/year. For the rest of the basin Bouwer further estimated a recharge rate of 5 mm/year. He concluded that 95% of the recharge of underground waters is derived from the hydrographic network. Outside the latter the rains fall in small storms, penetrate the soils to a shallow depth only and are subjected to evaporation during the days following the rains. In the beds of wadis, especially when the alluvial deposits are coarse-grained, and on the alluvial fans water can penetrate the soil to greater depths. These depths are sufficient for the water to be withdrawn from the influence of evaporation and for it to continue its progress towards the deeper horizons.

Balek (1989) proposed the hydrous equilibrium equation presented in Table 2.7 for a semi-arid basin receiving P = 420 mm/year, nothing of which will escape, and experiencing a potential evapotranspiration rate of 1760 mm/year. Introducing the pumping effect of the root system which, as we have to point out, draw water from two levels, viz. from close to the surface (around 0.1–1 m) and from a deeper level between 5–150 m, he arrived at the equation shown in Table 2.8.

Table 2.7. Hydrous equilibrium equation

P (mm)	Sheet-flow (mm)	Loss (mm)
420	8	412

Table 2.8. Hydrous equilibrium equation including pumping by roots

P (mm)	Sheet flow (mm)	Pumping by roots (mm)	Loss (mm)
420	8	8	404

Balek (1989) estimated furthermore that recharge may result from local rains or from hypodermic flow (just below the surface) even in arid regions. On dunes in Saudi-Arabia Duicer *et al.* (1974) observed the infiltration of a 20-mm-thick water layer to a depth of 7 m but did not state whether this water joined up with the deeper aquifers.

It is only with the help of isotopic techniques that we are able to date the recharge processes on regional scales and over historic times. Dating very old aquifers is affected by incertitude as ^{14}C is then no longer a good indicator. Airey *et al.* (in Balek 1989) probably found the longest storage time of 350 000 years in the Great Artesian Basin of Australia by measuring the chloride content, the minimum of which is related to the last glacial maximum and the maximum to the warm episodes of the Holocene and of the last interglacial. Since this basin is three times larger than France, it has been exploited by several hundred wells down to 1000–3000 m. The water is pumped up at a high temperature close to boiling point (Loup 1974).

The 1 : 1 000 000 hydrologic map of the Hoggar and the Tassili in Algeria (Gribi *et al.* 1992), where precipitation ranges from 100–200 mm/year depending on altitude, reveals water resources contained in the underflow of wadis in the basement and the Tassili as well as in the porous fractured Cambro–Ordovician and Lower Devonian sandstones. The underflow in the wadis is increased by the filling of the beds with coarse Quaternary alluvial deposits, the hydrological interest of which is directly related to their thickness and to the floods which alone assure their recharge.

The nature of the terrain plays an important role according to their permeability. Comparing four soils from the Negev, Hiller and Tadmoor (1962) could show that sandy soils take up moisture to deeper levels than soils from stony slopes, loess or clayey soils and exhibit the maximum storage capacity, highest infiltration and lowest evaporation rates.

2.2.6.2
Discontinuities in the Recharge of Aquifers

As the recharge of aquifers in drylands is always tied to floods, the periods of resupply are short and occur at long intervals. When rains are very weak or when the retention potential of the soils is too large, the phreatic layers are not recharged at all or are even depleted. The elevated retention capacity of sands and other unconsolidated formations extends the periods during which the aquifers will not be recharged. In fractured or jointed solid rock these periods are shorter and evaporation does not play an aggravating role. The trend for discontinuities in charging the aquifers with water grows with the diminution of the hydrological zone of aeration and the increase in its permeability. On the whole, from the dry to the humid regions, the recharge passes from a very discontinuous regime through periodic regimes in the Sudanian-Sahelian zone to a more regular one in wet tropical environments.

The Kalahari, a vegetated sandy desert, receives 150 mm/year in precipitation in its dry southwestern part and 650 mm/year in northeastern Botswana. Potential evapotranspiration exceeds 2000 mm/year (Thomas and Shaw 1991) which leads to a humidity deficit even during the rainy months from October to April. This deficit may be ascribed to the texture of the sands and their thickness of up to 400 m. Recharge will thus be possible only where sand cover is thin, along the edges of the desert, around outcrops of rocks or crusts, under bodies of water like the *pans* described above and along lines of runoff.

The Kalahari does not possess any perennial flow. With the exception of the Okavango and Chobe basins all its water resources are underground, requiring pumping from wells to gain access. The uncontrolled multiplication of wells has become critical. Yeager (1989) has shown that even a pumping station with a meagre yield will lead to raising more livestock than the surrounding pastures can support. Around certain wells Martens (1971) counted more than 2000 heads. In order to prevent the degradation of the environment Jarman and Butler (1971) proposed a tolerance threshold: wells 8 km apart can supply 400 head of livestock.

The discussion of the present recharge of the underground aquifers of the Kalahari has not ceased since it started some 80 years ago. Until 1968 the hypothesis of a lack of recharge had been commonly admitted because of the elevated evaporation and the vegetation cover. De Vries and von Hoyer (1988) estimated that the roots of trees reach down to the horizon at 141 m, representing a deep level for roots. With the aid of the isotopes ^{3}H (tritium), ^{13}C, ^{14}C and ^{18}O Verhagen *et al.* (1974) and Mazor *et al.* (1977) have shown that in the centre and south of the Kalahari since the beginning of the 1980s there was an accelerated recharge during years with above-average precipitation, where the thickness of the sand cover did not exceed 4 m. This finding did not apply to areas where infiltration was strong anyway, such as fault lines, areas of bioturbation, limestone outcrops, and drainage lines. For the Makgadikgadi basin de Vries (1984) estimated a recharge of less than 1 mm/year.

2.2.7
Degradation of Soils by Water Erosion

The specific hydraulic features described above as controlling the water deficit exposed the dual role of rill wash: as the elementary stage of the runoff and as a mechanism of erosion. We shall now analyse this second, constraining role of water for man and his activities.

There will be erosion of a soil when the mechanical loss from its uppermost layer exceeds its formation rate in volume. Of the four prongs of degradation, viz. mass movement, aeolian erosion, erosion by water and removal in solution, the last three will exhibit specific features in drylands. Soil erosion is the most traumatic and frequently irreversible stage in the degradation of lands. Because of the different degrees of its intensity an estimate of its extent over the surface of the Earth is rather difficult. Steep slopes and loess areas are the most vulnerable sites in drylands (Photo 2.4).

Photo 2.4. Anthropogenic desertification in the badlands of Somalia. Somalia has in the coastal area a system of massive red aggregated sand dunes along the shores of the Indian Ocean. Until a few decades ago this landscape was covered by a tree steppe with *Acacia tortilis*. This tree has been overexploited, both as a source of charcoal exported by boats to Saudi-Arabia, and by overgrazing by the numerous herds of cattle which were also exported to the same country. By 1982 the fixed dunes near Merca had turned into uncontrollable badlands presenting a threat to this town itself (Photo M. Mainguet)

The *erodability of a soil* itself and the *erosivity of a rainfall* depend on the degree of vulnerability of the soil against erosion by water:

- The erodability depends on the infiltration potential and the degree of moistening of the soil, on the swelling action of certain clays, on their drying-out, on the degree of water saturation and on compaction by trampling. When the vegetation cover disappears, even only on a temporary basis, erosion will increase considerably.
- The erosivity is the aggressive action of rain on the soil on which it falls, and its capacity to dislocate soil particles and transport them. It depends on the direct impact of the raindrops and on the rill wash started by the drops on the soil surface. Because it is proportional to the size and density of the drops, the erosivity of rainfall depends on the kinetic energy of its drops, i.e. the product of their mass and velocity, on the duration of the rain, on its frequency and on the total amount of precipitation. Lal (1988) remarked that the determination of erosivity is not yet reliable and more work will be required to define any indices permitting the es-

2.2 · Hydrological Constraints in Drylands

tablishment of a relationship between the energy of the precipitation and its intensity; this applies especially in tropical dry environments where storms produce precipitation of high intensity.

An experiment carried out by Hudson (1963) in the former Rhodesia illustrated the effect of the intensity of precipitation: Two experimental plots were prepared in the same way and one was covered with gauze at a distance of 10 cm above the ground whereas the other was left unprotected. On the first one the kinetic energy of the rain drops was dispersed which was not the case on the second one. Between 1953 and 1956 erosion from the protected surface amounted to 2400 kg/ha per year against 2900 kg/ha per year from the unprotected one.

Although Fournier's aggressivity index of precipitation was criticized, it was retained by the FAO/UNEP Commission (1978) because of its simplicity as a method of estimating the degradation of soils:

$$\sum_{1}^{12} \frac{p^2}{P},$$

where p = monthly precipitation and P = annual precipitation.

The *Universal Soil Loss Equation* (USLE) was introduced in the US during the first half of the 20th century by the agricultural soil conservation services. Hudson (1971) combined the various contributing factors in a diagram (Table 2.9). The equation is:

Soil loss $A = (0.224) R \times K \times L \times S \times P \times C$ (kg/m²/s).

In this equation the factor 0.224 is required when using the metric units (kg and m) instead of feet and acres.

R is a factor for the erosivity or the aggressivity of the precipitation relating the intensity of the precipitation to their kinetic energy. The most useful formula for it is the one proposed by Wischmeir and Smith (1960): $R = EcI_{(30)}$ with R = index of climatic aggression during the 30 min of maximum intensity of a rain. K is the erodability factor, a value for erosion measured on experimental plots of 1/100 acre

Table 2.9. Factors contributing to soil erosion (after Barrow 1991)

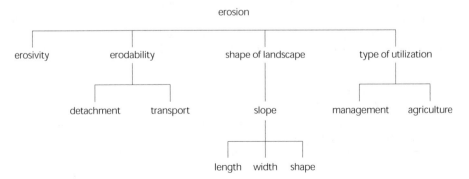

(2/500 ha), i.e. an area of 1.83 × 22.6 m. *L* is the length of the slope, in this case equivalent to the length of the experimental plot (22.6 m). *S* is the slope value, the relationship between the soil loss from a certain field and the loss from a 9% slope with soil types and length of slope being equal in both cases. *P* describes methods of soil protection, e.g. terraces or furrows etc. *C* is a correction factor.

The complexity of the factors mentioned serves as an illustration of the difficulties encountered when applying such an index.

A vegetation cover is still the best protection against erosion by wind and water and it also plays a role in the distribution of nutrients and in the humidity in the soil. Any activity of man exploiting the vegetation cover entails as a consequence the exacerbation of the erosion. In addition to its role in moderating erosion by reducing the effects of wind and precipitation, the subaerial and underground apparatus of a plant is responsible in part for the microclimate below a plant and around it. The subaerial part of a plant favours the accumulation of moisture in its vicinity and modifies the kinetic energy of the rain drops by changing their size distribution.

When there is a vegetation cover, rain will have three ways to go: one part of the water (which may amount to one-third) is intercepted by the foliage, a second part falls directly onto the soil whereas the third and smallest part runs along the branches or the trunk of the plant, favouring the humidification of the surrounding soil (Table 2.10). In the region of Zaragoza (Spain) Gonzalez and Hidalgo (1991) measured that 30–40% of the intercepted water ran off along branches and trunks.

The water drops possess their highest kinetic energy on hitting the ground when the vegetation is tall, as in oases, along gallery forests or in the tree steppe. This is because the drops form by coalescence of droplets, and the drops become larger in mass when they are falling from higher up. In contrast, when the vegetation is low (bush or grassland), the intercepting effect is increased because the water droplets become fragmented and fall to the ground at a distance which is not enough for them to regain their final falling velocity before impact.

The shadows cast by plants on the soil around them lead to a reduction of evaporation, of the intensity of luminosity and of temperature and thereby create a microclimate. However, we do not yet know the real effects of the plants: an increase in the gramineous cover, increase or reduction in the biomass and in the growth velocity. It is furthermore possible that the effects change with the type of plant species.

Williams and Calabry (1985) have shown that soil degradation is low in the arid areas of Australia, whereas in semi-arid areas erosion related to human activities is severe. Gifford *et al.* (1986, in Thornes 1994) investigated soils covered by *Agropyron desertorum*, one part of which had been under grazing for 20 years whereas the others had not been subjected to grazing at all. They found that the capacity of infiltra-

Table 2.10. Interception of precipitation by the foliage of a vegetal cover during 17 storms in semi-arid NE Mexico (after Navai and Bryan 1990)

Precipitation (mm)	Interception (%)	Direct hits on soil (%)	Runoff along branches (%)
230	27	70	3

2.2 · Hydrological Constraints in Drylands

Table 2.11. Comparison of soil and water losses at a station in the dry Nordeste of Brazil and in the dry parts of West Africa (after Leprun 1989)

Parameter			Station Caruaru	Station Linoghin
Soil type			Litholic	Brown eutrophic
Mean annual precipitation (mm)			709	800
Annual mean temperature (°C)			24	28
Erosivity (Rusa)			260	310
Erodability (K)			0.10	0.15
Gradient (%)			12	1.5
	Soil loss (t/ha)	av.	7.4	3.7
		max.	32.6	5.8
		min.	0.04	0.7
Under maize	Water loss (mm)	av.	43.6	165.3
		max.	126.5	229.8
		min.	2.8	67.7
Yield (kg/ha)			2993	2950
	Soil loss (t/ha)	av.	21.43	13.9
		max.	–	35
		min.	–	6.7
Barren soil	Water loss (mm)	av.	9.0	286.3
		max.	–	516.4
		min.	–	16.4

tion in spring was twice that in summer and three times higher in the non-grazed than in the grazed areas. They confirmed the results of Walker *et al.* (1981) who had analysed, also in Australia, the stability of grazed areas in the semi-arid savanna and observed sudden changes in the soils and an acceleration of the erosion. Grazing changes the grass species on a grass cover with some bushes and trees, the soil is compacted, infiltration is reduced, the perennial grasses recede and the forest comes back. The decrease in infiltration and humidity favours deeper rooting of bushes and trees but reduces their productivity.

With the overexploitation of the vegetation cover in the Sahel of Niger and in particular in the Hausa sand sea active dunes of the seif type with sharp crests become locally superimposed over a hummocky system of fixed dunes. Similar changes took place in China around the oases in the southern part of the Taklamakan desert and Tsoar and Möller (1986) described them from the Sinai as a consequence of the overgrazing caused by the concentration of the Bedouins after 1948.

Comparing the intensity of erosion by water in the Brazilian Nordeste and the Sahel, Leprun (1989) noted a rather surprising natural difference in behaviour and resistance of the soils under otherwise equal conditions (Table 2.11). The reasons for those differences are not yet understood.

Comparing erosion at the experimental stations, Lingoghin in Burkina Faso at the southern edge of the Sahel and Caruaru in Pernambuco (Brazil) revealed that under maize as well as on barren soil erosion is two times higher in Burkina Faso than in

Table 2.12. Comparison of soil losses and sheet-flow rates at stations in dry West Africa and the dry Nordeste of Brazil (after data of Leprun 1989)

	Kounkouzout (Niger)	Sumé (Sertao)
Mean precipitation (mm/year)	450	450
Soil type	Brown subarid	Brown non-calcic
Soil loss (t/ha)	0.76	0.40
Rate of sheet flow (%)	21 – 60	5

Brazil. At Caruaru with a slope of 12% erosion as estimated from the Wishmeier equation approaches 30 t/ha per year which is eight times the value on a slope of 1.5% in the Sahel. Rill wash under maize is one-fourth and on barren soil one-thirtieth of the figure at Lingoghin. This difference was explained by Leprun (1989) by a better state of the soil surfaces in the Nordeste which are virtually without crustings. Leprun (1989) held that

> "The bulk of the data compared above proves the impression of naturalists that the natural or agricultural environments of the Brazilian Nordeste are much better preserved and resistant than the strongly degraded and fragile ones of the Sahel. It appears in particular that the Sahel is subjected to a 'normal' erosion taking place in the natural non-cultivated environment and is much higher than that observed in highly discrete manner in the Sertas." (Table 2.12).

> "Favoured by the lower erosivity of its rains, the good properties of its soils especially its top soils, by its dense plant cover with a good regenerative potential, by the low degree of mechanization of its cultures and the lack of itinerant agriculture the physical environment of the semi-arid Nordeste possess qualities permitting it to face until now the processes of erosion and accelerated rill wash much better than is the case, unfortunately in the Sahel. An abrupt increase in the cultivated land surface and a strong mechanization in the Nordeste could question this equilibrium ... which has already been destroyed in the more humid 'serras' with their steep topography and high population densities where pronounced gully erosion has appeared." (Leprun 1989)

The effects of the wind, representing constraints supplementary to those of the water will be dealt with in the next section.

2.3
The Omnipresent Wind in Dry Ecosystems

Any form of land management in the drylands should pay greatest attention to the degradation of soils by aeolian erosion, and as a rule great care should be taken to prevent potential damage induced by the wind. The most degraded areas are located on the same dryland diagonal from the Sahara to the Chinese deserts and are subject to aeolian erosion. China has realized before Africa or even the West that it was necessary to consider the battle against aeolian erosion as a priority, and such erosion is seen as the principal cause for degradation in the drylands of the northern parts of this country.

The wind, especially as a transporting agent for sandy particles counteracts the preservation of the vegetation cover and the soils in drylands. This mechanism is dangerous but seemingly non-critical during its initial phase of activity, as the damage caused by it is then barely detectable. This changes when the degradation is already advanced and sometimes irreversible. During the 1980s the FAO stated that of the total area of the African drylands some 16% were affected by water erosion, 25%

suffered from salinization and 45%, by far the greatest proportion, were subject to traumatic aeolian activity.

On the whole it may be said that, although in humid and temperate ecosystems wind action is restricted to the littoral areas, it may also affect continental areas in the periglacial regions, but its effects, and here especially those of deflation, are reduced by the humidity. This humidity facilitates a certain cohesion of the entrainable particles and in addition transport is slowed down by the vegetation cover. In the drylands, however, the omnipresence of wind action may be explained by the absence, the discontinuous nature or the open vegetation cover. In the analysis of the interconnections between man and drought it is thus justified to place great emphasis on the aeolian mechanisms even though they are not restricted to any particular climatic zones.

Graded according to the decreasing severity of the degradation, the socio-economic consequences of aeolian erosion are:

- Famine, the cause of forced human migrations from the degraded areas and, in the worst case, ecological refugees, forced to migrate due to an ecological catastrophe.
- Costly technical intervention necessary for rehabilitation which would not set in naturally.
- Introduction of a programme of natural rehabilitation by total restriction of any use of the area for 3–5 years, allowing neither grazing nor agriculture.

Until the 1980s scientific geography took note of aeolian action only because of its morphogenetic effects. The aeolian mechanisms are envisaged here as one component of the vulnerability of the lands. An outline of the principles of aeolian dynamics and a classification of dunes will be presented here to bring the analysis up to the control of mobile sands.

2.3.1
Wind – Definition and Basic Principles of Its Dynamics

Vertical movements in the atmosphere are referred to as ascendances or descendances, the term wind being restricted to horizontal movements. Wind depends on pressure, variations of which lead to movements of air. The atmosphere is a mixture of weakly bonded gas molecules continually in motion. This produces collisions between them and against adjacent surfaces. When a pressure is exerted, the molecules will recombine. An omnidirectional pressure results from the combination of the occurring forces.

In the lower layers of the atmosphere the molecules are close to each other. This results in a multiplication of collisions and, because of this, to higher pressures. In the upper layers the gas molecules are farther removed from each other, collisions become rarer and pressure drops. In warmer air the molecular motion increases and with it velocity of the molecules and number of collisions, leading to a higher pressure. However, air will expand on heating which reduces its density and thereby also the pressure. These relations are referred to as the gas laws.

Solar radiation is the initial motor in the genesis of winds: Unequal heating of land surfaces and of the atmosphere caused by the radiation – the cause of horizontal pressure variations – is responsible for movements of air masses. Wind actually represents an attempt by nature to even out the unequal distribution in atmospheric pressure, the aeolian currents flowing from the high-pressure towards the low-pressure areas.

The direction of the wind is determined by the equilibrium between the following three forces:

1. *The pressure gradient* which causes the wind to move at right angles to the isobaric contours.
2. *The Coriolis force* which is responsible for a displacement of the air parallel to the isobaric contours. Any mobile unhindered object in motion will be deflected towards the right on the northern and to the left on the Southern Hemisphere. Because of this wind will be deflected at right angles to the pressure gradient and will tend to flow parallel to the isobaric contours. When the wind flows parallel to these contours it is referred to as geostrophic. The Coriolis force (Cr) is a deviating force, the horizontal component of which is proportional to the wind velocity and is at right angles to the wind vector:

$$Cr = 2\omega U \sin\varphi ,$$

where ω = angular rotational velocity of the Earth; U = wind velocity; and φ = latitude of locality in question.

The Coriolis force is proportional to the sinus of the geographical latitude and is thus zero at the equator and highest at the poles. In its absence, under the sole influence of the force G resulting from the pressure gradient, the wind would blow along the line of steepest gradient on the isobaric relief. However, the rotation of the Earth introduces a force of inertia, viz. the Coriolis force.

3. *The friction force*, controlling an intermediate wind direction at an angle across the isobaric lines. This friction is only developed in the lower parts of the atmosphere in a layer limited by an altitude of 1500 m – the boundary layer. It exerts a dual effect in slowing down the wind and modifying its direction. Above this altitude most winds exhibit a geostrophic flow.

The velocity of the wind is controlled by the pressure gradients: at a low gradient the wind is slow and at a higher gradient rapid. Winds tend to be more rapid above the boundary layer. The strongest tropospheric winds are encountered at medium levels as what is usually referred to as the *jet stream*, or at lower altitudes in the more localized thunderstorms or hurricanes. At tropical latitudes between the equator and 20–25° N and S as well as in the polar areas the air circulation is from the east and outside these zones mainly from the west. The tropospheric circulation may be envisaged as a closed system divided into seven zones arranged symmetrically along either side of the equator (Table 2.13).

At the interface between soil and atmosphere the lowermost air layer is characterized by a lower and an upper zone. In the lower zone flow is laminar and at its very base the velocity is zero over a thickness equivalent to about 1/30 of the soil particle's diameter, whereas in the upper zone flow is turbulent. The wind velocity close to the soil varies according to its roughness which itself depends on the topography. Turbu-

2.3 · The Omnipresent Wind in Dry Ecosystems

Table 2.13. Distribution of main air pressure areas and winds

Pressure areas	Winds
Polar high pressure	North polar easterlies
	60° N polar front
Northern subpolar low pressure	Northern westerlies
	30° general subsidence
Northern subtropical high pressure	Northern trade winds
Convergence fronts	
Doldrums or equatorial westerlies	
Southern subtropical high pressure	Southern trade winds
Southern subpolar low pressure	Southern westerlies
Southern subpolar high pressure	South polar easterlies

lence is caused by the roughness resulting from vegetation cover and topographic micro-, meso- and mega-obstacles (Fig. 2.4). The wind is slowed down by a friction effect in contact with a sandy surface or unconsolidated particles. Its velocity is reduced over a sandy surface with a fine-grained texture and a loose particle structure. In contrast to this it will grow over a coherent surface or over coarse unconsolidated material (a reg or a pebbly surface) or over material with a heterogeneous texture. Turbulence has dynamic causes tied to variations in the pressure field and may be static resulting from fixed obstacles or thermic from reheating of barren soils under the influence of insolation. Depending on its colour a soil will absorb different proportions of the incident energy. A dark humid soil will absorb more of it than dry lighter coloured and more reflecting soil. The top of the turbulent layer correponds to the top of the boundary layer.

2.3.2
The Effects of Wind on Vegetation and Soil Humidity

The wind velocity starts to fall off from a height about ten times the top of the canopy of the vegetation. It influences the thickness of the boundary layer on the leaves, raises air convection and lowers the temperature of the plants. Its increase raises evaporation from the soil surface and transpiration by plants (their water loss), both being referred to together as evapotranspiration. Through the action of the stomata in their leaves, each of which may be opened or closed by guardian cells, plants are able to control their exchange of gases. The closure of the stomata represents a temporary response which, however, is not very efficient for conserving water as it takes place only after a considerable loss of water. It reduces, however, the degree of gas exchange (carbon dioxide) and with it photosynthesis and growth.

According to Ash and Wasson (1983), sand becomes mobilized when the rate of vegetation coverage is below 35%. When the vegetation cover has degraded in a sandy area and the wind velocity is high enough to lift particles and displace sand, vegetation will be rehabilitated only with difficulty, because of the traumatic impacts by the particles in motion:

- On a microscopic scale the saltating sand will create lesions on the shiny cuticles of the epidermis of the leaves which will lead to higher evaporation. These damaged leaves dry out more rapidly than undamaged ones.
- The winds dehydrate the floral buds and thereby stop flowering.
- Pollination is influenced by strong winds desiccating pollen and pistil.
- Furthermore, pollen and grains may be blown into areas in which their value is reduced or even destroyed.
- Fruits may also be damaged by material in saltation; when wind velocities exceed 6–7 m/s fruits will be scarred, whereas winds with 10–12 m/s will cause fruits to drop. Turbulent winds with velocities above 12 m/s may flatten cereal fields reducing productivity by up to 40%.
- Grain plants or small adult plants may furthermore become covered by sand or, inversely, uprooted; grasses are susceptible to winds with velocities of about 6 m/s whereas woody plants become damaged when the winds reach 12 m/s.
- Experiments by Armbrust and Paulsen (1973) have shown that winds also affect the chemical activities of plants.

As nutrients and grains are concentrated in the uppermost centimetres of a thin soil, erosion of this upper layer represents a destabilizing mechanism reducing fertility as well as the capacity for infiltration and water retention. In the alfalfa steppe of Tunisia the action of wind combines with that of water in excavating the tufts to form pedestals. The reproductive capacity of the steppe is reduced as the soil is less fertile and less permeable around the tufts and does not allow germination to take place and thereby diminishes productivity.

On sites of aeolian erosion, transport, and deposition various indicators of aeolian activity may be recognized:

- Changes in soil texture resulting from deflation and winnowing
- Disappearance of the A-horizon, bringing the B-horizon to the surface
- Exposure of the root system of trees and bushes
- Residual soils shaped by *ripple marks*
- Residual sheets of coarse grains and nebkas (sand arrows)

The wind can remove great quantities of soil, 1 mm carried away amounting to 10 –15 t/ha, a value considerably above the amount replaced by pedogenesis in drylands. Above a threshold of 4 m/s (15 km/h) on average, a wind may start to lift mobile particles to different heights, depending on their size. This phenomenon of *deflation* leads to *winnowing*, the sorting out the finest particles, transforming the soil texture and reducing its humidity and amount of nutrients. In the Hausa land of Niger winnowing has been so intense since the beginning of the 1980s that all particles below 0.4 mm have been carried away, resulting in a residual soil fraction of coarse sand which is virtually sterile, reducing the production of millet.

2.3.3
Aeolian Processes in Sand and Loess Drylands

To prevent aeolian erosion we have to understand the aeolian processes affecting the areas to be protected. These processes may be subdivided into three groups:

1. *Aeolian erosion*: ablation, deflation, winnowing
2. *A second group* encompassing the transport of particles and the genesis of transporting dunes (barchans and linear seif-type dunes)
3. *The mechanisms of sand accumulation*, the sand sheets, the dune fields and the sand seas

2.3.3.1
Aeolian Erosion Proper on Unconsolidated Soils

Influence of Soil Condition
Middleton (1990) subdivided the variables of an aeolian system at the soil level according to the two dominant parameters which act together: erosivity and erodability (Table 2.14).

Entrainment of Particles
A study of particle entrainment is necessary if we are to better evaluate and make provisions for the onset of aeolian erosion. A particle is subjected to four different forces:

1. F_c: *forces of cohesion*: forces of attraction between the various grains
2. F_g: *gravity constant*: directed against the forces of vertical movements

A wind blowing over a horizontal surface exerts two other forces on resting grains:

3. F_d: *pressure of moving air* on the windward side of the grains
4. F_l: *the buoyancy force* acting vertically upwards

Table 2.14. The parameters *erosivity* and *erodability* (after Middleton 1990). The extend of erosion is *reduced* when the value of certain variables increases (+) and that of others decreases (–)

Erosivity		
	Aeolian parameters	Velocity (–), frequency (–), duration (–), surface area (–), disection (–), turbulence (–)
Erodability		
	Sedimentological parameters	Particle size (+ –), height (+), orientation (–), abrasivity (–), transportability (–) organic matter (+)
	Surface parameters	Vegetation: residue (+), height (+), orientation (+), density (+), fineness (+), degree of coverage (+), soil humidity (+), soil roughness (+), length of surface (–), slope of surface (+ –)

Motion of a particle will start when $F_d + F_1 = F_c + F_g$, i.e. when the sum of pressure and buoyancy are equal to the sum of forces of gravity and cohesion.

The buoyancy force results from the gradient between the underside of the grain and the soil surface where the wind velocity is zero and the higher velocity on the upper side of the grain. A high static pressure is established below the grain and a low pressure on its upper side. The grain will be lifted up when the difference in static pressure is larger than the inertia force which is proportional to the weight of the grain.

It is generally accepted that a wind blowing over an unconsolidated sandy soil will exert three types of pressure on the sand grains:

1. Pressure resulting from the wind movement on the windward side of the particles
2. Pressure resulting from the viscosity force on the lee side of the particles
3. Static pressure on top of the particles

The combination of these three pressures creates an uplifting force as a result of the difference in pressure between the windward side and the lee side of the grains and the one acting between their top and their base.

Einstein and El Samni (1949; in Pye and Tsoar 1990) introduced the following take-off equation:

$$F_1 = \Delta p A = \frac{CL \rho U^2 A}{2},$$

where F_1 = lift force; Δp = pressure difference between top and base of grain; $A = d^2 \pi / 4$ for spherical particles with d = diameter; CL = coefficient of lift; ρ = fluid density, for which Chepil introduced the value $CL = 0.0624$; and U = velocity of a fluid measured for a grain of 0.35 mm diameter.

In his experiments Chepil observed a value of $c = 0.85$ for spherical grains and a mean of 0.74 for grains with a diameter of 3–5.1 mm.

2.3.3.2
Aeolian Transport

In the battle against aeolian erosion a prerequisite is a scientific assessment of the modes of aeolian transport and of the particles displaced, using remote sensing and sedimentological methods. Two international workshops on the *Physics of aeolian lift-off and displacement of sand* given at the University of Aarhus in Denmark in 1985 and 1990 facilitated a better understanding of the various problems.

Aeolian transport takes place in two ways:

1. Movement of particles grain by grain: *traction or creeping, reptation, saltation and suspension*, to be defined below
2. Movement of sand in the collective form of dunes
    - Migration of dunes: barchan-like bodies and barchans
    - Elongation of dunes: linear dunes (or seifs)

The dynamics of the two types of dunes (barchans and seifs) will be discussed separately to allow a better understanding of the protective measures against the dangers which they represent.

The Modes of Aeolian Transport

There are four types of aeolian transport which have been redefined by Anderson *et al.* (1988, 1990). These are *traction and creeping, reptation, saltation* and *suspension*.

Traction and creeping occurs when particles too large to be lifted are mobilized by creeping over a surface. This mode of displacement may probably also move pebbles and blocks by sliding.

Reptation, intermediate between creeping and saltation, entails a brief jump over a short distance.

Saltation is a mode of aeolian transport which may be started by direct pressure on the grains or, when the grains are already in motion, may be supported by impact from grains in saltation. This saltation lifts the grains obliquely and transports them before they come to rest nearly vertically. The grains exert bonds and at the same time eject grains from the soil surface. Before being uplifted the grain vibrates in the stationary position and the critical velocity of detachment must be above the force of friction between the grains. The detachment velocity (u^*t) or fluid threshold velocity is lowest for particles in the range of 0.08–0.1 mm (Fig. 2.7) explaining the vulnerability of loess deposits, the size of which is exactly in this range.

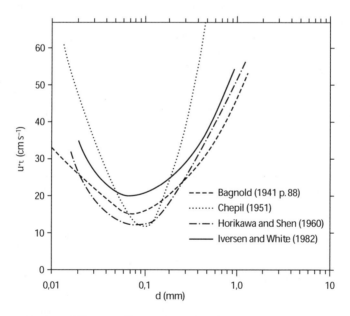

Fig. 2.7. Curves for the mobilization of loamy/silty and sandy particles of 0.01–2 mm vs. wind velocity. On the *abscissa* is the particle diameter (d) in mm. Except for the curve of Chepil where the particles are more homogeneous, all others were of mixed sizes. On the *ordinate* is the wind velocity (u^*t in cm/s), indicating the speed of mobilization of the grains by the wind. N.B. The curves agree with each other and show that grains of a diameter of about 100 µm require a minimum effective wind velocity

According to Bagnold (1941):

$$u^* t = A\left(\frac{\varphi_s - \varphi}{\varphi} gD\right)^{1/2} ,$$

where φ_s = density of air; φ = density of grains; g = gravity constant; D = equivalent diameter of grain; A = empirical coefficient depending on the function of the Reynolds number u^*Dv; u^* = threshold friction velocity and v = kinematic velocity of air.

A grain lifted up vertically by moving air encounters a resistance in a direction opposed to that of the moving air. During saltation a grain moves by rotation and comes to rest at an angle derived from Bagnold's equation:

$$\frac{V^*}{V} = \tan \alpha ,$$

where V^* = terminal fall velocity of a grain (m/s); V = air velocity; and tan = tangent.

Saltation is influenced by the texture of the substrate, and will be favoured if the latter is coherent and retarded when it is unconsolidated. The cloud of saltating grains reaches a height of 0.1–1 m and even 2 m on pebble surfaces, depending on the roughness of the surface. Its maximum density occurs at a height of around 9–60 cm, but mostly below 25 cm from the ground. In this layer the abrasive power of the wind on plants is most severe and the lower parts of plants are the ones most endangered. On dunes, the dry leaves at the base of grasses exert a protective influence.

Suspension, this mode of transport of dust, permits fine-grained terrigenous particles to attain heights of 3000 m and to flow over intercontinental distances.

Transport and the Influence of Topography

Applying the *laws of Bernoulli and Boyle*, according to which *in fluids in motion the product of pressure and velocity is constant,* leads to the Venturi principle. This implies that in an air current any increase in velocity corresponds to a drop in pressure and vice versa. Thus when the streamlines of air converge they behave as if they circulate across a spatial restriction (neck) with velocity increasing and pressure dropping. The opposite takes place when the streamlines of air diverge (diffusor) and they behave as in a Venturi valve, with pressure increasing and velocity dropping (Fig. 2.8). The air flow model according to *Bernoulli's law* is

$$P + \varphi \frac{V^2}{2} = C^{te} ,$$

where P = pressure; V = wind velocity; and φ = density of air.

Applied to a movement of air, this law entails that any increase in air velocity leads to a diminution of its pressure.

It has to be kept in mind that this model also applies to obstacles in relief like hills, mountains, dunes, unconsolidated material, slopes etc. and also to "negative" obstacles like depressions. Although it appears paradoxical, that in the line of travel of wind the opposite slope of a depression may represent the site of an acceleration by the wind, this incontestable situation results from Bernoulli's law.

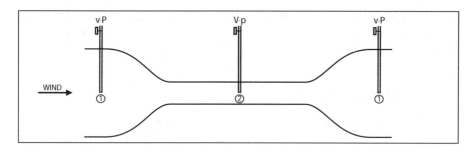

Fig. 2.8. Venturi valve. *1* On passage through a region in which the air is expanded, the streamlines spread from each other and diverge, pressure (P) increases and velocity (v) decreases; *2* on passage through the narrow portion the streamlines become compressed, pressure (p) increases and velocity (V) decreases. Velocity is accelerated by a convergence and slowed down by a divergence

Because of this the value of the gradient of the windward slopes of an obstacle plays such an important role in the aeolian transport of sand.

Case 1. Imagine an obstacle with a steep windward gradient. The velocity of the wind is accelerated as it approaches the obstacle as the air streamlines are strongly compressed and the pressure drops. At the base of the obstacle, on its windward side, an area of deflation or erosion makes its appearance. On the lee side of the obstacle the streamlines have enough space to separate from each other and as the pressure increases the velocity drops and a deposit of particles is produced at the edge of the turbulent area created by the obstacle where the wind velocity becomes nearly zero. The maximum effect on the windward side is created by an impermeable vertical obstacle like a wall or palisade oriented at right angles to the wind direction.

Case 2. Imagine a permeable obstacle possessing a porosity of 50% (windbreak). This obstacle slows down the wind without modifying the direction of the streamlines. Sedimentation by wind on the windward and leeward side depends on porosity and shape of the obstacle. For a porosity of above 50% the profile of the resulting sedimentary deposits is long and low, but shorter and higher at a porosity below 50%.

Although these data result from theoretical considerations, they should nevertheless not be forgotten during the assessments of a windy area with a differentiated topography.

2.3.4
Aeolian Deposits

These may be subdivided into three groups: loess sheets, sand sheets and dunes.

2.3.4.1
The Loess Sheets

Deposits of loamy grains, with diameters of 10–70 µm (Pye and Tsoar 1990), calcareous in composition and deposited in entire horizons, result from the fall-out of dust transported in suspension through an atmospheric layer. A description of this has been presented in the first part of this book. We must, however, recall that such loess deposits are most attractive for agriculture and also the most vulnerable ones in the face of aeolian erosion.

2.3.4.2
The Aeolian Sand Sheets

The surfaces of aeolian sand sheets are flat or only slightly undulating, being sometimes covered by ripple marks. The aeolian sand covers of the Selima along the frontier between Egypt and the Sudan cover discontinuously some 100 000 km^2 of Quaternary alluvials (Breed *et al.* 1987). According to Pye and Tsoar (1991) aeolian sand sheets cover some 1 520 000 km^2 in the warm deserts, representing about 38% of the aeolian deposits. They are also observed in cold climates where they imply a periglacial phase. Sand sheets were deposited during the Pleistocene periglacial over a vast belt from northern France to the former USSR and North America (Koster 1988).

When considering potential future utilization of these areas it is important to know that most of these sand covers possess a bimodal grain size distribution (100 to 300 μm and above 650 μm), being made up of poorly sorted sand of low fertility with a poor water retention capacity.

2.3.5
A Proposal for the Classification of Dunes

Depending on the degree of mobility of the sand, we may distinguish between active dunes, fixed dunes and dunes with a vegetation cover.

The term active dune is restricted to mobile aeolian edifices or to mobilizable material irrespective of its grain size. There are dunes made up of clayey or silty particles and of fine to coarser grained sand or a mixture of all four fractions, irrespective of the mineralogical nature of the grains.

A sandy edifice loses the characteristic features of an active dune when, e.g. the particles form aggregates with iron oxides or silica or are encrusted by gypsum or carbonates. A fixed dune may also result from winnowing, which leaves a pavement of coarser grains on the surface with a size above the aeolian competence.

The third group consists of dunes stabilized by a vegetation cover. These are e.g. the dunes of the Sahel and the southern Sahara, and those of Rajasthan. After climatic changes or degradation because of overexploitation a process of reactivation may affect the dunes.

In areas of continental dunes the border between active or fixed dunes and dunes covered by vegetation corresponds to the position of the 100 mm isohyet when precipitation falls during the cold season and the 150 mm isohyet when it falls during the warm season, as in the Sahel from June to September. There are no naturally active dunes in areas where the rainfall is above these values. They are then the result of degradation caused by man and increased by droughts. Their sand can become mobilized and these previously fixed dunes are transformed into reservoirs of sand which may be dispersed by the wind and pose a potential threat to human activities in the downwind direction. The invasion by active dunes is considered as the main danger resulting from wind effects. In order to combat the different types of threat posed by them, it is indispensable that we understand their dynamics. To make the solutions more obvious we shall introduce a classification of dunes according to the various parameters controlling their dynamics:

- Dunes of catchment of aeolian particles by trapping with vegetation cover: *nebkas* (sand arrows) and *rebdous* (sand accumulation in a tamaris bush).
- Dunes intermediate between transport and deposition: linear dunes or *seifs*; barchans and barchan-like structures.
- Accumulation dunes: transverse structures; barchan-line chains and transverse chains when sand is available in abundance. Pyramidal dunes or star dunes (ghourds).
- Residual dunes after erosion: longitudinal or parabolic dunes.

2.3.5.1
Classification of Dunes According to the Aeolian Regime

In his dynamic classification of dunes, Aufrère (1931) distinguished:

- *The conjunctive winds*, a regime corresponding to one dominant wind direction. It is called unimodal condensed by Fryberger (1979 when in a unimodal wind regime the wind directions are concentrated in a narrow range, or unimodal dispersed when the spread is wider, as in the case of the trade winds. A monodirectional wind regime may encompass winds of various directions, in only one of which the winds are strong enough to lift particles.
- *Opposing winds*: a regime with two dominant directions diametrically opposed to each other, like the *chergui* and the *saheli* of southern Morocco, the trade wind (*alize*) and *sirocco* of southern Algeria and Tunisia, the trade wind and the monsoon of the Sahel.
- *Incident winds*: these are bidirectional regimes or monodirectional regimes locally subdivided by topographic obstacles. Fryberger (1979) called them sharply or bluntly bimodal.
- *Multidirectional winds*: these are complex regimes encompassing at least three dominant directions.

2.3.5.2
Classification of Dunes According to the Sedimentary Balance

The concept of the sandy-aeolian sedimentary balance introduced by Mainguet (1984) and Mainguet and Chemin (1985) is a useful tool for defining the type of dune and from this the strategy of combat to be adopted against the degradation by aeolian processes:

- *Positive sedimentary balance* (BS+) when in an area subjected to winds carrying sandy particles deposition quantitatively is higher than removal.
- *Balance in equilibrium* when supply and removal of particles are equal.
- *Negative sedimentary balance* (BS−) when in an area subjected to winds carrying sandy particles deflation and removal are higher than deposition.

Table 2.15. Sedimentary balance and depositional type

Sedimentary balance	1. Positive: supply > removal Accumulation dominates	2. Negative: supply < removal Deflation dominates
Aeolian regime	*Type of dunes*	*Type of dunes*
One dominant wind	Barchanic edifices Barchan Barchanic chain Transverse chain	Longitudinal dunes (on reg-type pavements) Parabolic structures (forming in environment with vegetal cover)
Two dominant winds	Linear dunes (seifs)	
Three or more wind directions	Pyramidal dunes (ghourds)	

The concept of the sedimentary balance allows us to subdivide all aeolian structures into two families (Table 2.15):

1. *Depositional dunes* when the sedimentary balance is *positive*; this group of dunes entails three subfamilies: (1) barchans and transverse dunes in a monodirectional wind regime; (2) linear dunes or seifs in a bidirectional regime; and (3) pyramidal dunes in a tri- or multidirectional regime.
2. *Erosional dunes* when the balance is *negative*: in this family of deflation edifices there are two subfamilies: (1) longitudinal dunes; and (2) parabolic structures.

Whereas the genesis of the dunes that we have called depositional by accumulation is universally accepted, this is not yet the case with erosional dunes resulting from preponderant deflation in a sandy area.

When combining sedimentary balance and aeolian regime, all dunes may be assigned to three families (Table 2.15).

Two dune types deserve separate treatment as their transport characteristic present a severe danger for human infrastructures: the barchans, sand crescents moving as a whole, and the seifs or linear dunes, edifices along which the sand migrates as along a transport rail.

2.3.5.3
Dynamic Analysis of Barchans

Barchans and barchanic edifices (ovoid dome, barchanic shelter, barchanic dihedron) make up a family of *transverse* dunes (Fig. 2.9), i.e. oriented at right angles to the dominant wind. Such structures are formed in a monodirectional regime in an environment with a positive sedimentary balance (BS+). A barchan is a crescent-shaped dune the convex side of which faces against the wind whereas the two wings on either side of the symmetry axis are drawn out into horns with a concave leeward edge. The windward side may be considered as a slope of accumulation and transport of sand with a slope of 6–12° and the leeward side as the slumping and sliding face with a slope of 22–33°.

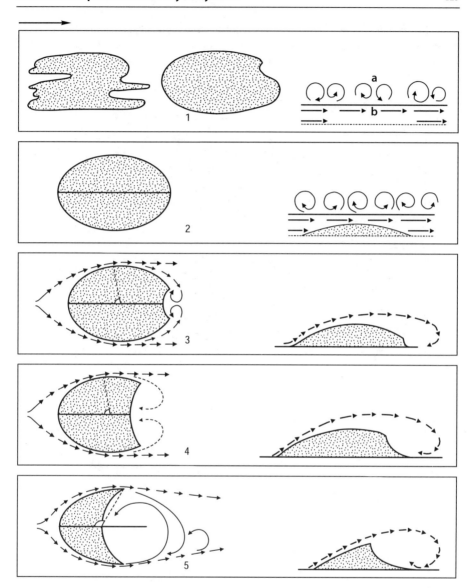

Fig. 2.9. The various barchanic edifices. *1* Irregular sand veneer (thin initial deposit); *2* dome; *3* barchanic shield; *4* barchanic dihedron; *5* barchan; *a* turbulent layer; *b* laminar layer; → dominant wind

When we understand the conditions of formation of a barchan we will be able to control such structures. The determining factors are: a monodirectional aeolian regime, a source for sand with a diameter of 0.125–0.350 mm, a flat topography, a coherent substrate or one made up of particles coarser than the aeolian competence and a missing or only sparse vegetation cover. Barchans occur isolated or like a flight of ducks or lined-up barchanic chains when the wings touch each other or transverse dunes in sand seas with a positive sedimentary balance. The Chinese sand seas of the Taklamakan and the Baidan Jaran are the most representative examples.

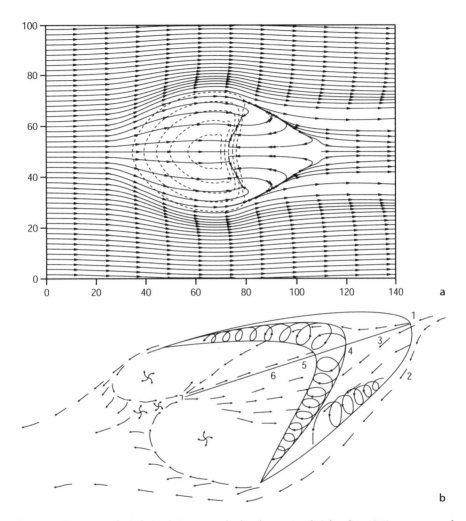

Fig. 2.10 a,b. Pattern of wind circulation around a barchan. **a** At a height of 1 m (Wippermann and Gross 1986) the wind blows from left to right. **b** Field observations (Mainguet 1991) with the wind blowing from right to left. *1* Division of current at the windward foot; *2* the pressure decreases to the sides and the velocity increases, leading to smoothing. The more the velocity grows, the more ovoid the dune will become; *3* part of the current climbs the reverse side and transports sand; because of the windward slope the velocity increases and the transport capacity reaches its maximum at the crest, limiting the growth of the dune in height, as erosion takes place on the summit; *4* because of the change in slope at the crest a detachment of the streamlines develops. The dune is overlain by an air layer in which the velocity is lower. This drop in velocity is accompanied by deposition of sand at the upper part of the leeside front increasing the gradient to the angle of repose preceding slumping; *5* recirculation to the foot of the front, explaining the absence of sand here; *6* this recirculation sweeps the sand towards the front and one may even observe small eddies climbing up the front from its foot

Fig. 2.11. Linear or seif dune. *1* Sinuous crest of structure; *2* dominant winds; *3* sand-blocking barrier

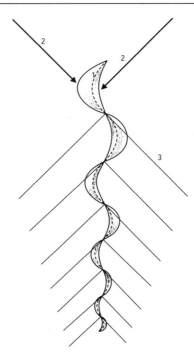

Around a barchan wind circulation is rather complex: The wind is split up into two lateral branches supported by a median circulation. In these three circulation areas turbulence and complicated return circulations are developed as shown in Fig. 2.10.

Pre-barchanic structures in the dome or shield stage are not yet mobile but present a potential danger as a source of sand which may develop further into edifices with an angular crest, barchanic dihedra and barchans proper which themselves are mobile indeed. It is agreed that the migration velocity of a barchan is inverse proportional to its volume. Watson (1991) modified this rule somewhat: A barchan tends to slow down its velocity during the growth stage and accelerates its migration during the ablation phase.

The advance of a barchan was envisaged by Bagnold (1941) using the following formula:

$$C = \frac{q}{\gamma h},$$

where q = quantity of sand to be displaced, γ = cohesion factor and h = height of the dune.

2.3.5.4
Dynamic Analysis of Linear Dunes or Seifs

Linear and longitudinal dunes are frequently confounded, with regrettable consequences for the control of each type. The former ones actually are depositional edifices forming obliquely against the wind. Clos Arceduc (1969) was the first to demonstrate scientifically their oblique nature. They result from a positive sedimentary balance, whereas longitudinal dunes are erosive residual structures oriented parallel to the wind with a negative sedimentary balance. The former are dangerous as they grow longer during phases of accretion whereas the latter, typical mostly of arid environments, are not so, except when disturbances of their surface lead to the removal of sand.

Linear dunes or seifs (Fig. 2.11) are elongate dunes forming in a bidirectional aeolian regime or on the lee side of an obstacle which divides a monodirectional regime into two branches at the convergence of which a sand structure is formed. They extend at an oblique angle against the two dominant winds like the sail of a boat and because of their advance they may reoccupy or invade human infrastructures. There are three types of these dunes: sand arrows with two symmetrical slopes, sand arrows with asymmetrical slopes and sinuous structures extending over dozens of kilometres.

2.3.5.5
Ghourds or Star Dunes

Such dunes form in a wind regime with three dominant directions or without any dominant direction but with a strong vertical component. The ghourds are sand pyramids from the summits of which several sandy arms with a dihedral profile diverge. They are the highest sand structures on Earth, reaching heights of over 400 m in the Chinese deserts and in the Lut Desert of Iran, and more than 320 m in the Issaouane sand sea of Algeria. In the *Grand Erg Oriental* the highest ones measure 320 m and their diameter in the south of this sand sea (30° N, 8° E) may achieve 3 km. Such less frequent edifices are located at the edge of sand seas or close to topographic obstacles except in the *Grand Erg Oriental* where they account for more than two thirds of the surface area.

The ghourds form dune fields hostile to human activities. This is caused by their diameter (500–1000 m), their height (50–400 m) and the mobility of the sinuous arms diverging from them which makes the installation of routes of communication and of centres of activity between them rather difficult.

2.3.5.6
The Dynamics of Parabolic Dunes

Parabolic dunes (Fig. 2.12) are models of aeolian erosion features or deflation dunes with U- or V-shapes opening towards the wind. The most typical ones possess a concave windward slope with active sand and a lee side slope fixed by a vegetation cover. There are three types of these dunes: isolated ones like the huge ones on the NW coast of Jutland/Denmark, coalescing ones with their bodies aligned either parallel or obliquely in relation to the dominant wind (Rajastan) mainly in coastal areas, or in sand seas and vegetated dunes where the vegetation cover has been locally destroyed by overgrazing. The dynamics of this type of dune have been described by Pye (1982) and Robertson-Rintoul (1990; Photo 2.5).

2.3 · The Omnipresent Wind in Dry Ecosystems

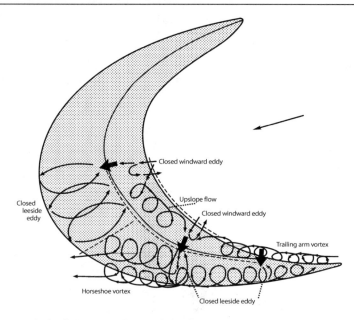

Fig. 2.12. General circulation around a parabolic dune (Robertson-Rintoul 1990). → summit jet; → general wind direction; - - - line of separation

Photo 2.5. Aeolian erosion in Texas. In this oblique aerial photo taken in the region of Plains, Texas, in 1981 the scars (in white) resulting from deflation in an overgrazed sandy hilly area give rise to parabolic dunes. With precipitation slightly above 200 mm/year, simple defence measures allow the vegetation cover to mend itself within three to four years (Photo M. Mainguet)

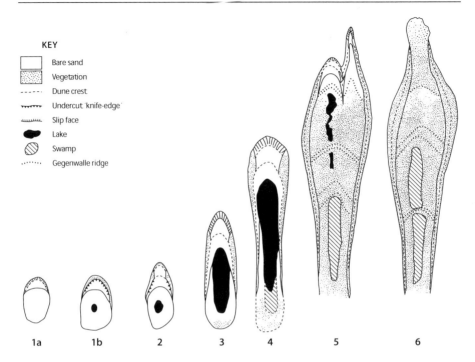

Fig. 2.13. Evolution of a parabolic dune by elongation (after Pye 1982). *1a* Parabolic dune with unvegetated slip face; *1b* parabolic dune with a slip face covered by vegetation and a groove on the windward side; *2* parabolic dune with erosion of the dune body and an active slip face; *3* dune with drawn-out wings; *4* increase of dune volume in height and length and "increasing convergence of the wings"; *5* reduction in height and "increasing convergance of the wings"; *6* eroded dune front and sandy tongue in groove. In this monodirectional regime the wind blows from the bottom to the top of the figure

The formation of a parabolic dune progresses in four successive stages (Fig. 2.13):

1. From scars in the vegetation cover on the windward side of sandy hills *blow-outs* will appear.
2. Accumulation on the lee side slope of the sand swept out from the windward side.
3. Deflation on the windward side and deposition on the lee side take place concurrently.
4. The edifice develops its characteristic U-shape by thinning out its wings and migration of its body in the wind direction. When the body has disappeared by erosion the two residual arms will develop into longitudinal dunes. This represents a possible mode of formation of the second subfamily of dunes, the erosive dunes.

The danger of these dunes lies in the deflation during their initial stage when they are still immobile and then in their migration after they have become mobile.

2.3.5.7
The Sand Ridges

Sand ridges may be defined as erosive residual dunes elongated in the direction of the wind in an area with a negative sedimentary balance. Their axis runs parallel to the dominant wind, their cross-section is convex and sometimes symmetrical. They are separated from each other by corridors of deflation the origin of which is not yet fully understood.

They are the only ones to deserve the descriptive term longitudinal given by Madigan (1936) to dunes with axes parallel to the dominant winds, and properly referred to as sand ridges. The parallelism between the longitudinal sand bodies and the wind was described from the Thar Desert by Blanford (1876), from Egypt by Beadnell (1910, 1934), King (1918), Hume (1925), Ball (1927), Bagnold (1931, 1933) and Kadar (1934), from the central Sahara by Aufrère (1930, 1931) and from Australia by Madigan (1930, 1936, 1938).

Longitudinal dunes are the most frequent sand structures in the arid and hyperarid regions of our planet and Jordan (1964) estimated that they make up 72% of the sandy areas of the Sahara. Sand ridges are described from all deserts except those in China where they are rare, and accompany the great trade wind systems in Australia, Saudi Arabia and the Sahara where they are even deflected into a nearly zonal direction in the Sahel to the south of the desert. In the area of Niangay (Niger) around 16° N and 3° E a system of NE–SW trending sand ridges converges with another system of larger ridges (2.5 km wide) oriented 88° ENE–WSW. This unique situation results from the encounter of the NE–SW trade winds which blow directly from the Sahara towards the north of the Sahel without suffering deflection with a typical trade wind of the Sahel whose direction has rotated to nearly east-west.

The sand ridges of the arid and hyperarid regions are frequently fixed by a pavement of coarse particles and many also be fixed or stabilized by a vegetation cover in semi-arid, subhumid or humid areas. Describing the sand ridges of the Simpson Desert in Australia between Marshal River at 23°S and Lake Eyre at 28°S, Madigan (1936, p. 213) wrote: "It is an ocean of spinifex-covered sand waves." Grove (1958) described those of the Hausa land in the Sahel of Niger: SE of Termit, touching the 150 mm isohyet they become covered by an increasingly denser Sahelian steppe as one moves more to the southwest. Tricart (1974) could prove the existence of dry phases in the Quaternary of Amazonia by discovering longitudinal dunes below the forest. The analysis of sand deposits in Mali east of the Niger bend south of the 150 mm isohyet revealed the presence of longitudinal dunes reaching down to the banks of the Niger at 14°N.

2.3.6
The Theory of the Global Aeolian Action System (GAAS)

It is usually assumed that in drylands the winds represent a danger essentially because of their capacity of forming mobile dunes. Much in contrast, the real danger of winds in sites of erosion – source areas for sand and its transportation – in transport areas and in sand seas with a negative sedimentary balance is still neglected. In order to better understand and correct these dramatic effects we proposed in 1992 the theory of the "global aeolian action system" or GAAS.

2.3.6.1
Definition of a Closed and an Open GAAS

A GAAS combines a source area or area of deflation in which the wind lifts its charge of particles depending on their competence, with a transport area along which the transfer of particles takes place and in which dunes of the barchan- or seif-type may already form. In addition, an area of accumulation may also be present. The latter, in turn, may become an area of re-exportation, i.e. a source area.

A GAAS is a dynamic, open or closed system (Mainguet 1992):

- In a *closed GAAS*, which may be qualified as endorheic, the particles are imported and accumulated, and exportation is negligible. The best example of this is in the Taklamakan (China), a deep basin surrounded by high mountain ranges, which are the source of its sedimentary material (Mainguet and Chemin 1988).
- An *open GAAS* is a system which after importation and accumulation may be subjected to re-exportation. The Sahara–Sahel system is the largest example of this. For a long time the Sahel has been an area of accumulation of aeolian material and of reworked fluviatile alluvions from the Sahara. However, since the last drought crises of 1968–1985, the effects of which were exacerbated by degradation resulting from human activities, the Sahel has turned into a potential area of exportation.

A GAAS possesses an upwind and a downwind side. In the Sahara–Sahel system the upwind area is located to the north of the desert with the large north-Saharan sand seas – the *Grand Erg Occidental* and the *Grand Erg Oriental* – whereas the downwind area is made up of the vegetated sand seas of the Sahel. The sand seas are arranged in chains along the aeolian currents defining the harmattan. These are traced by the axes of transport in which saltation is discontinuous in time and space and which show deviations and subdivisions around mega-obstacles presented by mountain massifs (Fig. 2.14).

The Saharan GAAS is divided into currents discovered in 1972 during a study of the kalut system east of the Tibesti (Mainguet 1972) and could be confirmed by the study of Meteosat images taken between May 20, 1972 and January 25, 1973. According to these documents, the currents which are conformable to the harmattan split up on encountering mega-obstacles, mainly the Eglabs around which the current flows across the Igidi sand sea to the west and the Chech sand sea to the east, the Hoggar-Tassili, the Tibesti and Djebel Marra, other mega reliefs which also divide the wind flows follow the continental NE–SW trade winds in the centre of the Sahara. The direction of the wind flows turns ENE–WSW in the Sahel at the latitude of the Tropic of Cancer and then becomes E–W with the decrease in the Coriolis force near the equator.

The westernmost current in the Sahara blows along the Atlantic coast from SE of Cape Juby near Tarfaya to the E of Cape Blanc. At a total length of 800 km it is made up of a mobile sand veneer and of barchans up to 15 m high (average 6–10 m), and a width of up to 75 m (average 15–35 m). The barchans occur in groups or are isolated (Querroum 1990). Over a rocky floor the barchans migrate on average some 25 m per year. The modal size of their sand grains is 220 µm. According to Oulehri (1990) the

2.3 · The Omnipresent Wind in Dry Ecosystems

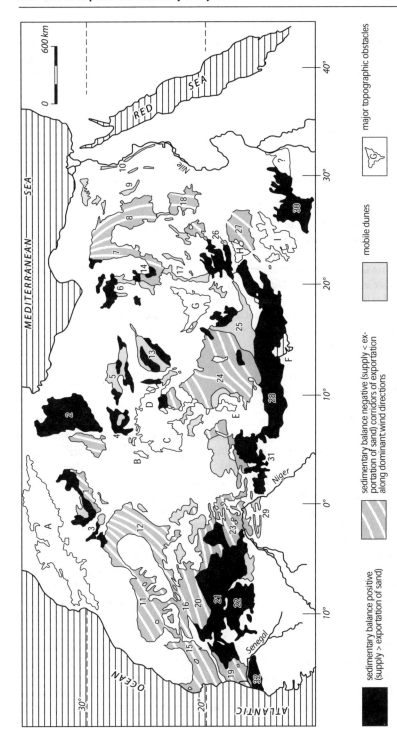

Fig. 2.14. The global aeolian action system (GAAS) of the Sahara and the Sahel. The ergs of the Sahara and the Sahel and their sedimentary balance. *A* Atlas; *B* Mouydir; *C* Hoggar; *D* Tassili N'Ajjer; *E* Aïr; *F* Lake Tchad; *G* Tibesti; *H* Ennedi. *1* Grand Erg occidental; *2* Grand Erg oriental; *3* Er Raoui Erg; *4* Isaouane N'Irarraren; *5* edeyen of Oubari; *6* edeyen of Marâdah; *7* Calansho Erg; *8* Great Sand Sea; *9* Farafra Erg; *10* Gharb Abu Muharik; *11* Iguidi Erg; *12* Chech Erg; *13* Mourzouk Erg; *14* Rebiana Erg; *15* Magteir Erg; *16* Ouarane Erg; *17* Libya/Tchad Ergs; *18* Selima Erg; *19* Trarza Erg; *20* Ijafene Erg; *21* Mreyye Erg; *22* Aoukar Erg; *23* Azaouad Erg; *24* Bilma Erg; *25* Man-ga Erg; *26* Mourdi Erg; *27* Goz Kordofan; *28* Haussa Erg; *29* sand ridges of Niger and Burkina Faso; *30* Sudanese Goz; *31* Talak; *32* Ferlo

wind transported 55–220 m³/m per year over a width of 7.5 km between 1980–1988 at Layyoune, corresponding to a volume of 0.4–1.6×10^6 m³. Sarntheim and Walger (1974) estimated that 93 000 m³/year were displaced in the dune field NE of Cape Blanc over a section of 80 km width, the amount of sand transported by saltation being 50–100 times higher.

2.3.6.2
Dynamic Units in a GAAS on the Synoptic or Regional Scale

A GAAS may be subdivided into three successive areas in the direction of the wind: the source area, the area of aeolian transport and the accumulation area.

The Source Area
There are a number of sources of the sand, which may be transported in steps over longer periods and great distances:

- *Marine*: along the coast of Baluchistan over 600 km and along the Atlantic coast of the Sahara or at Essaouira in Morocco the sand is swept up from the shores and transported into the interior of the continent.
- *Fluviatile*: by winnowing from alluvial beds and from the sandy material of wadis coming under aeolian influence, as is the case for a large part of the sands of the Sahara.
- *Glacial or peri-glacial* from moraine material.
- *Lithological* by ablation or corrasion of rocks in the respective transport areas, in particular over a substrate of sandstones.

The Area of Aeolian Transport
According to Dresch (1982) a grain with a diameter of 200 µm covers a distance of 800 km within a century. At the soil–atmosphere interface the wind transporting the sand organizes itself like water into hierarchic runoff channels which allows one to think of rivers of sand. However, sand flows differ from runoff as, despite being channelized by topography, they do not develop any banks, and in addition they may climb uphill even where the slopes are steep. Wind flows exist over the Himalaya massif (P. Cour 1997, pers. comm.). The roughness of the topography may increase along a transport area and reach such a threshold that the former ground level may turn into a depositional area. The opposite may also be true and the best example for this situation is the Kaouar escarpment in the centre of the Ténéré sand sea. The Ténéré sand sea stops on the windward harmattan side of this topographic high, is blown over the sandstone escarpment, and settles after several kilometres on the leeward side of a blow-out depression which extends along the lee-side of the escarpment.

The wind may or may not transport sand along the same current within a transport area. The transporting medium and the material transported have to be considered separately. In the case of the Saharan winds (continental trade winds or harmattans) the material deposited along the edges of the deserts is not necessarily of the same origin further along. It may be taken up all along the course of the aeolian current.

The sand of the northern Saharan sand seas was probably derived from the decomposition of the crystalline rocks and sandstones of the Atlas reworked by riverine action during pluvial phases in the lowlands of the Sahara flowing from the north out of the Atlas mountains. They were therefore of fluvial origin during their initial transport phase but were reworked by winds and shaped within the framework of the sand seas. The Saharan sand seas are areas of accumulation but also of resting, with the sands arriving and, after a certain period, departing again. The successive phases of remobilization rework the same material. Thus the sand seas of the southern Sahara and of the Sahel may be derived from three different sources:

1. Aeolian and allochthonous from the Sahara
2. More autochthonous fluvial alluvions coming from the alluvial fans of wadis flowing out of mountain massifs such as the Hoggar, Tibesti, Aïr and Adrar des Ifoghas
3. Minor amounts of deflation material derived from local decomposition products of areas like those underlain by the crystalline rocks and sandstones north of Burkina Faso, for example.

The Accumulation Area

This may represent the end point of a GAAS but may also form wherever the wind encounters a local feature slowing it down. If it runs against a slope steep enough to counteract its vertical component, the transporting wind will deposit its sand on the slope if this is not too steep, or it will deposit its sand in the form of an "echo" dune at some distance from the obstacle when the slope is steep. This echo dune grows in volume and develops into a climbing dune resting on the slope. It eventually overlaps the top of the obstacle and extends onto its lee-side, forming a hanging dune which will turn into a linear dune (or seif) on further growth.

A sand drift which is discontinuous in time and space will develop over a long period along these currents. Over time this drift will be more efficient during dry climatic phases when the winds are stronger. In space it will be favoured by rocky areas without sandy cover, whereas it will be slowed down by sand seas and dune fields. We have used the model of Reichelt (1992) to clarify their position over time:

- *Stage 1*: maximum extension during the upper Pleistocene (17 000–15 000 B.P.) in the Ogolian desert which was larger than the present Sahara
- *Stage 2*: humid period during the Holocene which favoured pedogenesis fixing the dunes and the development of a grass steppe with trees
- *Stage 3*: return of the drought around 7000 B.P. when the fixed Ogolia sand seawas reactivated with an acceleration resulting from anthropogenic pressures

In this climatic framework we may place our own theory of the sedimentary balance: during stages 1 and 2 the sedimentary balance was positive, turning to negative during phase 3. This actually represents for us one of the best expressions of the desertification of the Sahara for those looking for proof for this idea.

2.4
Conclusions

Life, limited by the deficit in water, is more arduous in the drylands of the African continent than in those of Asia, because the latter are situated at the foot of great mountain ranges crowned by glaciers furnishing water. This lower availability of water in the African drylands where, furthermore, the flat topography of the continent favours the establishment of aeolian currents abrading the soils and desiccating the vegetation, explains the dramatic developments of the last three decades on this continent. These developments result from the geographical peculiarities and from difficulties of the respective societies to adapt to this situation.

A correction of the ecological constraint is possible but only with stringent organizational efforts, at particularly high costs and with a rigorous care in selecting the most suitable methods. Managing the rare resources of the water-courses for their own sake, knowing the existence of sometimes fossil underground waters under the plains, exploiting them with increasingly more complex technologies in order to avoid their exhaustion, preventing erosion by water and wind action and finding short-term remedies against the drought as well as long-term strategies: all these aspects force man to bring forward new proof of his particularly creative genius. The various forms of his ingenuity and of his creativity are the object of the third part of this book.

CHAPTER 3

Human Genius: The Search for Water and Its Management – Battle Against the Wind

In Western cultures water presents a fundamental wealth. With regard to the Sahel Benoît (1984) remarked: "Although it may surprise, one has to note that here water is not perceived as a wealth in itself but as a means to achieve the wealth represented by pastures." For a nomad priorities rest in his herd – his fortune and his bank account – a visual difference of impression explaining the shift when a choice becomes necessary for management methods and for the extent to which concerted action by those using the pastures and those using the land is required.

The drylands impose constraints on the users of the land: the ever present aeolian action, a deficit in surface water because of precipitation erratic in time and space, strong evaporative pressure resulting from high temperatures and insolation and a scarcity of water-living organisms. Up to a certain minimum of precipitation, ranging from 300–400 mm/year depending on other environmental factors, the vocation of land utilization should be only pastoral. With higher precipitation rain-fed culture becomes possible, but precipitation will turn out to be insufficient as far as the needs are concerned and insecure with regard to the yields. Because of this man has devised numerous irrigation systems for using the available surface waters or taken recourse to the costly exploitation of increasingly deep aquifers.

Overcoming aridity and the necessity of utilizing surface waters parsimoniously has historically led man to prove his special and surprising capacity of adaptation and invention to put into operation techniques for the optimal management of water resources. Man would not have survived in the drylands were it not for his knowledge of how to increase the quantities of immediately available water according to his requirements.

We shall review the systems of exploitation in regions without irrigation, – rain-fed agriculture, nomadism or semi-nomadism – then the search for water from the mastering of water by the great hydraulic civilizations of the past to traditional village hydraulics and modern techniques of gathering and storing water (Photo 3.1). In this third part dedicated to human genius, we shall also reserve a special space for describing the art of combatting the ill effects of wind.

3.1
Non-Irrigated Agricultural Systems in Drylands: Their Difficulties in the 20th Century

Drylands are naturally infertile environments with limited potential. Toupet (1992) graded the modes of exploitation of these spaces according to their "degree of rooting":

Photo 3.1. Flood cultures along the Senegal River. Along the Senegal River north of St. Louis the river regime with high floods and sandy–loamy banks permit the planting of cereals in flood-irrigated cultures, in particular rice and dryland cereals like millet (Photo M. Mainguet)

- "*Pure nomadism* all encampments follow their herds and live from their products.
- *Semi-nomadism*: the herdsmen always follow their herds but restrain their movements and, at harvest time, they raise their tents near fields of millet and sorghum, cereals which represent an essential element of their food.
- *Transhumance*: livestock breeding remains prominent but men and women stay near the fields whereas the herds are entrusted to shepherds.
- *Rain-fed culture*: this is haphazard and the "farmers" frequently practise itinerant culture.
- *Subsidence culture*: this is less haphazard; villages, protected from the floods, become important and the resources provide not only sorghum but, because of strict utilization, also rice. Fishing and livestock represent additional complements not to be neglected.
- *Irrigated cultures in the oases*: these are actually referred to as "phenico-cultures", i.e. date palm cultures, and gardening. Oases have remained stationary for centuries, they no longer resemble villages and are sometimes even like small towns."

3.1.1
Itinerant Rain-Fed Agriculture and Fallow Periods

Rain-fed agriculture after fire clearing, always itinerant in nature, exists only along the edges of the drylands in subhumid dry areas. The fallow lands may be used for livestock breeding during these times. These two forms of extensive land utilization yield only meagre returns. When itinerant agriculture settles on virgin ground the harvests depend on the fertility of the soils, but such lands are only rarely still available and most areas have not been utilized for the first time and their fertility thus depends on the length of the fallow period.

The requirements of such a system depend on the conservation of the root system of the ravaged vegetation from which the plants may regrow and on the conservation of nutrients and organic matter in the soil. The main counter-indication is represented by mechanisation. As manual labour has to overcome the resistance of the roots, the surface is only scratched. This system therefore is not able to sustain a strong population increase. Any demographic pressure leading to a shortening of the fallow periods and of the chances of rehabilitation of the soils and to a decrease in the quality of the pastures on dry lands will result in overexploitation. In less dry areas this overexploitation leads to the conversion of the fallow system to permanent mechanized cultures which Shaefer and Kehnert (1987) consider to be disadvantageous for the development of small farms. Is there any solution?

Whatever the mode of exploitation or utilization of the ground in drylands, demographic pressure is the main difficulty facing agriculture. Economists are of the opinion that a change from subsistence farming to cash crop production represents a great progress. But are there any markets for these cash crops? We cannot believe this as e.g. the irrigated cotton culture of the Aral region had dramatic effects on the environment and, with the preference for man-made fibres, the expected profits have not been realised.

In addition to the annual rain quantity, the variability of the precipitation and the length of the dry season, the possibilities for agriculture depend on the timing of the rains (during summer or winter) which controls the ratio of precipitation to potential total evaporation (P/PTE). In semi-arid northern Africa where the rains fall during winter, rain-fed cereals (wheat, barley) are successfully cultivated with a mean annual precipitation of 350–400 mm. In the Near East, where 60% of the precipitation falls from December to January, these cereals are cultivated with success from 300 mm/year onwards, but they require 500 mm/year in the Sahel where the rains fall in summer. In North Africa and in the Near East the variability of the rains is lower than in the Sahel where it is itself lower than in the Brazilian Nordeste. Consequently the Nordeste, despite a similar, even a higher rainfall to that of the African drylands, is not able to furnish similar yields of cereals (Le Houérou 1989).

In the past decades a different approach to the understanding of water scarcity and to living with drought has developed: taking cognisance of differences in the resistance of the rural production against the lack of water. According to the insights gained in the Brazilian Nordeste since 1958 productive activities are grouped according to their vulnerability under a deficit of water:

- Food crops with a short cycle, especially maize, which require a continuous precipitation lest they perish.
- The culture of *moco* cotton, a xerophilic plant which in the adult stage sends its roots deep into the soil to compensate for the absence of rain.
- Cattle breeding, a less vulnerable activity, for which water and fodder may be brought to the livestock.

From this classification follows: "... that the reorganization of the economy of the semi-arid zones of the Nordeste implies a specialization on growing xerophilic plants and on livestock breeding and a reduction of the subsistence sector ..." (Pessao 1989), a comment that is particularly valuable in the light of the much talked-about need for self sufficiency in food production.

In the Brazilian Nordeste traditional rain-fed agriculture relies on maize, beans and cotton. Research on short-cycle cultures and especially of beans leads to an improved utilization of the poorly distributed rains. Sisal and castor oil plant are retreating and in contrast to this, spineless cacti as cattle fodder together with maize and cotton were rediscovered after the last drought at the start of the 1980s. The first experiments are being carried out with sorghum and millet, cereals which are better adapted to drought.

There is, however, one contradiction: the activity most interesting for land-owners is livestock breeding, whereas subsistence farming is the choice for agricultural labourers without land, share-cropping farmers (*parceiros*), tenants (*moradores*), land occupants without titlehold (*posseios*) or salaries and sometimes for the owners of smallholdings who are the most vulnerable in case of a drought (Cavalcante 1989).

With regard to the Brazilian Nordeste Coelho (1989) "outlined the agricultural practices which are to be recommended for maintaining the maximum humidity of air and soil, for maintaining or improving the biostructure of the soil and for reducing the impact of droughts."

Dry farming is adapted to many areas not suited for irrigation, and there are a number of techniques to improve its results:

- Using organic matter as fertilizer to improve the biostructure of the soil.
- Executing the so-called "*cultivo-minimo*", i.e. not subjecting the soil to working methods which lead to a compaction of agricultural soils.
- Avoiding runoff and holding back the water in small trenches cut so that its infiltration into the soil becomes possible.
- Practising "dead cover" or *mulching* which entails the placing of plant debris on the soil which inhibits erosion, retains humidity suppresses the growth of weeds and, after the harvest, becomes incorporated into the soil as humus.
- Providing shade for the pastures and certain cultures by a tree cover.
- Re-afforestation.
- Practising silo storage for conserving fodder.

Coelho (1989) is highly concerned that irrigation poses technical problems, especially the risk of soil salinization, economic problems because of investment and production costs, and social problems because of its low potential for creating new employment compared to the number of employment facilities destroyed by the opening of the surrounding areas.

3.1.2
The Move Towards Irrigated Agriculture

Rice, planted in both rain-fed and irrigated cultures, marks the transition between two types of land utilization. Managers too often are of the opinion that only permanent irrigated agriculture represents the solution for the development of drylands. We shall see, however, how this method leads to failure, especially because of the degradation of the environment.

The strategy of food production in the Senegal is based on the cultivation of rice. In Senegal with its population of 7.3 million and growth rate of 2.8% (or even 4% in the urban area of Dakar which will grow to 2.5 million inhabitants by the year 2000), the agricultural sector accounts for 22% of the economy, which is however less than industry (29%) and services (49%). "Cereal production represents only 40% of agricultural production as a whole, the remainder of about 50% coming from export crops like peanuts, cotton and tomatoes. "The mean cereal production in 1987–1989 was 1 million tons of which 60% were millet, 15% rice paddy, 13% sorghum and 12% maize. The country is nearly self-sufficient in dry cereals (millet, sorghum), but consumes 80 000–140 000 t of wheat, all of which is imported, and 0.35–0.5 million tons of rice of which it produces only about 15% (Diagne et al. 1993).

The hydro-agricultural management along the Senegal River forms the centre of the debates about development. In Mauretania the right bank of the valley has been opened up by private investors for planting rice, whereas Senegal has not yet become ready for this change which would require liberalising the transformation and commercialization of this cereal. As the consumer is more attracted by the taste of imported than of the local rice, a policy protecting local rice has been introduced. Rice gradually replaces the rain-fed cultures millet and sorghum despite the – albeit small – price difference and the general efforts in favour of changing these cereals.

In Senegal about two-thirds of the rice fields rely on rain and less than 25% of the harvests come from less haphazard irrigated fields. Less than 20 000 ha of rice paddies along the river are irrigated, where there is a potential for about 200 000 ha. Diagne et al. (1993) pointed out that "the potential for expansion in reality is rather hypothetical, taking into account the problems of profitability". In contrast to this, increases of 30–40% have been achieved at reduced costs in the lowland rice fields of Casamance.

> The production of rice in Senegal cannot compete with imports from the world market except at the place of production. Dakar has therefore shown interest in turning towards the latter for its supply. In the extreme case, the semi-intensive planting of rice in the wider vicinity of the delta will not be competitive even at the sites of production. The costs alone ... show ... that rice became more expensive in the irrigated areas than in the neighbouring countries (30–40 CFA more than in the officially rehabilitated parts of Niger and Mali and 20–30 CFA more than in Mauretania).

Official imports into Senegal comprise only 25% of the food production and only 10% of the rice. Imports of rice amount to one-third of the total trade balance. According to the "cereal plan" of 1986 the agricultural policy with regard to cereals expects a strong increase in the production primarily of paddy rice and to a lesser extent of millet and sorghum. At the beginning of the 1990s the per-capita consumption of rice in Senegal was 65 kg/year, compared 49 kg/year in Mauretania, and 26 kg/year in Mali. This "cereal plan" does not seem to work especially because of the failure of the policy of opening new areas as a result of a lack of finance as the donors fear a poor economic viability. On the other hand, as the local rice is subsidized, albeit on a small scale only, any increase of the volume by local production will thus be slowed down.

This analysis of the situation of rice in Senegal was presented in order to illustrate the difficulties which an irrigated culture in drylands and especially in a developing country face against worldwide competition. Similar difficulties have appeared also in the case of cotton.

3.1.3
Livestock Breeding, Nomadism, Pastoralism, and Transhumance

Nomadic herdsmen, including agropastoral farmers and breeders, represent the best known form of nomadism (Photo 3.2). Some 30 million in number, their activity is based on the domestication of herbivores, representing their means of production and consumer goods (Bourgeot 1994). It is thought that the lands utilized for herding cover twice the area used for agriculture. Most of the former are drylands in the developing countries and possess only a low productivity. Africa and Asia – the Old World – differ from the New World – the Americas – in their type of livestock breeding. Prior to the arrival of the Europeans, with horses, cattle, sheep and goats, the New World possessed only the llamas and alpacas of the high Andes as domestic animals. The lands became private property and were fenced in only after the arrival of the Europeans.

In the Brazilian Nordeste, cattle breeding practised in the natural environment of the Caatinga is barely economic as the carrying capacity is one head for every 20 to

Photo 3.2. Nomads in the Algerian Sahara. The photo, taken at the southern edge of the *Hamada du Tinrhert* in a hyperarid intensely sand-covered area, shows three nomads with their turbans and traditional wide summer garments of white cotton, which conserve the refreshing sweat, and also with the winter garment of brown wool (*right*), as the nights are rather cool (Photo M. Mainguet)

30 ha. For Cavaille (1989) any improvement would require irrigated fodder production, which casts doubts on the economics of this method. Goats and sheep are less adapted to the natural vegetation cover but the recent introduction of forage trees like *Prosopis juliflora* and *Leucaena leucocephala* have yielded good results.

In the Old World herding has taken place over several millennia on unfenced collectively owned lands. Droughts and their consequences, viz. famine, livestock epidemics and wars maintained a fragile equilibrium between nature, man and his herds. Colonialization has reduced the epidemics without interrupting the tribal wars, without finally stamping out famines or the competition between private and collective interests for the pastures or the overgrazing more rapidly developing in drylands. It is generally held that privatization of the land represents an assurance against the degradation of the environment. However, the nomads who are, as we have seen, best adapted to the natural arid environment, have actually been deprived because of this of a large portion of their grazing areas and concentrated on the poorer areas and it is difficult to be convinced of the justification of such an opinion.

In the dry areas south of the Mediterranean, where depending on latitude and altitude cultivation is not possible, land usage is dominated by transhumance. In grazed forests, on high-altitude grassland (matorral) and on the dryland steppes, sheep seemed irreplaceable gatherers of vegetal biomass and suppliers of wool and meat. There are three forms of herding which have been referred to already in general:

1. Over large distances (above 100 km).
2. Over more modest distances in the case of transhumance: one shepherd leads the herds to the summer or winter grazing areas.
3. Sedentarism with the daily movement of the herd and its return every evening to the shepherds station or to the enclosure. It is difficult for the animals to find fodder during those periods when the lands have nothing to offer.

Over the entire Maghreb double transhumance is practised with a descent in winter into lower-lying areas like in the Middle Atlas and to the cereal-growing areas to the north after the harvest. This double transhumance becomes simplified where the low-lying areas are cultivated. In Algeria the *achaba*, the transhumance towards the cereal areas in the north, is losing in importance and only some 2 million sheep are moved north every summer. The other more or less stationary livestock is fed with concentrates supplied by the state at low prices.

Since the start of the second half of the 20th century the world production of animal proteins has grown considerably, whereas in contrast to this the contribution from the traditional regions of extensive herding in the tropical dry zones has fallen. "The herdsmen were able to maintain animal production high enough for their own needs until the start of the 1970s, despite the demographic growth .... It has fallen ever since because the carrying capacity of the environment was exceeded" (Breman et al. 1990). This became obvious after the drought of 1968–1985.

In order to safeguard this source of protein and foreign exchange governments have called for the development of water-supported pastoralism and of veterinary ser-vices. The growth of pastoral production could only be achieved by large stock numbers requiring larger areas, which hurt the farmers who also had to extend their fields in order to increase or even only maintain their production. The actions of the farmers hinder the movement of the herds mostly during the dry season even when the animals utilize the residues of the cultures. This explains why the pastures deteriorate on the areas where the animals stay during the dry season, and mainly where vegetation – perennial grasses, shrubs and trees – resists moderate droughts (Photo 3.3).

There is general agreement on overexploitation of the natural environment as a fact, not because of ignorance but by necessity, resulting from overpopulation and the lack of alternatives. And finally, the difficulty rests in the lack of land and "competition for space in which the herdsmen are nearly always the losers" (Pélisier 1989). A remedy would be an increase of the productivity of the pastures and an intensification of agriculture. The quality of the pastures is mainly controlled by the relation between water and nutrients and in the northern Sahel where the precipitation is below 300 mm/year water is the limiting factor. In the southern Sahel the increase of biomass with precipitation becomes gradually smaller and nutrients become the limiting factor. The quality of the pasture drops forcing the cattle to find its fodder increasingly closer to the savannah. This situation may be the only justification for mobile or semi-nomadic herding, the most rational adaptation to a variation of the quality of fodder in time and space.

Experience in the northern subhumid dry part of Nigeria has shown that for the nomads the problems of access to pasture and water make themselves felt in different terms not associated with each other. The north of Nigeria is underlain by an arte-

Photo 3.3. Regrouping of herds in a Sudanese wadi. Several herds of cattle come here to drink at pits dug into the bed of a wadi near Faro Burungo in the Sudanese Sahel in March 1985. View across valley with the poorly marked bank bordered by a degraded riverside vegetation. The wadi is dry and the pits tap its underflow (Photo M. Mainguet)

sian aquifer with free-flowing wells spread over some 75 000 km^2. Prior to the utilization of the water resources of the region the herdsmen had to rely on shallow pits with manual pumps. The herds were concentrated along the Yobe River, a tributary of the Chad, or they migrated to the south where losses due to trypanasomiasis were high. At the start of the 1950s artesian wells were drilled at intervals of some 15 km from each other, opening access to all pastures. It was noted that:

- Removal of the constraints presented by the earlier scarcity of water did not cause the nomads to increase the size of their herds during years of more favourable rainfall.
- The distance of 15 km between the wells led to overgrazing around each of them whereas the intervening areas were not utilized.

The first of these observations is an expression of the age-old prudence of the herdsmen which expects that rain-rich years will be followed by dry years, without taking into account the new conditions of more ample water supply. The second one results from the negligence of the herdsmen in not leading their animals farther away from the wells. This shows how difficult it is to influence mentalities. The second observation also underlines the limited distances which these animals travel. In Saudi-Arabia water is brought by trucks to supply cattle at some distance from the wells, a costly method but which facilitates the optimum utilization of the pastures.
Jahnke (1982) wrote that:

> the development of the arid zones does not mean teaching the herdsmen better methods of livestock farming … but making them able to look for grazing in other areas so that the drylands can be utilized according to their carrying capacity and continue to represent a valuable resource for the African economies."

This statement has to be toned down somewhat as the nomads have known for a long time that they have to migrate farther when a drought forces them to do so. During the period 1830–1840 the Tibati region of Cameroon has witnessed the arrival of a wave of Peul people from the drier Chad and northern Cameroon who established small sedentary states preserving some degree of mobility. Another wave, the Bororo from northern Nigeria, arrived thereafter but remained nomadic, taking out their cattle to graze, everywhere immediately after the appearance of the vegetation. In the Central African Republic the Foumban nomads who came from Cameroon settled in the drier parts of the country as the endemic nature of the trypanosomiasis transmitted by the tsetse fly kept them away from the humid areas (J. Hurault, pers. comm. 1993).

After the second insufficient rainy season in 1984 the herdsmen of Niger, according to Retaille (1989) tried to preserve their way of life, i.e. their herds, by following their traditional trails. An exceptional transhumance has brought the Tuaregs from the north of Koutous towards Nigeria. For the Peul this represented a return to their position at the end of the 19th century into a still remembered pastoral area. They have occupied uninhabited areas in the border regions which may be considered as a pastoral reserve. Occupation of the land by farmers is discontinuous and the Peul herds of Nigeria sometimes spend the winter time here whereas those from Niger more frequently stay here during the dry season around the many swamps and the inundated areas of the Komadougou. This displacement of pastoral peoples over large distances has also been followed by entire villages of Dagra farmers from the eastern part of the Koutous region, who have turned to livestock herding.

The migration of nomads to outside their original areas add a nuance to the opinion of Jahnke (1982). Although the menace of the tsetse fly is on the way to being eradicated, vast areas, which Ford (1971) estimated to cover at least 10×10^6 km^2, have been restricted for grazing because of flies carrying Trypanosomiasis. The later a control of the tsetse fly and the utilization of these lands by man and his animals becomes possible, the greater are the chances of developing suitable strategies for the conservation of the vegetation and the soils. Because of this one could be worried about the use of anti-tsetse substances which would give the nomads access to the humid areas from which they have been kept away so far by the fear of epidemics and livestock diseases (Hurault, pers. comm. 1993).

Gallais (1989) posed the question whether all this displacement would not lead to social disintegration:

> To the extent that the territorial structures support the social organization, abandoning the pastoral areas of the Senegal valley in favour of a return to the transhumance around the drillings of the Ferlo ... will possibly provoke certain conflicts. Seignobos refers to the difficult settlement in the lower Logone of the new Baguirmi pastoralists amongst the Fulbe and the Shoa Arabs. I have studied the deep troubles caused for the first time in 1913-1914 but accentuated since the 1970s by the Tuareg herdsmen in the highly regimented pastoral organization of the Peul of the interior Niger delta.

3.1.4
Demise and Mutation of Nomadism

Over thousands of years nomadism represented the most flexible way of life in response to the irregularities and scarcity of the rains, facilitating adaptation to any situation. The term nomadism, which until now was used mainly in the sense of pastoral nomadism, may also include the Bushmen hunters and gatherers of the Kalahari, the Southern Hemisphere equivalent of the Sahara, and the Aborigines of the semi-arid fringe of the central Australian desert.

Jansen (1986) defined nomadism in its new complexity as a way of life and an economic system. The group concerned (tribe, clan or family in the wider sense) is temporarily mobile and wanders with its herds, its main source of income, to assure its subsistence within a sociopolitically indifferent or only rarely benevolent context. All goods are carried on the migrations during which temporarily non-pastoral activities like agriculture, commerce, transport services, gathering of wood and resins, fishing and temporary agricultural employment are entered into. Herding, however, still remains the main activity. Although the herds are entirely private property, the land belongs to all, which makes rational management of the pastures so difficult.

Pastoral nomadism is still characteristic of the entire Sahel south of the Sahara, but it is also known in other steppe regions like East Africa and the Ethiopian massif, where the pastoralists utilize the pastures on the edges of the plateaus and on small high plains. In Somalia, Kenya and Tanzania the nomadic tribes (Rendile, Masai, Borana) preserve a strong cultural individuality, but their activities have become diversified. Those of the Sahel were the source of the Arab-Berber civilizations which are organized in hierarchic societies subdivided into castes and possessing slaves. This social complexity is found in all nomadic societies of Africa and Asia.

In north-western India the state of Rajasthan features a steppe cover similar in composition to the Sahel. Toupet (1992) recognized here forms close to nomadism and the same situation of transition between two opposed and at the same time complementary civilizations: those of India and Iran for Rajasthan and the Mediterranean Arab–Berber civilizations to the north and the Sudanese of Black Africa to the south for the Sahel.

The alpine and desert-like steppes of the Mongolian Republic also represent regions of nomadism. At present they contribute some 70% of its agricultural production and 30% of its exports. Nearly half the population of this country is rural and the revenues, mostly depending on the livestock, have dropped since the end of the socialist regime by 20% for agriculture and 26% for industry, but by much less for livestock holding, which presents the best prospects after the decline of the urban economies.

In this country 70 years of centralized socialist planning have replaced the traditional pastoral groups, i.e. the *khot ail*, a camp of pastoral nomads comprising 2–10 fire places and the *neg nutgiin*-han made up of 4–20 *khot ail* by camps and brigades of herdsmen. During the 1960s, the socialist system substituted itself for the ancient know-how, which had been improved over the centuries and had supplied the means for moving the livestock, food supplements and assurances against the risks incurred in these marginal arid and cold lands. The customary system has nevertheless not been lost completely as Meerus and Swift (1993) remarked, but there is the risk that it can no longer respond to the new needs created by the growth of the herds and against the menaces imposed on the environment.

We do not want to describe nomadism, but to outline the mutations caused by the mechanisms of sedentarism and to see how the regression of nomadism has led to a renewed interest in it.

3.1.4.1
Reasons for the Settling-Down of Nomads

Sedentarism represents conversion from a life dominated by mobility to an essentially fixed life. It is by no means a new phenomenon. It existed in Asia for several millennia and in Africa for several centuries, but became accelerated during the 20th century. Among the reasons for this regression of nomadism are the precarious nature of life in drylands, the fascination exerted by the sedentary life in oases, wars or raids leading to the loss of large parts of the herds, or their subdivisions because of inheritance problems, the rigidity of the frontiers and finally the droughts, to which cattle are less adapted than sheep and goats but from which they recover best. Camels are the least sensitive animals in this respect. During a drought animal birth drops by 50–85%, but in the year following a drought it will rise to 140% of the mean value.

Colonization also led to regression of nomadism. The pacification of the Sahara caused the Tuaregs to lose the duties which they levied on the settled inhabitants of the oases for protection against pillaging. The Tuaregs also possessed slaves:

- For cultivating the fields in drylands. These were emancipated at the start of the century.
- In households where emancipation took place after World War II.

Photo 3.4. Peul nomads of northern Mali. Peul women with their children in the semi-sedentary village of Diré in the north of the Sahel of Mali in February 1985. They wear their traditional costume of hand-woven cotton with the children being naked or nearly so. The black tent woven with goat hair shelters a magnificent wooden bed, the supports of which are carved for aesthetic reasons. The two kitchen utensils shown, the tin kettle and the wooden plate are common property (Photo M. Mainguet)

The loss of their slaves left the Tuareg nobility in disarray, and they moved closer to the oases and abandoned their nomadic lifestyle. In addition, transportation by automobiles led to the gradual disappearance of commercial caravans – a source of income for the nomads – and thereby, among other reasons, to the regression of nomadism.

Nomadic herdsmen are confronted with problems of the loss of land, conflicts, restrictions imposed by their displacement and limited access to social services, which leads increasingly to their marginalization. This applies as much to the Gujarat drylands of India as to the Sahel, where several states have introduced reforms of land tenure as e.g. in Niger and Ethiopia. In Nigeria nomadism is also declining. The majority of their herdsmen have become sedentary and are practising a limited form of transhumance alongside their other, mostly agricultural activities. In turn, the farmers are turning more and more to livestock breeding as an investment, for animal traction and fertilizer. Commercial livestock breeding is practised in the southern part of the country but is limited by the high cost of fodder and veterinary products. The 600 mm isohyet no longer represents the limit for livestock holding because of trypanosomiasis, as ground clearing and deforestation have suppressed the vegetation, the natural habitat of the tsetse fly.

The traditional way of settling which is still followed today as the most usual process, entails abandoning livestock holding in favour of agricultural activities (Pho-

Photo 3.5. Settlement of Peul nomads around a well in Mali. The well constructed from concrete rings in the Peul village Nioko in Mali, was photographed in June 1985. The extraction system using wooden pulleys is traditional. The well represents a place of social encounter for the herdsmen (Photo M. Mainguet)

to 3.4). In Somalia the nomads of the north had already started in precolonial times and at the start of the colonial period to install themselves on the fertile lands in the south of the country, where their found better climatic and hydrological conditions and a better vegetation cover (Jansen 1986). This example shows that the nomads are relatively easily prepared to restrain their migrations and even to take up agricultural activities and to settle down alongside an already sedentary population.

A nomad will settle in areas along his routes of migration which appear more favourable to him. The great nomads of the Saharan winter migrations settled down in the high plains of Constantine (Algeria) and the high plateaus of Oran (Planhol and Rognon 1970). The Anatolian nomads who migrated between the high plateaus and the unhealthy lower plains stayed on the plateau when they became sedentary as the demographic pressure increased the price of land too much for them to afford. In summer they are pushed back by the farmers into the mountains, in winter forced to rent pastures on the fallow lands around villages and as a result their ranges are continuously shrinking and they become more marginalized (Planhol 1958).

The most recent forms of sedentarism coincide with the installation of administrative points in the nomadic areas during colonial and post-colonial times. As water posts or watering points, centres of food distribution, dispensaries and as centres of commerce they have attracted the nomads and enticed them to settle for longer periods or permanently. The facilities supplied by the authorities to the herdsmen developed into points of sedentarization and for farming the land (Photo 3.5).

Any management of allochthonous water-courses, of swamps or of other bodies of water or lowlands, or ample forage will favour farmers and be disadvantageous for herdsmen, who will lose access to the temporarily humid areas where the beneficial effects of the rainy season extend into the dry season; the herdsmen find themselves driven back to less favourable areas.

> In the Senegal Ferlo the drilling of deep boreholes has accelerated and consolidated the march of the farmers to the east and the shrinking of the space of the pastoral Peul just as the erection of dams on the Sao Francisco River in the Brazilian Nordeste has contributed markedly to the retreat of the herdsmen into the Caatinga (Pélisier 1989).

The discontinuous droughts experienced by the semi-arid ecosystems also represented an incentive, as was the case during the droughts of 1968–1975 and 1982–1984. In general, all these droughts accelerated the process of sedentarization around watering points along the transport axes which are also trade routes, and around markets.

Jansen (1986) described the different stages of this process of settling in Somalia which is representative for the whole of Africa:

- Stage 1: a grass-covered hut made of branches is erected at a favourable point; at a cross-road or in an area with easy access to surface water. A nomadic family will sell tea or milk here and eventually warm meals.
- Stage 2: when these activities are successful, bush restaurants are opened and then small stores and repair shops.
- Stage 3: erection of a police station and of a mosque.
- Stage 4: whereas the small shops are opened up along the main route, the living houses or huts are built away from it along the side streets. To supply water to men and livestock staying around these agglomerations during the dry season, wells are dug and concrete cisterns are erected depending on the local natural conditions and on the financial capacity of the newly settled inhabitants. Finding construction material is also a problem. Many buildings in these new agglomerations are not lived in permanently and not by all members of a family. During the rainy season the men lead a nomadic life with the livestock. The smaller livestock is looked after in the vicinity by women, children or the elderly.

The changes become more pronounced when the people move farther away to the major urban centres or, as in Somalia, to the oil-rich Gulf states. The Toubou of the Tibesti have left for the Libyan oilfields. The spread of motorized transport in the nomadic areas has favoured migration towards the villages which experience rapid growth. The newcomers settle on the periphery where huts are built next to each other without any planning. This mixture of cardboard, tin or wooden huts without water and electricity grows into a *"cartonville"* in which the population density is highest during the dry season when life in the bush is more precarious. Some families keep a few goats as a source of milk.

The relationship between village and the open ranges persists as the settled nomads guard the large herds which are supervised by other members of the family. A few cows kept at the edge of the village may furnish some income from sale of fresh milk. The adult men all tend to look for work and gardening is much in demand. The Rendile nomads of northern Kenya are the most sought-after gardeners in Nairobi.

Bernus has shown that during the 1980s the herds of the Woodabé Peul herdsmen of Niger also included animals belonging to commercial breeders and village officials.

3.1.4.2
The Mechanisms of Forced Sedentarization

Forced sedentarization may be the result of the political will of a central government:

- This was the case in the central Asian republics of the former USSR where sedentarization of the entire nomadic population was introduced at the start of the 1920s before a change-over to a "softer" organization of semi-nomadism around fixed points took place (Wimer 1963).
- At the end of the 1920s the Shah of Iran attempted to settle the herdsmen of the Zagros Mountains with the aim of subjecting them to his political authority (Beaumont 1989).
- In Niger the political attitude towards the herdsmen during the 1980s entailed a policy of sedentarization against which the people reacted in a hostile way and most probably quite rightly so.
- In Somalia (Jansen 1986) where 50–60% of the 2.5–3 million inhabitants during the 1980s and early 1990s were still nomads or semi-nomads, this being the country with the highest proportion of nomads in Africa. Despite the severe drought of 1974–1975, referred to as the *dabadheer* and other secondary droughts as in 1983, the headcount of the livestock has not ceased to grow, supported by the numerous new wells, by veterinary services and by the increased demand for meat on the local market and in neighbouring countries like Saudi-Arabia. Despite these droughts the long-term growth rate of the herds amounted to 1–2% annually. In 1985 the national count gave a number of 40 million ruminants, equivalent to 8 per inhabitant, which represents the highest ratio in Africa and also worldwide. Despite the increase in livestock the government maintained before the civil war as one of its objectives for political reasons the sedentarization of as large a part of the nomad population as possible;the nomads were considered as troublemakers, whose mobility permitted them to escape any control. Furthermore, for the governing elites nomadism was an obsolete way of life incompatible with modern development. A campaign of public instruction was started in 1971–1975 with the admitted aim of preparing the mentality of young nomads for a sedentary life. This coincided with the drought of 1974–1975 which ruined some 250 000 nomads and formed part of the widespread campaign for sedentarization, which Somalia, following the Soviet example, has tried to achieve since (Jansen 1986).

In 1975 more than 110 000 nomads were forcibly settled far from their usual areas of life in the framework of three agricultural and two fishing projects (Labahn 1982) which, organized in an authoritarian and centralized manner, excluded the nomads from the various decisions and had their self-sufficiency as an utopian aim. The object was not achieved and the population had to be fed by the *World Food Programme* at the start of the 1980s and later by the *Food-for-Work* programme. Many of those who could not stand the changes in the way of life and in the physical and political conditions did not subject themselves to this system. Whereas women and children stayed at these places in order to benefit from free food rations and the educational and health systems, the men moved to Mogadisco or to the oilfields of the Gulf. This unhappy situation contributed to the destabilization of Somalia, further assisted since then by the world.

Forced sedentarization can only lead to complete failure as the proposed projects are too complex and too vast to be handled by the nomads themselves. National or international projects of sedentarization which did not take into account the ecological know-how, the economic needs and the wishes of the nomads have all failed. In many drylands the climatic and hydrological conditions and the lack of available arable land are mostly such that nomadism represents the only rational exploitation of these areas.

The ILCA (*International Livestock Centre for Africa*) with seat in Addis Ababa has presented in its 1983 report a comparison of three systems of livestock management:

- That of the East African semi-nomadic Borana in Kenya
- The modern ranching system of Laikipia, also in Kenya
- The ranches of the Northern Territories of Australia

The results were surprising and toppled numerous prejudices: the Borana adhere to a nomadic tradition with several objectives but, more than for consumption or sale of meat they practise the production of milk for their own consumption. They rather try to optimize the number of persons supported by a certain surface area than to look for economic advantages. With their system they manage to support 6–7 persons/km^2 of pasture, whereas a ranch in northern Australia supports only 0.002/km^2.

3.1.4.3
The Consequences of Sedentarization

The sedentarization of nomads will always lead to a deterioration of their lifestyle and their economic level at least for the first generation. The socio-economic situation of these freshly settled people is worse than that of the urban population as e.g. only a small proportion of their children attends school. Those abandoning herding and migrating to the cities are frequently illiterate and find it difficult to obtain adequate employment and, as a consequence, become marginalized. The lowest paid jobs are offered to them in construction, transport, and commerce. Once cut off from their ethnic group, it is difficult for them to reintegrate themselves.

The change from a nomadic life to a semi-nomadic one is less disadvantageous as it will preserve the connection with a group, permits self-sufficiency in food supply and sometimes even a small surplus. However, the change to semi-nomadism or sedentarization will inevitably lead to a greater concentration of the livestock on the pastures surrounding the new settlements and wells and leads to a degradation of vegetation and soils which is irreversible in the short term. From a more general point of view, the diminution of the mobility of the nomads constitutes a great danger for the ecological equilibrium in the areas where they become concentrated.

In the light of the failure which is represented by the desire to reduce pastoralism, the opposite trend is now becoming apparent. In February 1993, a conference in Addis Ababa on nomadic livestock breeding in Ethiopia came to the conclusion that nomadic herdsmen are always amongst the most marginalized citizens and denounced the authorities who are of the opinion that the production of the herdsmen is archaic and uneconomic. This critical attitude is difficult to abandon for the transitional Ethiopian government and other Sahelian governments. The conference recommended those responsible for development to renounce their plans of converting pastoral people to sedentary farmers or commercial herdsmen, and to reduce their political and economic marginalization. This implies assuring them of better land tenure rights and better access to social, veterinary and sanitary services, helping them resolve the problems of insecurity inherent in droughts, improving their traditional production systems and avoiding making them mainly suppliers for urban consumption or export markets.

Because of the risks already becoming evident from favouring sedentarization and in order to avoid the still larger risk of an exodus towards the towns, the states should develop a certain infrastructure in nomadic territories and support financially the establishment of small enterprises related to the production of milk, wool and skins. Networks of commercializaton capable of absorbing the animals offered for sale during droughts at reasonable prices will be created. In northern Kenya such a system was planned by UNESCO in the Rendile country to facilitate the direct transport of livestock from the production areas to the capital.

Is it not, however, essential to assure pastoralists of: A legal instrument for access to land and water, a guarantee for mobility, as under the new prevailing interests – absence of the owners of the herds and different types of legal access to land tenure – the customary institutions being no longer sufficient, and, finally, a progressive transfer of governmental duties in the field of management of the rural areas to local organizations of the pastoralists themselves? The population has to be mobilized to decide on halting the growth of the herds and to rehabilitate the pastures by protecting them.

Although it is clear that nomadism can no longer exist in its present traditional forms, because of the external pressures exerted on this type of life, it is also obvious that its brutal, forced disappearance has met with failure. Livestock breeding has been and should remain a primordial sector of the economy of the semi-arid ecological systems. However, in the Sahel as well as in the Brazilian Nordeste one must not ignore the transfer in animal ownership from traditional herdsmen to urban breeders and traders and the transfer of shepherding the animals from the urban breeders to the traditional pastoralists.

We shall conclude this section with Marty (1989) who, in attempting to clarify the aims of the pastoral strategies in the sub-Saharan and Sahelian grazing areas which had been most affected by the drought, rejected the three proposed "mirage solutions":

1. A return to the ancient livestock system with transhumance between dry-season and rainy-season pastures. This system which had worked for centuries is no longer possible today because of the shrinkage of the grazeable spaces and the natural resources, and because of the appearance of new ways of treating herds as demanded by the towns.
2. Sedentarization, the second mirage solution proposed by national authorities, has in most instances led to a situation of impoverishment and dependence on food

and has not developed into a starting point for a new dynamic exploitation of the areas.
3. Agriculture, which was visualized as the only remedy for the crisis of pastoral societies, represents the third mirage solution. This is wrong as in most cases it has resulted in the clearing of the last wooded areas. And furthermore, outside certain privileged areas agriculture is on its own not able to feed the families. "In the majority of cases it is only through diversification and thus complementary activities and sources of income that the future can be assured" (Marty 1989).

3.2
The Geohistory of Water Management: The Hydraulic Civilizations

Freund (1984) called civilization "the spirit underlying the way in which humans socialize with each other in a geological, climatic and geographical context." From our point of view the climatic aspect of drought is the decisive one, but we shall not lose sight of the other two.

The term "geohistory" is introduced with the aim of attracting the full attention of the reader to the fact that throughout our study of the triplet aridity–drought and development it appears incontestable to us that geography preceded history and has controlled its evolution and not the other way round. We have noted that climatic changes play a decisive role in the evolution of any civilization, especially when they fluctuate towards aridity. These considerations will represent the starting point for an analysis by which we will show how the hydrological constraints of drylands lead to the birth of agriculture and the great hydraulic civilizations, which are more a response to external climatic or internal demographic constraints than a stage in the inner social evolution of a group of humans.

3.2.1
Aridity, Drought and the Birth of Modern Man

The new archaeological hypotheses distinguish archaic man of African origin from modern man who, after an adaptation to a dry environment, viz. that of the Asian steppe, has then migrated towards the west (Otte 1994). Bonifay (1993) proposed a different hypothesis of "knowing that no hominoid source giving birth to the first representations the genus *Homo* existed in Eurasia before the African continent became populated." Otte (1994) distinguished archaic man from modern man by a new behaviour referred to as "upper Palaeolithic". This concept is of interest to us as, if it describes a historic and not an anatomic phenomenon, it reveals an evolution that is mainly tied to a steppe environment, i.e. a dry ecosystem.

Adaptation to a steppe environment finds its expression in weapons made of bone instead of stone and wood, both of which are rare in these environments. Referring to Hadyu (1976), Otte (1994) remarked that:

> the new tools reflect limiting the economic use of natural plant materials and replacing them by those of animal origin. This represents a profound difference to the preceding techniques which considered vegetal fibres as complementary only for handles or arms. This replacement by animal materials is even today an adaptation to an open tree-less environment, i.e. that of the central Asian steppes .... The ancient form of this process, referred to as "Aurignacian" by archaeologists, was dated at 44 000 years B.P. in Bacho Kiro and Temnata in Bulgaria (Kozlowski *et al.* 1993),

whereas the equivalent cultural expression dates from 32 000 B.P. at Ksar' Aqil in the Lebanon (Mellars and Tixier 1989). When following it geographically, this extension has been dated at 34 000 B.P. in Moldavia (Otte and Chirica 1993), in Russia (Boriskovski 1984) and recently in Siberia (Goebel *et al.* 1993; Abramova 1989; Derevianko 1989; Vasiliev 1992). The trail of these sites thus runs along the southern edge of the Eurasian plains in an otherwise homogeneous movement at the same latitude and in open vegetated areas (Otte 1994).

In the vast expanses directly connected to Europe numerous old and new discoveries have revealed the rise of a bone industry and a new stone-working technique with highly evolved ancient features (Abramova 1989). The new way of life adapted to the steppe here appears to be older and richer than in Europe, where the anthropological and archaeological features appeared concurrently. All this indicates that new ethnic groups with a way of life adapted to geographical zones which apparently were treeless and relatively dry penetrated into this part of western Eurasia.

The Aterian culture developed in Africa from Morocco to the Chad simultaneously, after a dry episode between 33 000–25 000 B.P., during the climatic optimum of the upper Palaeolithic (25 000–20 000 B.P.). Its disappearance coincided with the establishment of hyperarid conditions during the Ogolian (20 000–15 000 B.P.), which was accompanied by a general warming of the planet, leading from 15 000 B.P. onwards to the melting of the Würm ice cover (Schneider 1994). According to paleoclimatologists the humidity was higher during the glacial periods at least in the Northern Hemisphere, with grasslands and forest on most of the terrain not covered by ice.

3.2.2
Aridity, Drought and the Birth of Agriculture

The oldest agricultural tools, dated at around 11 000 B.P., were found in an area between the east coast of the Mediterranean and northern Iraq (Suzuki 1981). There is general agreement that at this time the last, the Würm Ice Age, came to an end because of a rapid warming of the entire planet. "The water thereby set free facilitated the Neolithic miracle with the invention of ceramics in the Orient as well as in the Sahara" (Schneider 1994).

Sanlaville (1981) has shown that between 12 000–10 500 B.P. a very favourable climatic optimum prevailed over the Middle East. The trees spread and wild cereals extended over the entire Mediterranean–steppe crescent. This observation is supported by climatic data from East Africa, where humid conditions are known to have occurred from 12 500–8000 B.P. with a maximum around 10 000 B.P., together with an intensification of the influence of the African monsoon and of the discharge of the Nile. Issar (pers. comm. 1994) for the same period reported a spreading of Lake Nissan, the present-day Dead Sea. The Nafutian period ended between 10 300–9500 B.P. in another dry period followed by an alternation of humid (9000–8000) and dry (8000–7600 B.P.) periods. This was succeeded by the Neolithic pluvial (7600–6000 B.P.) and by another dry period (6000–5000 B.P.).

"Between 10 000–7000 B.P. agriculture appeared … in a form which for a long time was difficult to distinguish from the gathering of wild plants, but it was accompanied by particular social structures" (Leroi-Gousham 1956). Two hypotheses for locating the cradle of agriculture have been proposed:

1. Its independent appearance at several places, South-East Asia and Central America

2. Its birth at only one place from which it then spread, as proposed by Sauer (1960) and Wirth (1968), for whom South-East Asia was the cradle of agriculture, which implies a tropical humid system to start from.

The studies of Lampert (1967), Golson and Hughes (1976), Golson (1977) and finally Sillitoe (1993) reconciled the two opposing views. They described how in the Wola Country of central New Guinea high and long rainfall combined with a steep topography to create conditions conducive to strong potential water erosion. However, the small Wola lands (five valleys) have found a way to conserve the environment and in particular the humus by extended fallow periods and by building small mounds of compost above decomposing organic matter, allowing them to obtain a yield of 20 t/ha from an area which would naturally not yield more than 7.5 t/ha under the mountain rain forest.

The interest in these studies lies in the discovery of archaeological indicators showing that these conservational practices have been inherited from 10 000 years ago and that the mountains of New Guinea were, at the same time – and in parallel with the *fertile crescent* and the Middle East – one of the cradles of settled agriculture and domestic plants.

For the Palaeolithic hunters the sphere of activity depended on the presence of animals and plants. During the Neolithic, man became a planter with an increased dependence on climate, topography and mainly the availability of water, which limited his activities as well as those of the plants cultivated by him. By necessity he became conscious of the seasons and of the vegetative rhythm of plants required by sowing and harvesting. In this context, it is understandable that a settled planter recognized the power of the sun and the moon, of rain and storms, of lakes, rivers and springs and that he created for himself a polytheistic pantheon. Agriculture thus gave birth to gods and their cults and in a larger context to the concept of space and time.

After 10 000 B.P. postglacial warming was accompanied by a climatic drying process. Some have referred to this as the "climatic optimum" or *Hypothermal Period* (Suzuki 1981), which on average lasted for about 5000 years until about 4600 B.P., according to the data of Schneider (1984) on the Chad Basin. During this period the temperature was a few degrees higher than at present. More humid conditions resulting from a more northerly position of the west-equatorial circulation zone prevailed over the Sahara and western Asia. Chu (1973) has shown that these more favourable conditions also existed in the Huang Ho basin of China.

The Mid-Holocene climatic optimum (9200–4600 B.P.) as defined by Schneider (1994) made itself felt over Africa by humid conditions with the establishment of the Mega-Chad and the rise of the Mid-Neolithic culture. Schneider ascribed to this rain-rich climatic optimum:

- The "important development of cattle and sheep breeding"
- The "rock paintings of the Sahara depicting the wild fauna and domestic animals"
- The "early transgression to a proto-history in several regions of our planet with the metallurgy of gold (Bulgaria), copper (Egypt) and iron (Nubia) and the invention of cuneiform script by the Sumerians"

Schneider (1994) adds:

- "around 5100 B.P. King Menes Marmer unified Egypt and founded the first, Thinite, dynasty.
- around 4650 B.P. the Minoan civilization of Crete started to emerge.
- a little later, around 4630 B.P. King Lugalzaggisi of Uruk created a Sumerian empire, the first of the oriental empires (it is quite apt to remember that the history of writing on our planet commenced at Uruk around 5100 B.P., i.e. in the Neolithic, with simple pictograms).
- and around 4320 B.P. Sargon established the Semitic dynasty of Akkad."

For Suzuki (1981) agriculture spread from the *fertile crescent* over the entire planet during the 5000 years of this climatic optimum:

- Around 7000–6000 B.P. at its birth place in western Asia there was already a quite diversified activity, as indicated by decorated ceramics from Iraq. Then irrigated agriculture descended from the narrow mountain valleys and became organized in the plains.
- One of the first waves arrived in Greece around 9000 B.P. and reached Vienna (Austria) around 7000 B.P., benefiting from the warmer and more humid climate, which was more amenable than the present one; this wave came to a halt at the edge of the ice-covered areas.
- Another wave advanced towards the Sahara with its grass-covered steppe plains and forest cover in mountain areas like the Hoggar and the Tibesti.
- An eastern wave flowed towards South-East Asia which was quite humid during the climatic optimum; in northern Thailand indications were found that agriculture here dates back to about 9000 B.P.

Around 5000 B.P. the world climate started to deteriorate and the postglacial climatic optimum was succeeded by a real aridification because of the southward retreat of the equatorial zone of westerly circulation (Suzuki 1981). In the Chad Schneider (1994) placed the start of this aridification at about 4600 B.P. Under the influence of the geological advance of the deserts, the farmers were forced to take refuge along the valleys of the large allochthonous rivers and major courses of water.

3.2.3
Aridity, Drought and the Birth of Hydraulic Societies

Is it not a fact that the first inhabitants of the valleys utilized these initial ecological refuges to create the great hydraulic civilizations by necessity? Is this not perhaps one of the causes for the appearance of slavery, a type of labour that facilitated the erection or the digging of canals, the *kiariz*, the construction of dykes and, at a later stage, of dams?

In response to the natural desertification, at a medium level of civilization and during the same era, followed the birth of the great riparian civilizations along the banks of the Nile, in Mesopotamia, along the Amu Daria and later the Syr Daria (at higher latitudes), along the Indus and the Huang Ho, the Yellow River. One could say that the great hydraulic civilizations originated around 5000 B.P. as a response to aridification on a geological time scale which undoubtedly was caused by a natural desertification together with a drop in mean temperatures of about 3 °C. These civilizations became furthermore endangered by a further drop in temperature accompanied by another, still higher aridisation.

Among the causes for this drop in temperatures Suzuki (1981) named the explosive eruption of the Santorin volcano in 3500 B.P. and the numerous eruptions which, according to tephrochronological studies, occurred in Japan at roughly the same time. According to Suzuki (1981), the ancient Huang Ho civilization declined around 3500 B.P. at the same time as those of Mesopotamia whereas the ancient Egyptian civilization declined around 3200 B.P. This decline is not observed along the Amu Daria as archaeological studies at the Tazabagjab site have shown that the complex irrigation system of Khorezm in the lower Amu Daria valley dates from the Bronze Age around 3500 B.P.

According to Vinogradov and Mamedov (1975), Dolukhanov (1986) and Szymcsak (1997), the Neolithic occupation of the Kyzylkum declined from the early Holocene (8, 7 and 6 millennium B.C.), with the hunting-gathering Keltaminarian culture when the Lyavialakan lakes were numerous, indicating a spring rainfall of 140 mm per year with the farming economy of the Djeitanian culture in the Karakum and the Hissarian stock breeding economy in Easter uplands. In the Kyzylkum, the Mesolithic-Neolithic sites are counted in thousands, and many sites endured for long periods. Later, in the 5 and 4 millennium B.C., the sites were less numerous and the settlements were of short duration.

In the third millennium B.C. in the Kyzylkum the sites are very poor, with one or two pottery sherds, indicating short-term camps. This decline can be considered as an indicator of an increasing drought with the disappearance of the lake surface water. Simultaneously, along the Amu Daria during the Bronze age human occupation evolved towards a promising hydraulic agriculture and civilization.

The question raised by the decline of the hydraulic civilizations is whether it resulted from a collapse of the equilibrium because of the climatic factor of increasing aridity or of the socio-economic factor of the decline of the empires due the invasions like that of the Hyksos during the Middle Bronze Age in Egypt around 1700 B.C. during the Middle Empire (XV–XVI dynasties) or due to a degradation of the lands correlated with it. The drought leads to hydraulic technology through a number of obligatory stages. Aridification forces the farmers to move closer to the sources of water, i.e. to relocate towards the valleys; the first users of the valleys will bring into slavery the ones arriving last; a hierarchy and possibly a civilization will be created.

Rain-fed agriculture is impossible in hyperarid, arid and Saharan–Sahelian zones receiving less than 250 mm/year in precipitation with an interannual variability above 30%. This same rain-fed agriculture remains a haphazard feature in the Sahel and even in the Sahelian–Sudanian zone with a precipitation below 600 mm/year. "All authors agree that there is a semi-arid domain defined by a slightly longer humid period (up to 4 months) which is, however, highly irregular. Rain-fed cultures are possible with hardy species, but not every year, total losses are frequent and non-irrigated agriculture is a veritable lottery" (Durand Dastès 1977).

The first cases of water management originated in the drylands and here developed the large "hydraulic societies", a term coined by Wittfogel (1975) for the first model of this type of society in China: despotic governments (the title of his book

3.2 · The Geohistory of Water Management: The Hydraulic Civilizations

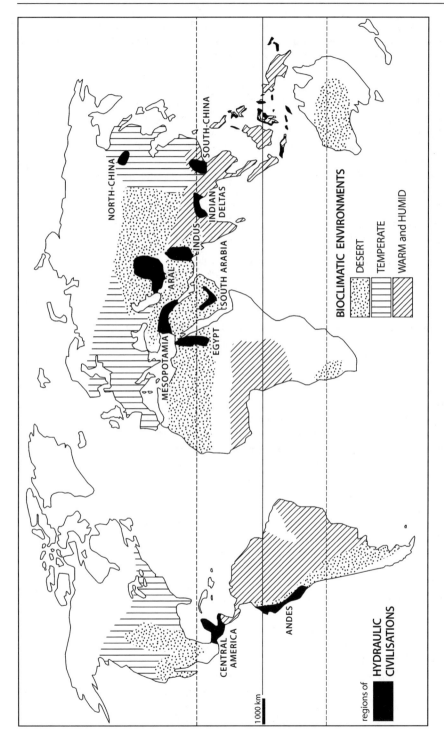

Fig. 3.1. Distribution of hydraulic societies and bioclimatic environments (after Bethemont 1981)

was actually *Oriental Despotism*) formed in pre-industrial societies in which agriculture was mainly by irrigation. Irrigation actually requires a strong organization for controlling water, its distribution, the construction and administration of the irrigation system, its maintenance, the allocation of costs and the vast pool of manual labour necessary.

Wittfogel's hypothesis that irrigation has as a corollary an authoritarian political organization has been rejected by many archaeologists. Adams (1974), Gibson (1974), Neely (1974), Vivian (1974), Farrington (1974), Mosely (1974) and Butzer (1976) are of the opinion that a hierarchical political organization preceded and facilitated the development of complex and even sophisticated irrigation. Spooner (1974) suggested that the scarcity of water or the difficulty of access to water or land influenced the social organization and its hierarchical structure.

Bethemont (1982) raised the question in somewhat different terms:

> The old fact has become rather banal: at the origin of the earliest civilizations there was frequently the control of water and it is here that the concepts of social stratification, political hierarchy and the concept of the state developed as a response to hydraulic problems. This correlation could be observed in Mesopotamia, in the valleys of the Indus and the Ganges, in Egypt, China, Peru and central America, all cultural areas revealing the categories defined by Wittfogel (1957): *hydraulic agriculture* (depending over vast areas on the control of water taken over by government) and *hydraulic society* in which the hydraulic and agrohydraulic structures play a fundamental role and are carried out by a particularly authoritarian government.

Remembering this question, the relationship between aridity, drought and the appearance of the various stages of irrigation and the great hydraulic civilizations will occupy our attention.

Archeologists like Cauvin (1981), Aurenche (1982) and Bethemont (1982) were among the first to ask themselves about the conversion of certain groups of humans towards hydraulic agriculture and, as a sideline, they raised the problem of environmental determinism, with special emphasis on sociocultural determinism.

The irrigated areas of central China and farther west the Asian hydraulic tradition of the deltas of the Ganges, Brahmaputra and Meghna and finally the irrigated areas of southern Mexico and central America with Guatemala and Honduras, all originated in the humid tropics. They shall not be dealt with here as there is no relationship between aridification and the appearance of successive stages of hydraulic engineering. We thus have to accept that hydraulic societies appeared in the dry tropics under different climatic conditions from those of the humid tropics, which would exclude any climatic determinism, were it not for the fact, as we shall see, that the hydraulic systems attained their grandest development in the drylands (Fig. 3.1).

3.2.4
The Great Hydraulic Systems of the Drylands

The most remarkable hydraulic systems of the Indus, Nile, Aral and Mesopotamia are all located on the same dryland diagonal of the Northern Hemisphere. Mesopotamia, about which we know more than any other hydraulic system in drylands, deserves a special position. All four areas reveal certain similarities and differences between themselves and the systems in humid areas. Along the dryland diagonal of Latin America which touches the Andes we observe that the hydraulic system of Peru, the only one on the Southern Hemisphere, made its appearance around 5000 B.P. on the piedmonts and in gullies where rill wash water controlled by small dams was utilized. Around 2000 B.P. irrigation in terraces became established on the mountain slopes with basins, aqueducts and a complex network of canals.

3.2.4.1
The Ancient Indus Civilization

This civilization, still referred to as the Harappa culture of India and Pakistan, extends in the west of the Indus plain in present Pakistan and east of the plain in the recent Rajasthan desert. From 4000–1000 B.C. three phases may be distinguished:

- *From the fourth millennium to about 2300 B.C.*, a pre-Harappa phase is characterized by an urban environment with mud bricks, pottery and cult objects and other, polychrome objects (figures and seals of terracotta), tools made of stone and bone and an increasing use of bronze and copper. Starting with the Mundigak culture of SE Afghanistan it extended into present-day Baluchistan and from there it gradually made its way down into the Indus Valley. It developed during the early Bronze Age concurrently with the by no means less grandiose Sumero–Accadian empire and the Old Kingdom of Egypt.
- *The second phase, from 2300–1700 B.C.* represents a more elaborate urban culture with mud bricks which may also be burnt; it is very homogeneous over a territory that is larger than present-day Pakistan and is characterized by large cities possessing defence systems, a sanitary organization and complex canalization networks. This civilization prospered because of a conversion to irrigated agriculture and then declined when probably natural events like an aridification led to the failure of this type of agriculture. Such events were also the invasion and domination of Egypt by the Hyksos during the Middle Kingdom (XV–XVI dynasty) and the biblical migration of Abraham from Ur in Mesopotamia towards Canaan.
- It is probable that such natural causes were violent floods or most probably as shown by the palynological studies of Bethemont (1982), a phase of aridification during which precipitation dropped from 830 mm/year during the 8th millennium B.C. to 200 mm/year in the 1st millennium B.C. This was accompanied by a degradation of the forest cover to steppe and then to desert. Such a considerable variability is only conceivable in an arid environment but this would imply gigantic upheavals. Bryson and Baerreis (1967) suggested a desertification resulting from an overexploitation of the environment because of overcultivation and overgrazing, accompanied by a *dust-bowl* effect reducing the condensation of atmospheric moisture. A high load of suspended dust in the air counteracts the formation of molecules as condensation nuclei.

The Harappa culture of the Indus is reminiscent of those of Mesopotamia and the Aral because of the same cultivated plants – wheat and barley – and by the same relationship with respect to the drought, all lying on the same dryland diagonal.

With the postglacial warming of the Northern Hemisphere the drylands grew in area and intensity and the dryland diagonal Sahara–Arabia–Thar deserts of former Soviet Central Asia–Taklamakan–Gobi reached its true desert dimension, although the Sahara should be considered as a much older desert as shown by Sarntheim, who actually discovered aeolian sands derived from a palaeo-Sahara in drill cores of Miocene sediments from the North Atlantic. Despite the great age of the Sahara as a desert, the hypothesis of the postglacial aridification remains valid. It can then be easily admitted with Suzuki (1981) that the Palaeolithic hunters who could apparently live by natural harvesting found themselves with growing aridity under the necessity of growing materials for harvesting, which led to the first agriculture. Agriculture thus originated in most cases from increasing aridity as was the case later for irrigated agriculture.

3.2.4.2
The Old Hydraulic System of the Nile

The biotope of the Nile valley is favourable for grasses and thus for cereals, which have been collected here since the 11th century B.C. (Bethemont 1981). The abundance of these plants and the ease with which they could be gathered led to attempts at domesticating these plants up to the early to middle Bronze Age or the fourth millennium B.C. Millet was probably derived from Black Africa (south of the Sahara), whereas wheat and barley came from the Middle East, probably from the Aral area. In the semi-arid fringe south of the Sahara Bernus actually found indigenous varieties of millet, sorghum and rice. Barley, dwarf wheat and most of the fruit-trees (peaches, apricots, plums) were derived from a different dryland area, the Aral basin and possibly Sinkiang.

Irrigation is younger and in its earlier stages it consisted of exploiting the waters of seasonal floods in swamps or in discharge-less depressions with a grassland vegetation. Irrigation from basins appeared between 6000–5000 B.P. and the erection of structures for the control and artificial retention of water between 5000–2000 B.P. According to Bethemont (1982):

> ...The Nile valley represents the oldest case of systematic management in the upstream and downstream parts by a regulating set-up made up of submerged basins arranged in chains .... This arrangement not only is evidence for a great technology, but it also shows administrative and political co-ordination. Other evidence ... for this high level of technology is the distribution of water at Bar Youssef and the management of the Fayum around 2200 B.C., which still belongs to the most remarkable engineering project of all time. One should take ... into account that the technique of continuous irrigation and with two harvests annually made its appearance in the delta during Ptolemaic times, a period that also knew the use of the saqieh.
>
> "...This testimony allows us to understand the high state of technology, possibly because the problems that had to be solved were quite different from those of the Middle East. We can ... conclude from this that the first utilization took place later here than in Mesopotamia but at a much faster rate .... On the other hand we observe in both areas the same evolution towards a centralized administration" (Bethemont 1981).

3.2.4.3
The Aral Hydraulic System (After Letolle and Mainguet 1993)

The earliest Palaeolithic industries on the foothills of the Khopet-Dag and of the mountains of the SE Turan date from 30 000–10 000 B.P., i.e. the upper middle Palaeolithic. Later, the Aral basin became a cradle of agriculture and of one of the major hydraulic civilizations. Archaeology allows us to follow the various stages and they deserve detailed description even if we might appear fastidious, as they are characteristic of the complete cycle of a hydraulic civilization in an arid ecosystem from its birth to its demise.

The alluvial fans and the alluvium in the river plains on the southern foothills of the basin below the Hindu Kush and the Pamir are the first sites. Aurenche (1982) confirmed that such sites were also utilized in Mesopotamia as well as in Iranian Kurdistan and Turkmenistan: "Gawra and Choga Mami in Mesopotamia; Ali Kosh, Sabz and Sefid in the Deh Luran; *Djeitun* in Turkmenistan ... , prior to the settlements in the alluvial plains itself (Eridu, Uruk, Queili in the south; Bendebal, Choga Mish, Djaffarabad in Khuzistan; Geoksyour in Turkmenistan). This multiple recurrence of the same phenomenon at several places cannot be fortuitous." Such a simultaneous development on a historic time scale actually implies contacts and thus major displacements. The prehistory and protohistory of the Aral basin is inseparable from that of northern Iran, Afghanistan, the Indus and of Sinkiang in China.

Wheat and barley were cultivated at the start of the *Namazga* civilization around 9000 B.P. at the *Djeitun* site in southern Turkmenistan, the cradle of soft and dwarf wheat, of barley, stone fruits and walnuts. This territory was occupied between 8000 and 7000 B.P. by pastoralists and farmers, and stone knives and round hoes have been unearthed there. Chlopin (in Andrianov 1990) has shown that the fields extended over 10 ha.

- The first stage of this civilization only required the action of individuals and led to only few changes in the environment.
- The second stage, the passing from the upper Chalcolithic to the lower Bronze Age in the Middle East, corresponds to the start of a local more humid episode called *Lavlakian* at the end of 7000 and the start of 6000 B.P. *Agriculture migrated towards the plains of the larger rivers*, the middle reaches of the Tedjen and the Mourgab. Floods were harnessed by dykes and canals. In the delta of the Tedjen remainders of complex irrigation systems dating from 6500–5300 B.P. were unearthed. Similar systems existed at the same time on the Mourgab, the upper Amu Daria, in Afghanistan and in Persia. In Turkmenistan, especially in the oasis of Geoksyour, Masson (1971) discovered the hydraulic civilization *Namazga II* which furnished the first copper tools and, between 6000–5000 B.P., a complex irrigation network, contemporaneous with the civilizations of Samarra and Obeid in Mesopotamia. The sites of Turkmenistan and Obeid furnished proof of the same social hierarchization in the form of collective buildings which are exceptional in size and decoration (Aurenche 1982).
- At the beginning of the seventh century B.P. the third stage, *Namazga III*, became estab-lished *on the valley floors and in the deltas of the allochthonous water-courses*. On these sites irrigated agriculture necessitated the control of floods by lateral levees and complicated gravitational irrigation systems erected by communal work and requiring a social organization, a strong state and a developed urban lifestyle.

Archeological studies by Masson (in Andrianov 1990) at Altyn Depe in Turkmenistan revealed a social organization around 5000–4000 B.P., a proto-town, fortifications, a complex irrigation system, developed commerce and a hierarchic society with diversified activities. Between 5000–4500 B.P. the domestication of the horse permitted the conquest of the steppes.

Around 4500 B.P., with *Namazga IV*, irrigation became even more complicated, with interconnected irrigated plots of land; Altyn Depe was a growing proto-town. Irrigated agriculture originating from the south of the Aral basin found its centre: the middle valley of the Amu Daria, the Surham Daria, the Mourgab, the Tedjan and the Atrek. At the end of the more humid Lavlakyan episode around 4000 B.P., derivations in the course of the Amu Daria towards the Aral were created.

From 4000–3000 B.P., during *Namazga V*, oases/proto-towns became established in the delta of the Mourgab. The *Souyargan* culture was established along the lower Amu Daria between 4000–3000 B.P., growing oats and wheat under irrigation, and living in contact with local populations of Neolithic hunters, fishermen and gatherers of the *Kelteminar* civilization. Several ways of land exploitation and degrees of development existed alongside each other. Contemporaneously with this during the Bronze Age (fourth millennium B.P.) the *Tazabagyab* culture on the Akcha Daria, the old course of the Amu Daria, was in contact with nomads from south of the Urals and the steppe north of the Aral. Here, one also finds long canal systems and flood reservoirs in river meanders and possibly the noria was invented here.[1] Irrigation works continued. Around 3000 B.P. the Oxus (Amu Daria) turned to the west and between 3300–2900 B.P. during the early Iron Age the *Aminabad* culture formed by the mixture of the Souyargan and Tazabagyab cultures resulting in complex activities: livestock breeding, agriculture and fishing in the canals became permanently established.

Around the same time the Bronze Age came to an end with *Namazga VI*, discovered on the Syr Daria in the Karasouk site (3300–2800 B.P.) where irrigation had not yet been introduced and agriculture still relied on natural floods around 3000 B.P.

Around 2700 B.P. the Zoroastrian epics of the *Azvesta* described the rivers, the canals and the battles between nomads and the settled population. One or two centuries later on the Aminabad sites where irrigation was more complex, possibly cotton was introduced and the building of dykes was discovered as protection against floods.

On the Jaxartes (Syr Daria) the irrigating civilization was launched later than on the Oxus (Amu Daria) and remained less developed. Between 2400–2300 B.P. the southern arm of the Yani Daria was managed and the Chisik Rabat fortress was built.

Hydraulic management persevered uninterrupted at Khorezm (in the Amu Daria delta), which became temporarily independent around 2400 B.P. Irrigation works were continually undertaken along the water-courses: In 73 A.D. at Otrar north-east of the Oxus. In the eighth century A.D., the canals were made wider and shallower to restrict their overflowing. In the ninth century A.D. the old canals of Kat, the ancient capital of Khorezm, were reactivated and the 150 km long Gavkor canal was built.

Around the year 1000 A.D. the oases of Turkestan and Timkent experienced another resurgence. The torrents of the Karatau were harnessed. On the Zerafzan a stone barrage was built, and filled two centuries later. During the 12–13th centuries

---

[1] *Chigir*: hydraulic machinery consisting of buckets fixed on a closed chain which enables water to be raised to irrigate fields located above the canals; the overturned buckets plunge into the water and are pulled up again filled with water.

agriculture grew with irrigation on the Jaxartes. The map by Al Idrisi in 1154 A.D. shows the Aral and the Oxus flowing to the Caspian Sea and also the great irrigation systems of Turkestan.

By 1220 A.D. Jengis Khan invaded the Aral basin and destroyed its towns and the irrigation system of Khorezm which thereafter were reconstructed one by one, that of Otrar on the middle Jaxartes e.g. around 1339 A.D. When Tamerlane invaded Khorezm in 1379 A.D., the new dykes along the Oxus were destroyed. The Karakalpak (black cap) nomads, who had settled on the lower Syr Daria, reconstructed the hydraulic systems. The stone barrage near Khodjend on the Syr Daria were rehabilitated during the 16th century.

The hydraulic works, as points of weakness for the settled population, were always targets for invaders. In 1795 the Khan of Bukhara invaded Turkmenistan, destroyed its capital Merv and the dam on the Mourgab, and in order to rid himself of the Karakalpaks he cut off the Yani Daria by a dam which was destroyed in 1848.

In 1872 the modern era of irrigation began with the construction of a deviation of the Syr Daria with the aim of irrigating the Hunger Steppe (*Golodnaya steppa*) southwest of Tashkent; construction, however, was stopped two years later. In 1886 the surface waters of Turkestan were declared state property, underlining the importance which Russia placed on this resource. In 1891 Grand Duke Romanov revitalized the project of irrigating the Faim steppe with a new canal which was abandoned again and then reactivated in 1899. It was only finished in 1915 and by 1917 it irrigated 35 000 ha. The value of the water and of irrigated agriculture for the Tsarist government is evident from a decision in 1897 which placed the agricultural services and irrigation under the authority of the general governorship of Turkestan.

In 1915, Russian "gigantomania" furnished the Morgunenkov project, the rerouting of the Amu Daria to irrigate Kyzyl-Arwat and the south-eastern coast of the Caspian Sea. The cotton crop grew until 1916 and then fell by 27%, to drop by 92% in 1920, whereas the cereal crop had declined by 47% already in 1916. With the decree of May 13, 1920 Lenin assigned 256 million roubles to the resumption of irrigation in *Golodnaya Steppa*, symbol of development of Central Asia and the decree of November 2, 1920 re-imposed the growing of cotton.

In 1924 the first 5-year-plan encompassed credits for putting the new lands under cotton. In 1930 the gigantic management project on the Waksh, a right hand tributary of the Amu Daria, was started. Between 1933–1937 the Zerafzan, the Mourgab and the Tedjan were managed, and the Kerki Dam constructed on the Amu Daria. In 1936 more than 95% of the agricultural land was collectivized. Despite the invasion by German troops in 1941 the second Fergana Canal was built in 1942, and in 1944, before the end of the war, the Farkhad Dam south-west of Tashkent was started to regulate the Syr Daria.

Central Asia has attracted particular attention because of its hydraulic projects, the aridity of its ecosystem, its topography, its remarkable extent and especially because of the experience acquired by the Russians with regard to canals over two centuries and under the Soviet government during the 1950s and 1960s. With Peter the Great as a driving force large canals connected Lake Lagoda with the Baltic Sea and with the Caspian ad the Black Sea via Moscow. A wide network of waterways was constructed thereafter, assisted by the numerous and vast bodies of water in the interior and the large rivers, a special feature of Russian geography.

When the Russians, driven by Krushchev, decided to populate and exploit the vast desert areas of central Asia, they committed the fatal error of transferring a technology acquired in cold humid areas to a hot and dry ecosystem and, aided by their "gigan-tomania", of digging reservoirs and thousands of kilometres of canals for irrigation, clearing, withdrawal and discharge, which were to lead to the ecological catastrophe of the Aral Basin (Letolle and Mainguet 1993).

3.2.4.4
Mesopotamia, Environmental Incentives and Constraints

"In the Near East one does not talk only about the importance of water – which is everywhere true – but about 'the water problem'. Scarce water limits occupation by man but, at least by stimulating his ingenuity, it does lead to these 'hydraulic societies of Mesopotamia … , i.e. to the starting point of our own civilization. Thus man either submits to a natural situation or he reacts" (Cauvin 1981).

The *fertile crescent* (Israel, Jordan, Lebanon, Syria, south-eastern Turkey, Iraq) does not receive on average more than 100 mm/year in precipitation, a mean value which, however, covers up contrasting realities:

- Precipitation above 1000 mm/year over the mountain ranges from the Levantine ranges to the Taurus and on to the Zagros.
- A pronounced aridity on the lee-side of these mountains.

Rainfall of the cold season, a long summer drought, strong insolation, elevated temperatures and strong evaporation force man to resort to irrigation for his survival. These are the characteristic features of the Middle East which is subdivided by the 200 mm isohyet into three biogeographic domains: the forests on the mountain ranges, the steppe on the foothills and the desert. The water used mostly comes from allochthonous water-courses and aquifers. Sanlaville (1981) wrote: "thanks to the size of their discharge the Tigris and Euphrates … are able to traverse arid expanses and allow humans to concentrate in densities which the climatic factors themselves would not permit."

Using only water as a criterion, Mesopotamia in the wider sense may also be divided from north to south into three geosystems, viz. ancient Assyria, Mesopotamia *s. str.* and the Shatt-al-Arab.

Ancient Assyria, corresponding to upper Mesopotamia, receives more than 200 mm/year in precipitation and offers the possibility for cultivating drought-resistant plants like olives. Assur, the ancient capital, lies on the Tigris.

The plains of the Tigris and Euphrates, which come as allochthonous rivers from the Anatolian plateau. In Mesopotamia the discharge of the Euphrates is $10-37 \times 10^9$ m^3/year and that of the Tigris $15-55 \times 10^9$ m^3/year. The spring floods, in May on the Euphrates and in April on the Tigris, are less useful than those of the Nile. Prior to modern river management, low water levels were pronounced, the floods abrupt and in damaging succession, because of the narrow topography, and particularly dangerous, because of the frequent changes of the beds: Baghdad has been devastated by floods several times in the past.

Mesopotamia proper, commencing below the confluence of the Diyala and the Tigris receives less than 150 mm/year in summer rains and as a desert area is unsuitable for dry agriculture. As a flat and low-lying plain with a negligible longitudinal gradient it experiences vast inundations during each flood. The beds of the rivers are at different elevations, they possess a dense network of channels, sinuosities and depressions complicating their management which is made still more difficult by the embanking of the valleys.

Mesopotamia has neither a climatic nor a biogeographic homogeneity:

- Because of the pronounced slopes, the runoff in the upstream areas is well organized; the rains with a still more Mediterranean trend fall in winter and allow wheat growing. It is the original area of distribution of the cereal plants *Hordeum spontaneum, Triticum boeoticum* and *Triticum dicoccoides* (Bethemont 1992).
- In the downstream areas the landscape is desert-like, the plains are areas of accumulation of an enormous mass of alluvions responsible for the talwegs raising and the site of inundations and channel blockages, in short an unfavourable environment. The vegetation consists of bushes and shrubs, tamarinds and willows in the north, reeds and date palms in the south.

Along the Persian Gulf, the present-day Shatt-al-Arab is an amphibian domain with abundant permanent water vegetation of phragmites (reeds). The date palms here represent the favoured culture because of the supply of water from the rivers. The swampy biotope is very favourable for utilization.

It is in such surroundings that agriculture started at the end of the eighth millennium B.C. in the Syrian middle reaches of the Euphrates at the border of the Mesopotamian Djezirah (Cauvin 1978). After rain-fed farming two millennia were then necessary for irrigation to come into its own: "the first traces of irrigation may be attributed to two civilizations, … that of Samara and of Obeid. And one has found again … the traces not only of a complex social organization but also of an apparently hierarchically organized society" (Aurenche 1982).

From the end of the sixth millennium B.C. an irrigation system with dykes, artificial draw points for water, regulating structures and canals was in existence in the Mandali region on the Samara sites, where six strains of barley, flax and irrigated Leguminosae (lentils, fat peas) were grown at Choga Mami (Cauvin 1981).

On the whole, this shows that the first agricultural attempts did not go beneath the 200 mm isohyet in drylands, i.e. they remained in the core zone of wild cereal where dry cultures were possible. They were no longer able to react during increasing aridity and to the lack of water in the more desert-like areas (Cauvin 1981). The Neolithic expansion due to irrigation facilitated the spread from semi-arid to arid areas where dry cultures had until then been impossible, but this expansion is also due to the choice of new cultivated species which are more demanding of water than the species cultivated previously. "The neolithization descended from the Zagros and its foothills … in Iraq (Jarms) as well as Iran (Ganj Dareh, Ali Kosh) with the first villages of settled farmers" (Cauvin 1981). Cauvin furthermore remarked that: "such an adaptive effort could have been a stimulant for human invention, the water problem acting as a challenge." May we thus not envisage that the birth of agriculture, of irrigation and of the hydraulic societies was also a challenge in the face of a natural weakness of the environment, i.e. aridification or of a human disequilibrium, i.e. demographic expansion?"

In conclusion, the diversity of the dry geosystems resulted in a large variety of hydraulic strategies. Bethemont (1982) gave several examples:

- In 5500 B.C. at Ras-Al-Ayma in the Babylon region the waters of an arm of the Euphrates were diverted into depressions, the submersion of which was controlled.
- On the alluvial fan of the Gangir upstream of Mandali piedmont in 5000 B.C. derivation canals permitted gravitational watering.
- during the third millennium B.C. the chadouf was utilized as shown by seals from Lagash.
- Nineveh was supplied by an 80-km-long canal. The Parthian rulers already knew how to construct retaining barrages around the third century B.C.
- Ancient pictograms show that the plough appeared in India during the third millennium B.C., an advance implying the domestication of cattle. According to the royal records, the cereal yields were 600–1000 kg/ha for wheat and 800–1200 kg per ha for barley (Bethemont 1982). In the Sahel recent yields of millet north of Maradi in Niger were 250–400 kg/ha and then only during years with good rainfall.

However, the study of hydraulic societies in the drylands has revealed a unidirectional model progressing from simple irrigation on the edges of the foothills and in lacustrine depressions towards more complex management structures along the rivers. Simultaneously, sedentarization by agriculture affected pastoral people and nomads, and the small isolated groups became replaced by organized nations. Although acceptable in its general outlines, this model has undergone a host of modifications. It is difficult to say whether cereal planting in Egypt of the fourth millennium B.C. and the subsequent hydraulic constructions of the sixth dynasty developed prior to or after the establishment of centralized institutions. Butzer (1976) and Cauvin (1982) support an earlier formation of the centralized structures.

This question of a chronological order which also covers the reflections of Marx on the development of material structures preceding that of social structures, is only one of the facets of this problem and becomes complicated by differences in the rhythms of evolution: irrigated agriculture is older in Mesopotamia than in Egypt where its development then took place more rapidly. In drylands in particular where resources are less abundant than in other ecosystems, the equilibrium between resources and population must not be disturbed. However, thanks to modern irrigation which created or opened an abundance of resources, a threshold has been crossed which allows other resources to be assured. Is it not that another threshold is presently being crossed with the calculated used of saline water for irrigation?

It is at an entirely different level that we would like to end this discussion of geohistory. Even if the model we have proposed is simplified, and in some aspects simplistic, it appears that the expansion and collapse of civilizations, waves of migrations, conquests and assimilations are so well synchronized with climatic changes that they are not just coincidental but the latter directly influence the former. The difficulty here is to define what a climate change means. It is probably not so much the continued increase in aridity but rather the slow accentuation of the aridity in stages which could be conveyed by the recurring drought crises?

3.3
Traditional and Modern Hydraulic Techniques

The battle against the chronic dearth of water and against the effects of the variability of precipitation in drylands has led to the application of a large variety of techniques. The projects – sometimes luxurious, sometimes foolish – which have been proposed and sometimes executed during the 20th century will be examined and then we shall try to inventarize the systems of parsimoniously handling water and to study traditional and modern methods of irrigation.

3.3.1
Modern Search for Water, Luxury Projects

3.3.1.1
The Inner-Sahara Water Plan

This old project, already proposed by the colonial services, was based on the assumption that a large evaporation area would become a cause of precipitation for the surrounding regions (Létolle and Bendjoudi 1997). This did, however, not take into account:

- The evaporation of the precipitation prior to reaching the soil surface
- The salinization of the water body during the course of evaporation; depressions like the Melhir would have turned within a few years into a body of water so salty that evaporation would be nearly non-existent (Janzein, pers. comm.).
- "Meteorology" which "furnishes an argument ... against an interior sea in the Sahara. Situated on the anticyclonic subtropical belt exporting water where rains are rare, it would tend to impede the formation of the summer thermal depression in the lower layers which pulls the intertropical front to the north and would thereby increase the drought in the Sahel zone without augmenting the rainfall in the area it would occupy" (Durand 1988).
- Lastly, the topography which would have required the costly digging of canals of unrealistic depth and length.

The idea was, however, revived by Hense and quoted by Rognon (1989): a canal 6000 km long should have connected the Atlantic and the Mediterranean, traversing Mauretania, the Niger bend, northern Chad along the Tibesti to reach the Mediterranean along the Libyan–Egyptian border. It would have furnished water to desalinization plants and seawater fish farms and would have required overcoming a topographic difference of 500 m. The project fortunately remained theoretical, no studies of the impact of such an outrageous undertaking have been performed and no provisions were made for an impermeable base layer to exclude the danger of salinization of freshwater aquifers.

3.3.1.2
Transport of Polar Ice Towards Drylands

Large chunks of ice from Antarctica towed in impermeable plastic covers towards the Red Sea and the beaches of Arabia would preserve the waters resulting from the melting ice. This method would be expensive even for Saudi-Arabia .... A first test had actually been carried out in Chile in 1911.

3.3.1.3
Artificial Rain

The results of the various studies are rather deceiving and the method is not sufficiently attractive to increase the water resources of developing countries.

3.3.1.4
Diversion of Rivers

Hydrologists of the former USSR have proposed changing the direction of the Siberian water-courses and have them flow to the Central Asian deserts. This original project of Davydoff (1948) had revived earlier ideas and was re-introduced during the 1980s. For Rognon (1989) the general use of this technology could present a solution for stemming the rise of the sea level and the submersion of coastal plains resulting from the warming of the climate which would lead to the melting of the polar ice caps. However, as the long-term impact of such gigantic projects on the environment is unpredictable, they are not worth any further consideration.

3.3.1.5
Desalinization of Seawater

Although this method is complex and costly, it is the most widely used by the countries along the Persian Gulf, distillation being the technology usually chosen. Distillation by boiling seawater and condensation of resulting vapour is an old method already used on sailing ships. The first plant for supplying drinking water to ships was erected 1869 in Aden (Hornburg 1967). The two presently most widely used methods are reverse osmosis and step-distillation. Hornburg estimated that in 1984 production amounted to 9.92×10^9 l/day, of which 60% were accounted for by Saudi Arabia alone.

3.3.1.6
The Great Man-Made River of Libya

Groundwater of the Sahara pumped from the Koufra and Sarir basins north of the oilfields is led from Tazerbo over more than 600 km by pipeline at a rate of $2 \times 10^6 \, m^3$ per day to irrigate the coastal strip of the Gulf of Syrte. Spaced out along this project are wells through which water is drawn for irrigation purposes. Considering that the aquifer in use belongs to Egypt, the Sudan and Libya and is fossil in nature, it is expected that it can be exploited for only 50 years or even less, putting an end to any ideas of a sustainable development.

3.3 · **Traditional and Modern Hydraulic Techniques** 171

Photo 3.6. Water collection in the Negev. In the semi-arid area of the northern Negev the techniques for gathering water are very old. This *liman* near Arad is an artificial body of water gathering rainwash. It was established with the aim of providing shade for Bedouins and for motorists using the nearby road (Photo M. Mainguet)

3.3.2
Parsimonious Management of Surface Waters

The drylands have seen the origin of controlled rill wash, as a result of human ingenuity, the modification of slopes to arrest erosion, the collection of water running off on slopes and the increase of water infiltration permitting the recharging of aquifers.

The technologies for recovering runoff water combat the short duration of the runoff by capturing the water on the slopes and by concentrating it in topographic depressions with a triple aim of:

- Establishing bodies of water from which pastures may grow after infiltration and evaporation
- Establishing areas of water concentration which may be cultivated
- Supporting the recharging of the aquifers

Artificial swamps, traditional rural hydraulic structures, are referred to as *limans* in the Negev (Photo 3.6), *boulis* on the Cape Verdes or *nadis* in Rajasthan, where they represent an ancient way of building up water reserves for man and his animals and where most villages possess their own nadi. They are also a widespread feature of the rural areas of Burkina Faso. Their digging fulfils two aims: to gather rill wash and to store the waters of the rainy season. Built on the mid-slopes, they are bordered by an earth dam open in the upslope direction. They possess the following advantages: simplicity of the concept and availability of the water from the first rains without delay, above all for the animals. Their disadvantages, however, are numerous:

- Low storage capacity for only a few months after the rainy season
- Strong evaporation
- High losses from infiltration
- Rapid filling by silt or sand, requiring frequent, sometimes annual maintenance

Hill lakes, abundantly found in Tunisia, are a modern version of this technique.

In Rajasthan the nadis have been developed scientifically. The Indian government recommended in 1988 a quantity of 70 l/day per herdsman for eastern Rajasthan, where the dry season attains a maximum of 300 days (Khan 1989). The value was derived from the following formula

$$V = \frac{CND}{100},$$

where V = basic daily water requirement in m^3, C = daily water consumption per person in l, N = number of persons and D = number of days without precipitation.

The calculations were carried out under the auspices of the National Water Board and a model nadi was proposed for 500 persons and their livestock: The nadi should possess a volume of 18 100 m^3 or 1.7 times the calculated requirements, thereby taking evaporation into account.

Some improvements in the concept of these traditional structures have been introduced (Rochette 1989):

- Construction of a network of stone dikelets on the uphill side to restrict the influx of sediment and to facilitate a better guidance of the rill wash
- Planting vegetation on the earth dam to reduce the influx of aeolian material, to limit evaporation and to control the approach by livestock
- Armouring the peripheral dyke for consolidation purposes
- Construction of a paved outlet canal
- Construction of small secondary reservoirs on the slopes which slow down the approach of the livestock to the main pond
- Covering the floor of the pond and its banks with small plastic sheets

Cisterns require a wadi topography or a feeding slope. The water is derived from the upstream side or laterally. Although invented originally for overcoming the seasonally insufficient supply of water in semi-arid areas, cisterns are found along either margin of the Sahara, viz. in the subtropical Mediterranean zone as well as in the Sahel zone. They are known in the former zone from Greece, Anatolia and the coastal areas of Libya and Egypt, are the *Harabe* of Bedouins of Negev in the 2nd and 1st millennia B.C. and are also the *maagurak* in the Hebrew of the Bible. Murray (1952) has estimated that there are 3000 of them between Alexandria and Sollum. From the centre of the Syrian desert Schlumberger (1951) described underground cisterns situated at the base of a slope and suggested that during ancient times these cisterns permitted the breeding of horses in the true desert, where precipitation varied from 100–150 mm/year. Cisterns are widespread in Asia and in particular in India where they benefit from the rich summer monsoons in the eastern part of the country.

In drylands cisterns are utilized in towns for collecting precipitation, benefitting from its suddenness and its concentration over a small number of days only. Their catchment areas may be as small as one courtyard or one roof. In 1921 Jerusalem possessed some 7000 cisterns with a capacity of about 0.45 million m^3. In the traditional Mediterranean or Near East urban environment, cisterns represented the only means of water storage for towns. About 2.5 m^3 per head annually were sufficient or 7 l/day per person, which would be equivalent to a catchment area of 13–15 m^2 per person at a precipitation of about 200 mm/year (Planhol and Rognon 1970).

Called *tanka* in Rajasthan, these quite traditional water reservoirs are used mainly to store drinking water and water for kitchen use. They are built close to religious places, schools or individual dwellings and may be reserved for the use of a single family. They are round and 3–5 m deep, their walls are covered by a clay mortar or in more recent cases by cement mortar and their tops are covered by branches of *Zizyphus nummularia*. More modern versions are covered by stone slabs. Their catchment area may be just the roof of a house or a specifically built peripheral area. It has been calculated that a tanka should have a volume of 21 600 l to supply a family of six for one year, equivalent to a demand of 10 l/day per person (Khan 1989).

In the Brazilian Nordeste water is by no means rare but still a limited resource. In the state of Sergipe in the early 1980s a programme for the construction of cisterns was launched, using roofs and other specifically prepared catchment areas to gather rainwater.

The management of water catchments for water collection by modern mechanical preparation of runoff lines (gullies and rills) and of flows like wadis, revives Nabatean system for gathering water created 2400–1500 years ago. The *khadin* method for gathering and concentrating runoff water on slopes resembles the Nabatean system, and aims at increasing the yield in the neighbouring valleys where water is stored by small clay dams. It has also been used in the 16th century by Paliwal Brahmans in the Jaisalmer district in India.

An experiment on these principles was carried out some 30 years ago in the Negev desert at the Shirta and Avdat sites, which receive only 80 mm precipitation per year (Lovenstein *et al. 1987*). A farm was established in the early 1970s in Wadi Mashash, some 20 km south of Beer Sheva and 60 km inland from the Mediterranean where the mean precipitation is 115 mm/year. The Thornthwaite humidity index is 40–60, representing a dryland with precipitation too low for agriculture. The intensity of the precipitation exceeds infiltration and leads to rill wash which may be concentrated into small cultivated areas with a soil cover thick enough to retain water and permitting deep penetration, thereby offering sufficient water reserves for the plants to survive a dry period of more than 6 months.

To improve the retention of water and its infiltration, a wall is built around the small cultivated area. Based on the same principle several such systems have been established in the area of the experimental farm:

- The wadis, not operating outside major floods were subdivided by terraces and were even cement-lined to receive the runoff from the neighbouring slopes.

- Macrobasins covering 0.3–0.5 ha were built to receive maximum rain-wash from the neighbourhood.
- Microbasins were established around individual trees.

The plants were selected on the basis of polyvalence, according to their potential for rapid growth, their response to trimming, in the case of forage or bushy plants, their adaptability to drought and to the rapid violent floods of these basins: *Prosopis* spp and *Acacia salicina* for forage and firewood, *Eucalyptus occidentalis* and *E. camaldulensis* for firewood, *Leucaena leucocephala* for food, forage, firewood and construction timber; *Atriplex barclayana, A, nummularia* and *Cassia sturtii* for fodder and *Phaseolus acutifolius* for food.

After selecting the most promising plants, mixtures of trees, bushes and annuals of different densities were studied. The plant mixtures took into account deep root penetration and the distribution of the water demand throughout the year. Because of their potential for intercepting sunshine and their ability to act as a windbreak, trees were planted together with vegetables for which they supplied shade. *Eucalyptus occidentalis* and *Acacia salicina* were not planted together because of the competition between their root systems for water, but they were mixed with sorghum and chick peas in different densities. These experiments were accompanied by measurements of water consumption and of optimum utilization of fertilizers. The keeping of goats, sheep and camels was combined with the system of cultivation in order to test the best forage species, the quantities of fodder produced and the requirements for their conservation.

It is surprising to note the tendency of the drylands to develop well nigh gigantic forms of concentrated water erosion. The technology of fixing such gullies and basic rills entails:

- Small transverse check dams behind which eroded soil may accumulate. By check dams and the deviation of water towards small banks inclined in the upstream direction and by plant cover, farmers have battled against erosion and benefited from the water concentrated in the gullies. At the ideal distance between two such banks there will be no surplus of water not infiltrating which might concentrate to cross the erosion threshold.
- Erection of dry bands of rock blocks spaced at increasing intervals from the upstream to the downstream side and at right angles to the line of flow. Several small bands are more efficient than one large band which is, however, more easily built.
- In the case of small wadis more than 5 m wide and 1–2 m deep the bands of dry rock blocks may be reinforced by *gabions*, referred to as a stabilization dyke by Rochette (1989). It should be anchored to the bed and the banks, possess an overflow structure at its top permitting the discharge of the larger floods and have a catchment basin on the downstream side.

More refined versions have been investigated:

- Earth dams consisting of several rows of *gabions* (two or three lines) and reinforced on the downstream side by several layers of big boulders in the bed.

- Dams with gates encompassing a bricked or smooth-set passage at their base. These are advantageous for reducing the impact of flood waves as they allow the excess to pass. Water flowing more slowly is less erosive and the water stagnating above the dam recharges the aquifer. The dam fixes the alluvial charge of the water on its upstream side. Rochette (1989) compiled a list of errors frequently encountered in such stabilization dykes or earth dams:
    - Under-design: too light, too high or with too narrow an overflow, the structure may be broken or breached from the side.
    - Insufficient anchoring; the structure may lose its lateral support.
    - Insufficient catchment basin.

In the larger wadis the banks are protected by groins or by dykes protecting the banks themselves. A groin dam is a structure found in all types of ecosystem, but in drylands where productive lands are rare it possesses a triple advantage:

- Protecting the banks between the individual groins and on their downstream side
- Promoting the advance of the banks and an increase in land area by deposition of sandy–clayey alluvial material
- Increasing infiltration between the dams

In drylands water and soils suffer the same destiny. The means for fighting for a better control of the waters also act as means for fighting soil erosion. The main aim is to improve the retention of water in the soil and to minimize surface erosion. The main techniques are: Anti-erosive levees along contour lines, techniques for combining impluvium with dug trenches, terraces.

Anti-Erosive Levees Along the Contour Lines on Slopes. These consist of tamped earth or of lines of upright or stacked rock blocks some 50 cm high arranged at a mutual distance of 25 m on a slope of 2° (Mietton 1989). The levee of tamped earth represents an impermeable obstacle whereas those made of rock can let the water pass. Rock levees are nevertheless more resistant than those made of earth and require less maintenance. They should lead to a damming of the earth coming from above causing the base of the block layer to become impermeable whereby the formation of water leaks can be avoided. Earth levees behave like trenches on the upstream side, and more frequently on the downstream side when made out of rock blocks. At their extremities the levees possess horns referred to as "wings" by Rochette. To allow the excess of water to be evacuated without damage, a diversion should be provided between two adjacent wings. Such a diversion represents a forced passage for the water. It is reinforced by large rock blocks, branches or millet stalks which prevent the formation of gullies. The earth levees are more resistant when covered by vegetation. Very long levees should be avoided because of possible degradation.

By slowing down the runoff the levees are a protective measure and at the same time they rebuild the cultivable ground and support the infiltration of water. They can improve the soil humidity and because of this the agricultural yields. According to Mietton (1989), the increase in yield was 30–60% for millet in the dry subhumid area of Kaya in Burkina Faso. They furthermore assure a better recharging of the wells on the managed sites.

In Sahelian Africa there are numerous areas for which levees could be envisaged and it is in Burkina Faso where they are utilized most. In Yatenga Province of northern Burkina Faso fields planted with millet, which are overlain by poorly permeable impluvium, have been equipped with filtering obstacles consisting of compartmented rock bands, i.e. double lines of ferricrete blocks, along the contour lines. Spaced some 20 m apart, they have increased water infiltration by 20%, led to an abatement of the floods and to their extension in time, at the same time reducing the erosive potential. Lamachère and Serpantié 1989 stated that:

> These installations facilitated a better establishment of the plant population, better penetration of roots, better fructification, increasing the grain quantity per ear and the production of dry matter on plots thus managed in 1986 and 1987. On the lower-lying parts of plots the increase in grain yield was 11% in 1985, 81% in 1986 and 31% in 1987 in favour of the plots thus modified, underlining the variability of this effect and allowing the expectation of a negative effect of this method when rains become weak and rare at the end of the cycle.

These authors estimated that this method prevents the strenuous ploughing of the soil on such slope situations. Improvements in water supply permit an increase in the production of non-restituted parts of plants being, however, accompanied by a more rapid exhaustion of the soils which then require the supply of fertilizer.

Techniques for Combining Impluvium and Dug Trenches. The water pocket method also referred to as *zai* (or *zay*) in Yatenga (Burkina Faso) represents a traditional method. Holes 10–20 cm in diameter, 10–15 cm deep and 0.5–1 m apart are dug during the dry season, enriched with manure covered by a thin layer of earth and then seeded during the first rains. In a zai, which represents a microbasin, water becomes concentrated and infiltrates, offering a moist soil to the seeds. The plant germinates, grows rapidly and establishes good roots. As a shoot it is protected against the wind. This method is also useful to avoid the formation of crustings.

The crescent method is a modification of the impluvium trench method. A hole or a counterslope is dug and the rubble is placed in a semi-circular embankment open in the upstream direction and running out into the contour lines. On the Cape Verdes these crescents, referred to as *caldeiras*, are covered on the downstream side by rock blocks. On very steep slopes of more than 60% inclination a method using small walls or *banquetas* is applied.

The Terrace Method. Terraces on slopes are walls made of rock blocks (formerly of dry rocks) built behind which an earth fill permits infiltration and proper storage of water. This method is employed on slopes with more than 8° inclination when highly inclined areas would become vulnerable to water erosion. They also constitute means for concentrating runoff to allow the crops to survive the dry season. The best terraces are found in the Yemen where they are still well maintained. When runoff on slopes becomes too violent the overflow of water is diverted onto the next lower terrace either directly or by a specially built exit pass. These structures are an efficient method for preserving water and soils in drylands.

According to Planhol and Rognon (1970), the terrace method formed the

> "source of irrigated agriculture in subdesert areas …. The original terrace of the semi-arid subdesert zone as a form of agriculture or irrigated agriculture probably developed only much later

for dry, agriculture tree plantations along the shores of the Mediterranean and in particular along its northern coastline."

Terraces did exist and still exist in the Turan (Zerafshan) and also on the mountain slopes of the Himalayas above valleys where irrigated rice growing is practised. The slopes are transformed into terraces used for dry agriculture or winter cereals (wheat and barley) in combination with orchards (Planhol and Rognon 1970). In a less elaborate form as simple earth ramps cut into arable lands they are also known in the Andes. In pre-Colombian America they made their appearance with the original agrarian societies on the high plains of the Andes and of Mexico.

All these techniques should be established only in co-operation with the local population. Mietton (1989) estimated after investigations carried out in Burkina Faso that the technology on its own was without any result. Even where the participation of the farmers in the construction of the embankments becomes a normal feature, this will no longer be the case during their maintenance. The degradation of these earth embankments is furthered by livestock tracks and footpaths and their cover by plants remains limited. This also applies to the Yatenga where south of Ouagadougou 10% of the structures have become degraded only 2 years after their construction. Are we not dealing here with one of the major problems throughout the entire African continent, the lack of proper training in maintenance? This comment becomes all the more crucial when we are dealing with irrigation as the following chapter will show.

3.4
Traditional and Modern Irrigation

According to the World Bank (1987), some 15% of arable land is irrigated, yielding 30% of the food production and using 65% of the water consumed worldwide, equivalent to seven times the combined demand of households and industry, the two next largest consumers.

Capot-Rey (1953) defined irrigation as: "any method which involves mobilizing water resources above or below ground to overcome the effects of insufficient rain and thus all methods of supplying water to and distributing it on the field level." Depending on the degree of aridity of an ecosystem one can distinguish:

- Irrigation compensating for the hyperarid or arid nature of climates (receiving less than 100 mm/year when precipitation falls in the cooler season and less than 150 mm/year during the warm season) where agriculture otherwise would not be possible.
- Complementary irrigation with a double aim of:
  - Alleviating the general or seasonal scarcity of precipitation; this is the case in the Mediterranean, Sahelian and Sudanian domains where the crops require a supplementation of the precipitation.
  - Alleviating the interannual irregularities in precipitation.

The slow evolution of traditional irrigation and its contrast with modern irrigation methods, which are audacious and sometimes rather inconsiderate, will be dealt with in this chapter.

Photo 3.7. Irrigation in a palm grove along Wadi Draa (Morocco). The Amezrou Oasis, 1000 years of age, owes its development to a system of traditional irrigation with water from the Draa Valley. The upper level of the oasis vegetation, a luxurious growth of date palms, towers over tamped earth walls subdividing the agricultural plots. To the *right*, between the two walls, flows a séguia dug into the ground itself. The closeness of the walls, the subdivision into plots, and an irrigation system difficult to modernize represent major handicaps for the development of this type of oasis (Photo M. Mainguet)

3.4.1
Traditional Irrigation

Following his definition Capot-Rey (1953) prepared a typology of the traditional irrigation systems encountered in the Sahara.

"Bour" Cultures. For this the Algerian souf is the type area. The *soufs*, about 15-m-deep excavations, are funnel-shaped holes dug to tap the original aquifer of the *Grand Erg Oriental* flowing from south to north and approaching the surface along the underground flow. Date palms are planted here. The cultivation of this crop is also practised in hollows along the edge of the *Grand Erg Occidental* in the Taghouzi and the Fezzan.

Irrigation by River Water and Water-Diverting Dykes. The course of a river is diverted towards a supply canal by a submersed dam. When the flow is not violent, the dam may be just a levee of rocks and branches which will be plugged by fine material during floods, or it may be a mortared structure. It is laid out at a slight angle to the current in order to prevent its being carried away. The Draa in Morocco, a wadi with a strong slope and a well-defined bed is easy to capture for irrigation. The water thus captured is passed into a network of 89 *séguias* (earth canals, Photo 3.7) classified according to Ouhajou (1986) into:

- *Melk*: the water is private property irrespective of the size of the land tenure; the water rights initially require a contribution towards the construction of the séguia.
- *Allam*: these séguias are not divided into individual property as this water is collectively owned and an integral part of the land which it irrigates. The amount of this water available is thus proportional to the size of the plot.
- "Mixed séguias" in which both previous types are combined.

This seemingly strict regulation actually covers up a permanent competition. "The control of the water frequently was the property of a socially dominant group" (Ouhajou 1986).

Irrigation by séguias is predominant above Agdz, on the Draa. Downstream of Ouarzazate, where the discharge of wadi Draa decreases, irrigation from the river is no longer sufficient and wells are the main source of water. Throughout southern Morocco the séguia system for irrigation is the rule; on the Rhéris wadi it permitted the development of palm groves at the Tafilalt.

Irrigation by Spreading of Floodwaters. This is practised throughout northern Africa and the Sahel, as in Mauretania, for example. It is based in part on temporary ponds formed during the rainy season and persisting for a short time thereafter, and in part on small temporary dams storing water.

Irrigation From Springs. Capot-Rey (1953) observed: "Ten springs at Tozeur, with a discharge of 500 l/s together with recent artesian wells permit the watering of a magnificent grove of some 400 000 date palms. On the other side of the Chott, in the Nefzaoua, there are hundreds of springs and many are connected to séguias. " Since the beginning of the 1990s, the springs have dried up from the multiplication of wells and motor pumps and all these oases are only able to survive by exploiting underground water resources.

There are also numerous springs in the northern Fezzan at Chati where, depending on their discharge, they may be used to irrigate an individual garden or the gardens of an entire village. Along the northern edge of the Tassili des Ajjer at Ghât, Serdeles and Djanet irrigation with spring water is also the rule.

Irrigation by Water from Aquifers and Wells. Capot-Rey (1953) distinguished:

- Wells with pulleys driven by animal traction (mules or camels).
- Wells with booms: a beam of palm wood pivoting on a traverse fixed to two upright wooden beams. A counterweight is attached to the end of the shorter arm.

Irrigation by Means of Foggaras. These underground galleries, as we shall see with regard to the development of oases (Chap. 4, Sect. 3.2.4), have spread throughout the world from Asia to North Africa and Latin America.

Irrigation by Artesian Wells. This has led to modern irrigation.

3.4.2
Modern Irrigation

Between the 350–400 mm and the 600–700 mm isohyets most semi-arid and dry subhumid countries with an agricultural economy utilize the potential of irrigated cultivation to gain at least a minimum of security in their food supply. The constructions for water control which have multiplied since the 1950s are subdivided into four classes:

1. Along perennial water-courses, catchment and diversion structures supply medium-sized irrigated areas nearby: a few hundred to a few thousand hectares
2. Along temporary water-courses storage structures of small capacity: a few million m^3 for small-scale irrigation, fishing and domestic or pastoral needs
3. Along more important temporary water-courses or small permanent courses storage structures of medium size, between 300–400 million m^3 permitting the irrigation of larger areas of 1000–10 000 ha and urban water supply
4. Along permanent allochthonous water-courses large storage structures for volumes exceeding 10^9 m^3 for major irrigation and the generation of hydroelectric energy

In the former Soviet Union where 60% of the territory was subject to drought, agricultural production was highly irregular, causing the authorities to utilize the abundantly available water resources for irrigation. About 20 million ha or 10% of the cultivated land was irrigated, yielding 30% of the agricultural production, and in percentage of land irrigated the former Soviet Union was only surpassed by China and India (Sullivan 1992). In addition to the technical difficulties induced by the quality of the installations, the yields of traditional irrigation and the viability of the existing installations posed problems after the shift towards market economy. In view of rising energy prices some pumping installations may no longer continue without higher financial aid. Simultaneously, the retail price of cotton and rice would have to be fixed within the framework of the market economy.

For rational irrigation the water requirements for cultivation have to be calculated from three parameters:

1. The water requirements of the plants, which may vary from 350–500 l/kg dry matter, depending on the species
2. The soil texture: soils of silty or loamy texture mostly possess higher water reserves than sandy soils
3. The climate, on which depends the value of evapotranspiration (PET).

Such rational irrigation consists of three stages, viz. capturing the water, transporting it, and distributing it:

1. *The initial stage of capturing water by canals* from springs, pumping of water from water-courses, establishment of reserves of surface water (retaining walls, large dams), extraction of groundwater (underground galleries, wells, boreholes).

2. *A transport stage* when the source of the water and the area to be irrigated are at some distance from each other. The transportation network may consist of open canals or buried pipes. Inherent to all programmes of water transport over large distances is their cost, which is not always justified from an economic view point but from the environmental aspect of utilizing dry areas.

Example 1, the Indira Gandhi Canal, also referred to as the Rajasthan canal, transports water over large distances from the state of Punjab to the dry regions of Rajasthan. The first phase of the project covered 200 000 ha to be irrigated between 1961–1978. Where irrigation had started in the early 1960s the desert landscape turned into an agricultural area because of the supply of water, the fixation of the dunes, the development of pastures and afforestation. The second phase should have irrigated 246 000 ha but raised questions of the availability of water, the increased selling price of each hectare and the inhospitability of the area for the new inhabitants (Biswas 1990).

Example 2, the programme of irrigating the *Imperial Valley* in southern California is one of the most successful ones in the USA. The transfer of water from the Colorado River via the *All-American Canal* facilitated the irrigation of 220 000 ha. Citrus trees, dates and winter harvests have become possible because of the warm climate in the Imperial Valley, the Coachella Valley and at Paloverde.

Example 3, the Karakum Canal and the former Lenin Canal, dating from the era of the large canals in the former USSR. Started during the fourth Five-Year-Plan 1946–1950. Work on the great southern Turkmen Canal (Karakum Canal) lasted until 1965. In April 1950 Krushchev decided to start a canal building programme, and in 1954, in view of the inadequate nature of Soviet agriculture, it was called *The plan for utilizing virgin grounds*. In the same year the section of the great southern Turkmen Canal between the Amu-Daria and Merv was completed. This first section used the dry bed of the Kelif Daria, an old course of the Amu Daria. A retaining dam with a capacity of 35 million m^3 was built, then a second one with 1 km^3 capacity on a side arm of the Tedjen and a third one on the river itself.

The Karakum canal has reached the lower Ouzboi by now – but not the Caspian Sea – and irrigates the foothills of the Kopet Dag. Excavated directly into sand without a cement coating, in the early stages it suffered tremendous water losses, which recharged ancient aquifers, lowered the salinity of the ground by percolation and formed lateral swamps over 200 km^2. Eventually a natural coating of mud reduced these losses. The excavation works, however, reactivated the sands over a width of 2 km, degraded the vegetation and caused the need of rehabilitating the vegetation cover in order to prevent sand storms.

The never executed North Turkmen Canal should have started upstream of Nukus and rejoined the Ouzboi. Conceived by Stalin against the advice of his experts, it was started in 1950 but was stopped several years after Stalin's death. It should have irrigated 1.3 million ha in the Khorezm and 500 000 ha on the southeastern shores of the Caspian Sea.

A third canal, the Karchi Canal, starts like the Southern Turkmen Canal upstream of Kerki, using the ancient bed of the Kacha Daria. Other projects for rerouting rivers of nearly as gigantic dimensions have been started in the valleys of the Syr Daria and the Zerafzan (Samarkand and Bukhara).

The consequences of these gigantic projects of river rerouting and withdrawal of water were rather varied: they were responsible for an ecological catastrophe, in particular the drying out of Lake Aral, which will be analysed in the fourth part of this book

3. *The distribution of water by the different types of irrigation* to be described below.

3.4.2.1
Irrigation by Submersion

This method of flooding a level area surrounded by dykes requires the costly installation of terraces.

Along the major course of the Senegal and other tropical rivers, alluvial levees surround depressions which are inundated during each flood. During years with normal rainfall these so-called decantation basins will be submerged for 20–60 days per year. The water arriving here is rich in clay which increases the water retention capacity of the floors of these basins permitting a natural irrigated agriculture still referred to as flood cultivation.

> The good results from cultivating these basins has led the agricultural service of Mauretania to think about reproducing this phenomenon by building barrages across the wadis in the south of the country .... At Mal a gravity wall holds back the floodwaters of the Guélaour wadi. The waters of this dam settle out and fertilize the already deposited alluvium. After one or two months of submersion the waters have seeped away and supply lakes downstream, which serve as watering points for the herds, whereas the floor of the dam will be cultivated in the same way as the basins along the Senegal (Durand 1988).

3.4.2.2
Irrigation by Gravity

This is employed on gently sloping ground. For this purpose trenches are prepared from which the water spreads over the fields. This method is still used for watering surfaces even nowadays.

In the Brazilian Nordeste the supply of water to man and his animals is of high priority, whereas irrigated agriculture is only second to this. The region does not have any history or experience as far as the latter aspect is concerned (Cavaille 1989). The battle against the dramatic effects of drought started after the great drought of 1877–1879 by storing water during normal years to account for the daily needs of human and animal consumption and for cultivation during dry periods. This has resulted in the so-called *açudagem* strategy with the word *açude* describing the combination of a barrage and a reservoir.

The first *açude (açude de Quixada)* was started in 1884, and had a capacity of 128 million m^3. Up to 1983, some 1121 barrages with a total storage capacity of more than 15×10^9 m^3 had been erected, of which 274 are public property and 847 mostly smaller ones private (Pessao 1989). To this have to be added all the other initiatives, and roughly 10 000 different structures have been built across the Sertao. However, despite the high potential, irrigation is poorly developed. Pessao ascribed this situation to the agrarian structure of the Nordeste with more than three quarters of the lands belonging to less than one-fifth of the rural landowners. The large, frequently absentee landowners are especially interested in extensive cattle ranching and therefore are more concerned with water reserves for consumption by their animals than with irrigation.

Because of this only the simplest gravitational irrigation methods of low efficiency are employed by the majority in the Nordeste. Cavaille remarked that "in the case of the poorest, methods are in use for capturing water *in situ* and covering the ground with straw and other agricultural residues and fronds of the Carnauba palm in order to limit evaporation."

Irrigation to extend the vegetative cycle of annual rain-fed crops from small retaining ponds on hills is largely scattered. Drop irrigation from perforated hoses or small basins for crops with longer cycles is also practised, drawing water from rather scattered, shallow wells dug into the alluvial deposits of non-active water-courses.

3.4.2.3
Irrigation by Infiltration

This method, which is rather costly to install, is carried out in several ways.

Irrigation by Furrows. Water is distributed through parallel grooves between flat areas. This method has to be installed meticulously so that it will allow higher discharges to flow in a thin layer over large areas. The flat areas have to be horizontal to assure a good spreading of the water, their length depending on the nature of the soil and e.g. reaching 400 m on heavy soils (Durand 1983). In the Turan experts recommend grooves in a zig-zag pattern to avoid excessive currents and to support infiltration.

Irrigation by Basins. This method is used mostly in orchards where basins are prepared around each tree. In Crete each olive tree possesses its own water supply pipe.

This more than 4000-year old technique of supplying water for agriculture has given way during the last 75 years to development and diversification allowing the constraints of topography and of abundant labour to be overcome. They have led to the replacement of gravitational irrigation by sprinkler irrigation.

Sprinkler Irrigation. This was carried out initially by rows of sprinklers which were moved manually. Because they required too much labour, they were replaced by automatically pivoting aspersion lines developed in the dry areas of the American west. A row consists of rapidly rotating pipes, and on each line rotating sprays branch out every 11, 18 and 24 m. Each row rotates for between 8 and 10 days before it returns to spray the same area again. In addition to pivoting rows and rows on wheels, giant sprays were developed in the form of automatic spray cannons.

In contrast to the older forms of irrigation, sprinkler irrigation does not require any preparation of the terrain. It is furthermore moderately economic as it allows to save 30–50% of the water. However, the equipment, rotating rows of sprinklers or water cannons distributing water as rain over the soil and plant foliage, are expensive. It is recommended to water only permeable soils by sprinklers.

> Its advantages are: no levelling required, excellent control of light water doses, low manpower requirements and better aeration of the water. For certain plants it permits the avoidance of early cultivation on earth ridges. On the other hand, the installation costs are high, maintenance is ex-

pensive, distribution of water during strong winds is problematic, working a soil soaked in water is difficult, compaction of the soil and the formation of scalding must not be neglected and the risk of sickness because of the dispersion of anti-pest products employed previously will certainly be higher. And finally, sprinkler irrigation by distributing large amounts of water may exceed the water capacity of the soil and may lead to its erosion when the discharge is poorly controlled (Durand 1983).

This method is suitable for an irregular topography and light soils and does not require much man power. However, it also has its difficulties: calculating the rate of watering, which should be below the absorption velocity of the soil concerned in order to avoid water erosion; spraying onto aerial parts of plants may lead to tissue damage by insolation.

Drop Irrigation. This is also called micro-irrigation, and has been installed in Israel and Australia. Perforated pipes, either suspended or lying on the soil surface distribute the water under low pressure of approx. 1 bar. This method is most economic in water consumption (2–10 l/h), can be highly automated and requires highly competent operators, but permits watering at closer intervals.

3.4.2.4
Underground Irrigation and Underflow Barrages

In drylands with their intense insolation irrigation and pumping benefit from solar energy. The irrigation system may become more integrated by the introduction of fertilizers into the waters. Among the main techniques of underground irrigation underflow barrages occupy a particular position.

Such underflow barrages are of particular importance in drylands. Their objective is to build up an accumulation of underground water in a river valley. Their advantages, in addition to the nearly total elimination of water losses by evaporation, have been summed up by Guiraud (1989):

- Building an underground reserve the volume of which may be high.
- Raising the top of an aquifer to a favourable site of utilization where the capillary rise can assure supply to plants.
- Reducing the salt content of the soil because of the virtually complete absence of evaporation; distribution of water from a network of dug-in canals enriches the soil in water from the bottom upwards and there is consequently no leaching.
- Great reduction of pollution by man and animals.
- Defined and reliable catchment structure.
- Lower construction costs when locally available manpower is employed. The walls may be made of concrete or cemented stones.
- Establishment of the level of perennial water and the possibility of securing the populations.
- Operating and maintenance cost low to zero.
- The structure may rise above the surface in order to retain floodwaters and to permit a more efficient recharging of the aquifer.

Guiraud stated that the "hydroclimatic domain" for preferably employing the underflow barrage principle is the semi-arid zone, but such structures may also prove their worth in arid zones where the geographical context allows surface runoff from time to time.

Guiraud gave a number of examples for this type of barrage: After the war such barrages were built in Algeria and Morocco, and in Kenya more than 30 such structures have been built. In the early 1980s they were built in southern Africa, Cape Verde Island, Somalia and again in Kenya.

For the construction of an underflow barrage one needs: An underground flow recognisable on surface by a line of bush vegetation in the bed of a wadi and a thick alluvial fill encased by an impermeable layer.

We should remember that Quaternary alluvial fills are found in the valleys generally throughout the world linked to the glacial episode 20 000 B.P. and in the tropical systems to the Holocene aridification with a degradation of the plant cover and an increase in water erosion and deposition. In both cases the lowering of the sea level by 120 m and of the lake levels was accompanied by an incision of the water-courses which on the renewed rise of sea level and lakes became the sites of thick alluvial deposits with pebbles, sand and silt with thicknesses up to several dozen metres. Such alluvial deposits are found in still functioning networks and in fossil networks like the one delineated in the Egyptian desert by El-Baz with the aid of Radar-SAT images.

3.4.3
Reservoir Dams

In arid and semi-arid zones the aim of water management is the same as in the other environments: to satisfy a demand for water from domestic rural and urban consumers, to control floods, to produce electrical energy, to regulate water-courses for navigation, and to erect water reservoirs for irrigation. However, the various projects are rather specific in nature as they always have to allow for the great variability in time (seasonal or annual) and space of the individual water-courses.

In drylands reservoir dams may be envisaged only for large hydrographic systems with permanent runoff or on seasonal water-courses with strong floods. Without going into the technical requirements of large dams we shall refer here only to the benefits which they entail for development.

Reservoirs dams were first used for irrigation in Persia and the oldest ones, near Persepolis in the Fars of southern Iran, date from the Achemenid period (sixth–fourth century B.C.). Thereafter they have been erected continuously and have spread out from the area of the Iranian civilization. Other structures, of the Sassanid period, also occur in the Fars. Around 260 A.D. Shapur I using Roman prisoners captured together with the emperor Valerianus constructed the first three dams with mobile sluice valves on the large rivers Chuchtar, Dizful and Paipol on the plains of the Mesopotamian foothills of the Zagros Mountains. From Persia the technology spread to Arabia: the Mârib barrage in the Yemen, the collapse of which in 570 A.D. marked the decline of the sedentary kingdoms under the onslaught of the Bedouins, the dam in the Batinah plain of Oman and the barrage east of Taif. It also moved west towards the eastern shores of the Mediterranean: the Harbaqua barrage in Syria west of Palmyra in 132 A.D. and the forest barrages of Belgrade near Byzance, built during the eighth century to supply water to the town (Planhol and Rognon 1970).

Iran again appears to have been the source for arched dams, the oldest examples dating back to the Mongolian period of the 13th century. The one at Qum on the high plateau was 5 m thick, 26 m high and measured 55 m along the crest. The technique possibly reappeared in the 16th century and from 1611 onwards with certainty in the western Mediterranean in Italy and Spain, but we cannot say whether this represents an independent discovery or the application of the Iranian principle. It spread from India to the USA via Europe (Planhol and Rognon 1970).

Intended to assure a reserve of water for permanent agriculture, they also lower the higher salinity of the floodwaters. During floods, the salinity of the waters may be high but when these waters become mixed with the water already present in the reservoirs, the salinity will drop.

Reservoir dams are becoming increasingly controversial: They are expensive to build and in drylands they represent large evaporating surfaces; they lead to destruction of the natural environment and to the agglomeration of cultivated areas. Among the most disquieting examples are the dams of Sélingué in Mali, those of Diama and Manantali in the Senegal, the Aswan Dam in Egypt and the Three Gorges Project on the Yangtze-Kiang in China, the advantages and disadvantages of which will be outlined below.

3.4.3.1
The Sélingué Dam in Mali and the Diama and Manantali Dams in the Senegal

Barth (1989) described the consequences of Sélingué Dam in the valley of the Sankarani, one of the main tributaries of the Niger upstream of Bamako. This dam, built in 1981, was intended to fulfil a number of objectives: storing water for irrigation, flood control, hydroelectric energy (with a generating capacity of 44 MW) and navigation. It immediately exerted an impact on the entire fluvial system. The natural floods reached Koulikoro in mid-September, Mopti by mid-November and Diré in December. After filling pools and lakes the river flooded its inner delta for 4–5 months. In a normal year large parts of the delta and numerous independent lakes in the local Sahelian regime held water over the entire dry season.

Evaporation withdraws 2000 mm annually or 45% of the flow, leaving a discharge of 1800 m^3/s at Niamey whereas Koulikoro receives 5280 m^3/s and Diré 2290 m^3/s. Regime and discharge of the middle reaches of the Niger depend on the precipitation in its upper reaches. We can then understand that the construction of a dam on the upper part will influence the hydrological conditions of the inner delta and why Barth said that the completion of the dam compromised the development of the entire region. Even if the contribution of the Sankarani is only 20% of the total, the dramatic effects of a series of years of loss like those in the first half of the 1980s have been aggravated by the dam.

In addition to fishing, which supports some 200 000 persons, the inner delta offers a high potential particularly for growing rice. Two projects, the *Office du Niger et Opération riz de Ségou et de Mopti*, working together on numerous hectares of land used for rice cultivation, benefit from the annual floods and their deposition. The traditional area of rice cultivation, depending on the floods, occupies some 100 000 ha (Barth 1989).

The initial object of preserving traditional rice cultivation has not been achieved. Because of the drought vast areas of the inner delta were no longer inundated, reducing the growth of the bourgou by more than 50%. The fishing industry collapsed in 1984. More than 1 million people were affected by the low flood levels and the construction of the dam benefited only a small urban minority at the expense of the vast rural majority.

The Diama Dam on the Senegal and the Manantali Dam on the Bafing, a tributary of the Senegal, concerns four countries: Niger, Mali, Mauretania and Senegal. *L'office de mise en valeur du fleuve Senegal* (OMVS) was formed to manage these dams. The difficulties here are similar; the irrigable lands created by these dams are still awaiting their optimum use and interested investors, several years after their completion. Nevertheless over 25 years some 300 billion CFA francs have been spent in French and international aid funds.

The Diama Dam which was built in 1986, just above the estuary of the Senegal, with the aim of preventing the intrusion of salt underground water from the sea flowing during the dry season for up to 200 km inland, proved to be ineffective. The Mantantali Dam, built in 1987, controls 50% of the discharge of the Senegal and should have increased the hydroelectric potential and satisfied most of the present electricity demand of Mali, Mauretania and Senegal. Furthermore, it was to transform the seasonal shipping on the Senegal into permanent navigation up to Kayes. Its objectives were thus fourfold: storage of water, generation of electricity, navigation and irrigation. Filling the dam led to the destruction of 10 000 ha of woodlands; the dam accelerated erosion and the sanding up of the valley and it holds back the mud from fertilizing the delta.

Regulating the discharge of the Senegal should above all have enabled the farmers to prolong the duration of the food supply for the first time. An irrigated area of 100 000 ha was foreseen for the Diama and 350 000 ha for the Manantali, with a total flow of 200 m^3/s. The overwhelming majority of this still has to be put into effect.

In the meantime the small farmers settling between the two dams have seen the loss of their lands and their traditional irrigation, whereas the livestock of the pastoralists, as was the case previously in the irrigation programme in the Awash valley of Ethiopia, no longer have access to the river for drinking and grazing and no longer can traverse the water-course. Resettling of the farmers and herdsmen requires continuous technical and sanitary assistance demanding subsidies in addition to the US$2 million maintenance costs of the dam (Derrick 1984). However, even the commercialization of production will not be solved and it is probable that small producers will be evicted in favour of larger companies, if this has not alreadyhappened. Furthermore, during the drought and when the lakes are no longer recharged, the situation will be more difficult than before. At present, only a part of the plots installed has come into production. "Of the 25 agricultural development programmes financed by the World Bank in the 1970s only one dozen may be considered successful" (French ONG 1993).

3.4.3.2
The Aswan High Dam

The high dam on the Nile represents a multi-purpose project and its official aim was the promise of new lands, the supply of water to urban and rural communities, an irrigated agriculture, protection against floods, river navigation, and the generation of electrical energy. The secret aims were of a political nature: giving an image of distinction to the socialist regime, rivalling the pharaonic traditions which would allow the population to forget its meagre present, and in particular to respond to the demographic explosion by an increase of the agricultural production. The demographic explosion resulted from a growth rate of 3%, equivalent to an increase of 1 million new inhabitants within 10 months.

In 1861 a system of modern irrigation along the Nile was started with a barrage constructed below Cairo. This was the start of controlling the delta, which led to five of the seven arms of the river in the delta drying up. Eventually a first dam downstream from Aswan was built in three stages (1902, 1912 and 1933) close to the First Cataract, to become enlarged in 1934. Several kilometres upstream, some 11 km from the town of Aswan, the 114-m-high and 3600-m-long High Dam was started in 1962 with Soviet design and finished in 1971.

A 500-km-long artificial lake, with a surface area of 5000 km^2, one of the largest in the world, reached a maximum depth of 70 m with a storage volume of 157 km^3, extending well into the northern Sudan. The fluctuations of the water level of the lake are about 30 m and in the flat desert country this leads to considerable lateral movements of the shore lines. Belal (1994) described the appearance of a seasonal vegetation which offers refuge to animals and which develops on the vast temporarily inundated areas on the borders of the lake. As these areas developed into attractive pastures, nomads have changed their routes of migration and started to settle here in semi-permanence. The areas thus developed an interesting economic potential because of their vegetation, the presence of fresh water and because of the mud deposits which led to fertile soils. However, as pointed out by Belal (1994), opening up this section is rather difficult because of the pronounced instability and vulnerability of this new ecosystem, the harsh climatic conditions, the wide fluctuations of the lake level and the low population density except in the immediate vicinity of the lake.

Among the benefits of the project are:

- The hydroelectric power installed: 2.1×10^6 kW/year are expected, with a production of about 20 TWh/year; at present more than half of this has been achieved; in 1966 the 12 turbines were operating. In 1988 production amounted to 88% of the potential, satisfying one-third of the entire demand of Egypt with a peak of 53% in 1974 (White 1988).
- The devastating floods have been harnessed; El-Baz (1989) mentioned the flood of 1975, the disastrous effects of which were avoided.
- The effects of the episodic droughts were nullified; according to El-Baz (1989) the droughts of 1973 and 1984 and the more severe one of 1987–1988 did not have the usual catastrophic consequences.
- Shipping was improved by regulating the floods and by decreasing the current velocities of the floodwaters.
- The cultivated areas have grown beyond the valley and the delta: the stored water offers a potential for irrigated agriculture on 800 000 ha in Egypt and 2 million ha in the Sudan (El-Baz 1989); White (1988) noted that at the time of construction of the dam Egypt possessed 6 million feddans (1 feddan = 0.419 ha) irrigated by traditional methods based on floods, gravity, and pumping. On the whole, White

estimated that the irrigated area has not really grown, because a total of 300 000 feddans of arable land between Cairo and the sea was used for industrial, commercial, and residential purposes. Formerly cultivated fields were abandoned because of salinization and waterlogging. The production of clay bricks annually uses 120 km^2 of land (White 1988). The land irrigated by the new Nubariya Canal could not be cultivated because of waterlogging and the newly irrigated ones in the Tahir (= Liberation) province were abandoned because of the wrong equipment.

- Three harvests have replaced the single one which until then profited from the annual flood of the river. Because of irrigation throughout the year the agricultural cycles have changed, winter crops, mostly clover, have spread over 400 000 feddans and summer crops, mostly rice and maize, over 2 million feddan. Submersion cultures have decreased but the yields of orchards have trebled (White 1988).

Among the main negative consequences of the project are:

- The destabilization of the riverbanks and the lowering of the downstream riverbed up to 100 km from Aswan necessitated the consolidation of the barrages at Esna, Naga Hammadi and Assiut, and along the navigable canals (White 1988).
- Simultaneously the lack of load which contributed to build up the delta and the Mediterranean coastline explains why the delta is in some parts receding by 100 to 200 m annually. In the upper delta, because its water is being held back at the level of the reservoir, the reduced discharge of the river is accompanied by a rise of the salt-water table, leading to damage to agriculture because of salinization.
- The loss of the legendary, albeit erroneous purity of the Nile waters because of the development of towns and industries along its banks.
- The loam and silt, otherwise spread over the sandy soils as a natural fertilizer and protective coating against aeolian erosion, are held back in Lake Nasser upstream of the dam. It is estimated that 100 million t of fertile mud brought down by the river during the annual floods are now blocked by Lake Nasser. These alluvium, which formerly had been used to make bricks and as natural fertilizer, now have to be replaced by artificial building materials and by expensive chemical fertilizers; moreover, they are withdrawn from the lands further downstream. The retention of these deposits by the dam leads to an impoverishment of the soils downstream and to a reactivation of the aeolian erosion of the sands, which had been lightly fixed by the annual spreading of these fine particles. The water deprived of its active load possesses an erosive potential downstream from the dam which has to be satisfied by newly eroded material which in turn will be deposited in sections previously without deposits.
- Sardines which until then had been abundant to the north of the delta have shrunk considerably in numbers. The diminution of fish in variety and volume and thus of the fishing industry itself may be ascribed to the impoverishment of the nutrient value of the littoral waters; this loss, amounting to 15 000 t out of a total of 18 000 t is barely compensated by the 22 000 t of fish caught in Lake Nasser, where overfishing and socio-economic problems of competition have started between the individual fishermen and the fishing companies much to the detriment of the former.

- In addition to the losses by evaporation exceeding 10×10^9 m^3/year, which are aggravated by the strong and constant winds over the lake and which had not been taken into account by the Soviet planners, there is a strong infiltration into the ground. The geological investigations had not predicted the presence of large amounts of evaporites in the ground, which may contaminate the infiltrating water with salts; such water is later pumped up for irrigation. In the early 1970s the salinity of the shallow aquifers had already risen to 36 g/l.
- Because of the permanent cultivation, the disappearance of fallow periods and the increased use of fertilizers, the drainage of the salts is no longer assured and more salt reaches the delta, leading to a salinization of the soils and the aquifer.
- The positive effect of supplying potable water to the settlements is counteracted by the demographic pressure requiring constant maintenance of the infrastructure.
- The changes in the position of the water table and in the water quality in the delta are considerable only close to certain agglomerations because of the discharge of untreated wastewaters.
- Among the difficulties, blockage of the soils and hydromorphism resulting from the rise of the water table which may exceed 1 cm/day or about 4 m/year; this ascent of highly mineralized groundwater contributes to the salinization of the soils and counteracts any drainage efforts; it is also responsible for the destruction of agglomerations, the deterioration of ancient temples and the flooding of archaeological sites. This is despite the relocation of the temple of Abu Simbel to a higher location at a cost of US $40 million, under the auspices of UNESCO and the Egyptian state, who each paid half of the cost. During this project the temple of Abu Simbel with its four colossal statues of Ramses II was transported piece by piece from its original position along the river 300 m to the new, higher site in order to preserve this testimony of a grandiose artistic and cultural past.
- Under the pressure of the irrigation water the flow of certain water-bearing horizons has reversed, mostly in the delta, facilitating the underground intrusion of salt water and thereby endangering areas which had been cultivated for thousands of years.
- Along the banks of the lake and in the irrigation canals a small gastropod, *Bulinus haematobium*, has proliferated; this gastropod is the intermediate host in the evolution of parasites of the genus *Schistosoma*, responsible for the bilharziosis or schistosomiasis endemic to the Nile valley. They have experienced, because of the demographic explosion which in the irrigated areas has led to population densities of 1000/km^2, a dangerous expansion. Some 70% of the rural population and 20 000 nomads, more or less concentrated along the shores of the lake, are affected by schistosomiasis. White (1988) suggested that the spreading of schistosomiasis is not so much due to the dam itself, but to the hygienic conditions and that this disease had decreased, while still remaining pronounced, since the Nasser government started a programme in 1952 of supplying potable water to 2000 rural health centres engaged in the detection of infested children, in health education, and prophylaxis.
- In Lake Nasser new varieties of zoo- and phytoplankton made their appearance as well as new stock of fish different in number and population. In the river itself the increased salinity is accompanied by an increase in the phytoplankton from 160 to

250 mg/l, requiring a stronger treatment of the urban water supply and a greater chlorination in the Cairo water supply;
- Among the still existent doubts those about drainage remain the most pronounced. Twenty-five years after the construction of the dam it is difficult to estimate the percentage of irrigated lands benefiting from improved drainage, but it is known that the major drainage programmes collapsed and that in many cases the material employed was poorly adapted to the task. It is also known that by reducing the number of days when the canals are closed from 21 to 7 the aeration of the soil layer above the root systems has been reduced. However, the effects of the reduced supply of mud or of the increased use of nitrogen fertilizers, phosphates or insecticides are not known. Kish (in White 1988) estimated that they have led to a loss in yields by 10% whereas White (1988) did not subscribe to this point of view. According to the statistics presented by him the production of maize, raw cotton and rice has increased, that of sugar and beans stagnated and that of lentils dropped. These results may be ascribed to the two annual harvests or, according to White, 1.9 harvests annually on average, whereas the original aim had been three harvests.
- White (1988) mentioned the necessary displacement of 50 000–60 000 Nubians who had to leave their homes to become relocated in eight new settlements on the new irrigated lands at Kom Ombo some 20 km north of Aswan. Not much is known about the 53 000 Nubians who had to leave their homes in the Sudanese part of the lake reservoir or about the inhabitants of Wadi Halfa.

The Egyptians are very preoccupied with the cost/benefit ratio of the dam, but no socio-economic balance is available. It would be interesting to know what the production would have grown to if the money spent on the construction of the Aswan dam had been used for smaller projects. We have to admit, however, that the objects of controlling the floods and generating hydroelectricity have been achieved. Three unforeseen but undeniable affects have nevertheless shown up:

1. A change in water quality
2. A too large reduction of the areas of raw material for brick making
3. A stagnation in size of the irrigated areas

3.4.3.3
The Dam Across the Yangtze-Kiang in China: Should Gigantomania Still Be Allowed?

The 7th National People's Congress on April 3, 1992, adopted a resolution for the construction of the Dam of the Three Gorges on the Yangtze-Kiang, the river separating the dry from the humid areas in China. Planned as a source of hydroelectric energy, the dam will be 185 m high with the eventual water level at 175 m. It will flood some 568 km^2 of cultivated land by the year 2009, lead to the relocation of 1.13 million persons, a figure never before reached by any other major dam project, and erase 40 000 years of culture. The problems of the impact it will exert on the environment and the social situation have not yet been resolved.

In conclusion, it can be stated that the necessity for constructing large dams particularly in drylands appears to be contestable more than ever, and the great priority placed on them must be discussed. It is erroneous to believe that they will be able to hold back major inundations on the downstream side as the overdimensioning of a dam to account for extreme floods occurring once in a century or a millennium would be too costly, and furthermore the very existence of the dam and the false feeling of safety from floods lead to great population densities in the valleys and thus to potentially higher losses in the case of inundations. Despite the progress made during the last three decades the risk of dam failure has not been eliminated, in particular in seismically active areas. Many reservoir dams in developing countries have been hastily erected without a comprehensive study of the area to be irrigated, the population densities, their real water requirements and their economic possibilities: the price of water has risen because of the cost of the dam. In addition to the direct effects of reservoir dams there are negative effects on irrigation itself which will be analysed in Chapter 4 of this volume.

3.5
A Response to the Aeolian Constraints –
The Art of Counteracting the Effects and Damage Done by Wind

Taking into account the omnipresence of wind in drylands any development programme in areas receiving less than 600–700 mm/year precipitation should include a section "Strategy for controlling aeolian erosion" based on a study of the potential vulnerability of dunes when they are vegetated in semi-arid areas and of the fixation of mobile sand and active dunes when the climate is still drier and the dry season longer. Any aggravation of the aeolian mechanisms damaging the natural resources will limit the possibilities for development and lead to suffering. *For efficient combat the factors causing damage should be distinguished from the processes of degradation and their effects.* It is, however, not of importance whether the former are climatic (i.e. natural) or human.

3.5.1
Basic Principles for Fixing Mobile Sands and Dunes

The two-edged battle, on the one hand against the reactivation of dunes and the mobilization of sand, which lead to the deterioration of pastures and arable land, and on the other hand against the effects on the harvests of droughts, which force farmers to move into marginal land, is not only of a technical but also of a socio-economic nature. It would be erroneous to believe that causes and processes of sand mobilization could be fought only by material means. The strategies for fighting against the ill effects of wind entail physical as well as biological methods of fixation. They will only be successful when supported by sustained cultural and social activities. Examples of such a successful battles may be found in Africa and Asia.

3.5.1.1
Physical Measures Against Sand Invasion

In hyperarid and arid ecosystems where movement and sources of mobile sand are nearly inexhaustible because winds are frequent and strong, anyone opening up of such lands will be confronted by a double danger: sand and mobile dunes. All counteracting strategies entail three courses of action based on the reduction of saltation by deviation, blockage and stabilisation:

1. *Deviation* during transportation in order to prevent the invasion of sand into human installations, by creating a deflecting dune with a palisade placed at an angle of 120–140 °C to the predominant wind
2. *Blockage* of sand in the source area or at any other stage along its transport course
3. *Stabilization* or fixation of mobile sands by mechanical, chemical but mostly by biological means.

At all times the basic rule is to inhibit the exportation of sand from or near the source and to prevent its accumulation over human infrastructure by allowing it to pass as freely as possible through the areas of migration. The erection of structures in the stream of sand should be avoided, as this will lead to accumulation. Where they are inevitable they should be designed in aerodynamic shapes. Any linear human installations (roads, railway lines, pipelines) should be laid out as parallel as possible to the dominant wind and should be placed on a raised bed so that the dominant winds have to rise and thereby accelerate, leading to transport rather than to deposition of sand.

Blocking Sand in the Source Area

In this case the delineation of the source area is of importance. The sand may be *autochthonous*, i.e. a reactivated dune, or *allochtonous*, i.e. coming from somewhere else. Because of the distance or the extent of the source area of the sand, it may be impossible or not economically feasible to intervene at this level. In the Egyptian desert the aeolian currents bringing sand to the Kharga oasis start in the Qattara depression some 700 km NNE. The source area may also be located in an ecosystem too arid to allow the establishment of a vegetation cover. Stopping the sand in the source area by mechanical barriers is of little use on the medium to long term unless accompanied by a programme of biological fixation, which, in order to be of any use, requires a minimum of precipitation or irrigation; in the latter case the question of cost is acute.

Under more favourable conditions (smaller source areas) palisades or mechanical barriers will lead to an artificial dune. At Essaouira in Morocco an artificial littoral dune was created during the beginning of the 1930s. Its maintenance at the present time requires constant trapping of the sand until the growth of a vegetation cover continues on the dune and in its hinterland, an aim that has only poorly been achieved so far.

Blocking Sand in the Area of Transportation

In the area of transportation any type of physical method which may reduce the wind velocity, provoke the accumulation of the sand and decrease wind erosion must be considered. Among the methods tested were deep ploughing, scarification or surface ploughing and earth ridging, methods which cannot be recommended in all situations, especially where the soil is made up of particles of a size susceptible being

Photo 3.8. Biological fixation of a dune in the Fergana (Uzbekistan). A dune of black sand in the Fergana valley, above the Syr Daria basin near the village of Kokand in Uzbekistan (40° N, 70° E) is fixed biologically by rows of bushes in 1981 (Photo M. Mainguet)

lifted up by wind. Covering the sandy surface or *mulching* has also been practised. Objects spread over a sandy surface will increase the erosive force of wind when the density of coverage is below a critical level, depending on the size of the objects and the velocity of the wind. Thus the coverage has to be dense in order to suppress deflation. *Mulching* may be carried out with any type of natural or artificial (plastic netting) material to replace a smooth with a rough surface to reduce the wind velocity. Chemical mulches like hydrocarbons spread in bands in the Fergana of the former USSR (Photo 3.8) and in Iran are combined with biological methods. The fixing layer formed by chemical products should not inhibit the regeneration of the vegetation and should be maintained until the re-establishment of the latter.

Two techniques have been employed to reduce the volume of particles transported: trenches or palisades. These are erected at right angles to the wind should lead to the accumulation of the aeolian sand in the areas to be protected. Digging trenches on the upwind side of human settlements in order to trap the sands is a costly method offering only temporary protection. In order to be really efficient, these trenches should be of a width which is larger than the distance of saltation of the grains and of such a depth that deflation can no longer pick up again the sand deposited at its bottom. Because of this the trenches have to be emptied continuously or doubled by parallel trenches. In the case of palisades similar problems will show up as they will be rapidly overflown. Blocking the sand with the aid of a mechanical technique represents a solution requiring indefinite maintenance.

3.5 · A Response to the Aeolian Constraints

Photo 3.9. Mechanical fixation of dunes at Chaotu in the Tengger Desert of China. In order to protect the Bejing – Lanzhu railway line against the active dunes and shifting sands of the Tengger Desert, the Chinese have invented a mechanical system for fixing dunes by a 20–30 cm high grid of pressed straw, which facilitates the second stage, the biological fixation. This method is rather laborious to install, but nevertheless has found worldwide acceptance and is encountered in particular where the fixation of sandy structures which endanger human infrastructure has become necessary (Photo M. Mainguet)

These mechanical techniques should be employed only in urgent cases like the protection of buildings, routes of communication or valuable agricultural land. Methods of blocking the sands are only palliative, the *Road of Hope* in Mauretania being a good example.

On mobile dunes remodelling methods (reduction in height, volume or profile) can aggravate the situation by increasing the migration velocity of these structures. Transforming a mobile dune into a sand mound is only a temporary measure as the mound will evolve very quickly into a newly mobile aerodynamic shape. In order to obtain a more durable result, the remodelling must be followed by planting.

In ecosystems receiving 300 mm/year or more in precipitation the stabilization of mobile sand may require a combination of mechanical and biological techniques (Photos 3.9, 3.10).

- *Stage 1*: erection of a grid of small low palisades rising 25–30 cm above the ground, spaced at 1 m intervals. A number of materials may be used for their construction: stalks of millet or any other harvesting residue, plastic strips or even cardboard.

Photo 3.10. Battle against the effects of wind by means of mechanical grids and biological rehabilitation in the Draa valley of Morocco. In the background are the houses and palm groves of Amezrou in the Draa valley of Morocco. Initially, a grid system made of palm fronds 1 m high and placed manually on an otherwise active dune protects the palm groves and the oasis against sand invasion caused by the *Saheli* wind. These measures for fixing sands should lead to a natural restoration of the vegetation. The photo shows a holding-back effect by grasses and some bushy specimens of *Calligonum*. But this installation has not led to sufficient biological rehabilitation and thus was not really successful (Photo M. Mainguet)

- *Stage 2*: development of a natural or planted vegetation cover, the only permanent solution for stabilizing dune sands. This "two stages" method was invented by the Institute for Research in Arid Zones (IRAZ) at Chapotu in the *Tengger Desert* of China.

The objectives of mechanically or chemically fixing sands should only be their immobilization for a period long enough to permit the natural regeneration or the planting of a vegetation cover able to survive without watering and, as a result:

- Avoiding the abrasive action of moving sand on the young shoots which have germinated during the rainy season
- Preventing the burial of the young shoots by moving sand
- Or alternatively preventing exposure of their root systems

Clearing an Obstructing Deposit of Sand

The wind itself may also be used to remove sand which has already been deposited, or to prevent its deposition; for this purpose its velocity around the deposit has to be increased by creating counter-slopes or by initiating turbulence.

3.5.1.2
Biological Methods for Stabilizing Sand, Dunes and Windbreaks

Mechanical methods are only short-term emergency measures when biological methods have become difficult in areas with less than 200–300 mm/year precipitation and in closed ergs like some of those in Asia or Iran, where there is a continuous supply of sand from glacial or periglacial sources. In any attempt at stabilizing sand preference should be given to biological methods of control, as these are the only durable ones on the long term. The development of a permanent vegetation cover can result from three processes:

1. Natural regeneration
2. Combination of natural regeneration and seeding
3. Creation of a vegetation cover by seeding

The methods of overall coverage may be combined with windbreaks and local biological barriers. An ideal windbreak is made up of three to four rows of trees:

- The first central row of trees will be highest, preferably consisting of rapidly growing trees.
- The second row, on the downwind side, uses trees of lower height.
- The third and fourth auxiliary rows on the windward side are made up of lower trees and bushes.

To obtain the maximum in slow-down of the wind, the windbreak ideally should be spaced at five times their height. However, such a spacing could be incompatible with agricultural methods and consequently, depending on wind velocity and topography, the recommended distance varies between 5–25 times the height of the individual rows, but there is no rule for predicting the best spacing. When the wind is turbulent and the topography highly differentiated, the spacing should be reduced to five times the height of the windbreak. The spacing of the barriers also depends on the gradient of the terrain in the windward direction when the streamlines are compressed and thereby the wind velocity accelerated. The distance between the windbreaks has to be reduced when they are oriented at 45° against the dominant wind or when they are placed on slopes exposed to the wind. In contrast, on down-wind sloping ground where the streamlines are spread out and where the velocity is lower, the density of the windbreaks may be reduced.

Such green barriers with several rows of trees or bushes are a classical technique in Asia, particularly in China. The species utilized are selected according to their resistance to drought, their growth velocity, the availability of water, and the depth of the water table. Trees such as walnuts, apricots and cherries are frequently planted in the Chinese oases. In the Turfan oasis of the northern Taklamakan 16 000 ha of vineyards are enclosed and subdivided by green barriers. A total of 1130 ha was planted between 1964 and 1989. Similar barriers have been planted at Hotan in order to protect the oasis against the winds in a bidirectional regime.

Other biological techniques applied are micro-windbreaks, like sugar cane or plantations with alternating rows of vegetables of differing heights.

3.5.2
Control of Barchans and Linear or Parabolic Dunes

3.5.2.1
Methods Against the Migration of Barchans

Many attempts to control the advance of barchans have proved ineffectual because:

- The source of the sand was too far away or too big to be controlled; in the line of advance of the barchans the trend to generate such dunes predominates, and the blockage of the supply of mobile sand proved difficult or ineffective because of the abundance of the sand source.
- The palisades at right angles to the axis of the dune on its lee-side were rapidly submerged and the trenches filled up.
- The volume reduction of the dune led to the opposite of the intended effect by accelerating the migration of the structure, as the velocity of a barchan is inversely proportionate to its volume.

The only possibly useful methods for controlling a barchan are its destruction or its fixation.

Destruction of a Barchan
This may be carried out by removing its material and will be of interest only where the sand may be put to other uses as the necessary technical means are expensive. We are here reminded of an old farmer at Kaouar (Niger) who, with a small enamelled metal basin as his only tool, deplored his helplessness in the face of a barchan some 20 m in height which invaded his garden and his wells.

Other possible solutions are:

- Breaking the crest of the structure as a convex dune is less conducive to mobility than a dune with an angular crest and steeper slopes, and sand in saltation is less difficult to control than the arrival of a large volume of particles in the form of dunes.
- Destroying the dynamics of the structure by cutting its wings and digging a median trench.

However, these solutions are only of a palliative nature. On the plain of the Draa in Morocco blocks placed on the crest of dunes create vortices increasing the velocity of the wind which then is able to export the sand. The sand edifice thereby becomes lower and eventually disappears.

Fixation of a Barchan
This control may be of a mechanical or biological nature, by complete revetment and replanting or by stabilizing strips arranged at right angles to the dune axis. As an additional measure palisades are arranged on the windward side of the dune across the approach route of the sand in order to prevent a further supply to the dune during the first stages of the fixation prior to the growth of vegetation.

3.5 · A Response to the Aeolian Constraints

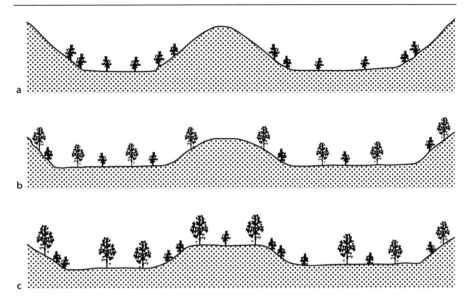

Fig. 3.2. Stages in the fixation of a linear dune (seif). Stage 1 Establishment of biological barriers over the interdunal corridors and on the lower half of the sides. Stage 2 after 2 to 3 years, planting on the lower to middle parts of the slopes (the aim of these two stages is the natural lowering of the dune). Stage 3 Planting of the entire eroded structure

The principle of any mechanical stabilization supported by a chemical adhesive is the establishment of a cover of agglomerated grains. This is done by spraying solutions or suspensions of chemical products, a large number of which have been tested (Mainguet 1992). Tests have also been carried out with clay suspensions, which may be employed where precipitation is low enough, so that the clay skin on the dunes will not lead to water erosion in gullies, as was the case in the fixed littoral dunes at Brava in Somalia. Certain chemical products also containing fertilizers have been tested, to be used during the rainy season to support the growth of any plants and to serve as accelerators for germination.

3.5.2.2
Methods for the Control of Linear Dunes

Here again, the three methods diversion, blockage and stabilization are employed. But the only durable one is biological stabilization, carried out in three stages (Fig. 3.2):

1. Establishment of a plant barrier at right angles to the two dominant winds and thus obliquely in relation to the structure itself
2. Planting of pioneer trees in the corridors between the dunes
3. Planting on the lower third to half of the dunes.

When the crest of the dune is left unprotected the wind will cut off the summit part of the edifice by deflation, and the dune may then be fixed by plants.

Experience has shown that from a precipitation of 200–300 mm/year upwards on a material rich in fine-grained sand, simple exclosure may lead to a natural rehabilitation of the steppe, which then fulfils its role of trapping particles.

3.5.2.3
Methods for the Control of Parabolic Dunes

The control of parabolic dunes results from the battle against the causes for their genesis and is based on an attempt to naturally rehabilitate the plant cover by enclosure or by artificial rehabilitation with seeds or planting. It can also result from the destruction of the dynamics of the dune itself by cutting a trench along its axis, in which the wind velocity is accelerated. The increased deflation here progressively carries away the body of the dune, leaving only its two wings which then behave like longitudinal dunes.

3.5.3
Strategies for the Control of Aeolian Erosion on the Level of the Various Units of the Global Aeolian Action System (GAAS)

The combat against sand mobilized by wind and against mobile encroaching dunes requires different strategies, as the disturbing effects of mobile sand and those of a mobile dune are by no means the same. Unfortunately, the control of aeolian erosion is undertaken only when sand invasion takes place and mostly entails only the stabilization of the dune in the wider sense, whereas it should cover the control the movement of the particles from the source to the transport area and eventually in the depositional area.

For the source area, which represents an area of erosion, one has to know its location, the processes of erosion, the the extent of the eroded area and the causes and mode of exportation of the sand. It is frequently cheaper and more efficient to undertake the control of the sand in the source area and to inhibit its exportation, than to try to fix the dunes in the area of accumulation.

The objective of such a control is protecting the soil against winnowing and deflation. This may be attempted by:

- Reducing the wind velocity through the erection of windbreaks made of inert material, or preferably of living plants, where the precipitation or available water make this possible
- Increasing the cohesion of the soil materials by planting a vegetation cover
- Fixing the soil with various chemical products, by porous plastic covers or any type of nets, etc.

In the transportation area the aim will be to try and reduce the wind velocity by increasing the topographic roughness. Forced deposition of sand in selected areas could reduce the menace of deposition in areas of human settlements. Watson (in Goudie 1990) compiled measures for reducing the volume of sand in motion:

- Forced deposition of sand in trenches or behind barriers
- Planting of vegetal barriers to trap the sand

- Increase in speed of the aeolian currents

In the accumulation area a sandy cover or dunes may form. The shape of the deposit depends on the quantity of sand arriving, on the wind velocity and its duration, on the aeolian regime and on the local topography.

In conclusion, any reasonable control strategy of the aeolian dynamics on the scale of a GAAS should include an investigation of the sedimentary balance (BS). The characterization of the sedimentary balance is derived from the typology of the sandy edifices occupying the respective areas. A region with transverse dunes, like isolated or juxtaposed barchans or transverse chains of the transgressive type, is always indicative of a positive balance (BS+). Linear dunes or seifs at an angle to the dominant wind are also indicative of a BS+. In contrast to this, erosive, parabolic or longitudinal dunes result from a BS–.

Each type of sedimentary balance has its own specific strategy:

- Areas with a negative balance (BS–) export large amounts of sand, so that the creation of human settlements on the downwind side of such areas should be avoided. Within the areas themselves the loss of soil particles should be prevented by any means of slowing down the wind (windbreaks, re-afforestation, planting of adapted species, etc.).
- In areas with a positive sedimentary balance (BS+) the risks are the overaccumulation of sand where human activities create obstacles (structures) or increase the roughness of the surface (agriculture, axes of communication). Any new management of an area should allow for the circulation of the sand by avoiding blockages of the particles and their concentration in deposits. Two approaches are possible to combat the covering of existing human infrastructure by sand: forcing the accumulation of sand on the upwind side of the area or initiating the dispersion of the deposits also upwind of the new infrastructure.

3.6
Conclusions

The third part has reviewed innovations which man's creative ingenuity has put into effect to enable him to overcome the constraints of handicapped ecosystems. All human attempts at life in drylands have led to his advance and included the spiritual level: is it not in such environments, that the Western monotheistic religions were conceived (Suzuki 1981), and have taken shape the philosophical concept of a possible continuous progress, each generation advancing over the preceding one? Is the 20th century, however, not one of a reduced certainty and does not objectivity lead to the question of whether any sustainable development is possible in the drylands? Considering the gigantic extent of the works produced and the gravity of their sometimes disastrous consequences, are we not witnessing the human genius running out of control, and is this not the prelude to a disquieting decadence? The question then will be how to assure continued progress in development. We shall see in the fourth part how difficult it is to assure such progress and maintain development on a stable level and we shall ask: has the 20th century not perhaps experienced the acme of a phase of development and the start of a decline?

CHAPTER 4

From Ingenuity to Decadence: Geohistory of an Actual Decline – Grounds for Hope?

In the third part we have seen what a wide field the fight against drought has turned into. However, it appeared that attempts at utilization or development of this environment accelerated its degradation because of a failure to foresee all the factors involved, and that such attempts have unleashed certain mechanisms as precursors to a possible decadence. The first section of the fourth part will cover the definition of this decadence and come to a conclusion which will allow us to explain it, before we analyse the changes in the oasis environment and the degradation of its soils in four case studies involving irrigation and the specific problems of salinization. Desertification will be analysed under this perspective before grounds for hope are given.

4.1
Proposals for a Definition of Decadence

The term decadence is used "to describe any process of degradation and even the mere diminution of the intensity of an activity or of any manifestation or even a plain loss of vitality ... , the concrete process of dilapidation of a material object ... or the fact of ageing" (Freund 1984).

In his analysis of decadence and of its significance Freund proposed that

> ... there may be decadence in relation to an earlier state of blossoming referred to as an apogee, which itself is the result of a more or less rapid rise from a situation of relative mediocrity .... Decadence generally represents the third phase of movement of the same society, the other two being the rise and ... the peak or blossoming .... I believe that it is barely necessary to point out that each of the three phases may entail dangerous tensions, ... momentary backsliding and rapid progress. The movement of rise and fall is not harmonious.

Combining the terms *geography* and *decadence* places them in a spatial context which in a first stage was favourable for the birth and progress of a civilization, when the equilibrium between man and his environment was advantageous to man, and in a second stage led to decadence, after the collapse of this equilibrium. There thus appears to be a linear pattern from development to decadence:

- Development → saturation → overstepping → attrition → decadence

In an ecological system with a particular complexity, in the absence of any possible compensation, a breakdown will arise by modification, beyond a certain tolerance level, of one or more parameters contributing to an equilibrium. Because of this we intend to propose an inventory of possible principles for the explanation of decadence.

"The menaces for the environment are certainly growing more rapidly than the desire or the ability to control them." With this observation Barrow (1992) aptly described the whole problem of the threats due to human activities, their rapid aggravation during the 20th century and our modest ability to restrain them. The growing hold of man on his environment and the menace to his safety emanating from the degradation of his framework of life are the hallmarks of a decadence which we shall analyse with the aid of ten explanatory principles. We will also analyse the changes to the oases, the degradation of the soils and eventually the controversial question of desertification.

4.2
Ten Principles Explaining Decadence

We have chosen to describe decadence in the form of one of its expressions, viz. the degradation of the environment. As its explanation is by no means unequivocal we shall present ten perspectives.

4.2.1
The Naturalist Perspective

This involves a natural determinism with the following opposing poles:

- The environments and their more or less pronounced instability or vulnerability as a result of topography and geographical position; the most vulnerable ones are areas with slopes, low-lying or littoral areas subject to tidal currents and/or submarine slumping or to flooding after exceptional rains; depending on the geographical position these are zones with fragile soils, areas exposed to hurricanes, earthquakes or invasion by locusts. All these cases make up a category of endangered areas.
- The climatic factors: aridification, repeated drought crises, strong and frequent winds, all parameters characteristic of drylands and climatic factors unavoidable within the framework of our present knowledge.

This naturalist perspective is invoked by those who would like to exonerate man from any responsibility for the degradation of his environment and it is contested by those who do not evade a human responsibility.

4.2.2
The Neo-Malthusian Perspective

According to this the degradation of the environment is caused by demographic growth of human and animal populations, representing a pressure on the environment which has passed the tolerance threshold and is beyond regeneration. This raises the notion of carrying capacity. This idea is criticized by the anti-Malthusians and especially by those who prefer global means and arguments (the surface area of Africa and its resources, especially water, in relation to its population which could be much larger than it is ...) which do not take into account the spatial heterogeneity of these resources.

4.2.3
The Marxist Perspective

In this perspective the wealth of the developed countries stems from the exploitation of the countries formerly referred to as underdeveloped and now as developing:

- *Agricultural resources*: export crops also referred to as cash crops and criticized as having taken over the supply of the population from food crops.
- *Mineral resources*: accused of having distracted the populations from the care of the land and their traditional activities for the sole benefit of the developed countries, which because of this are responsible for the increase in poverty of the Third World, in itself the cause for the degradation of the land:
  - Either through insufficient or maladjusted exploitation
  - Or through overexploitation: overgrazing, lowering of the water table by pumping or waste of water, and chemical pollution of air and waters.

4.2.4
The Anti-Colonialist and Neo-Colonialist Perspective

The first critics of the former colonial powers accused them of having neglected the overall development of their colonies in favour of a more sectorial development in the fields of sanitary infrastructure, hospitals and services fighting the major endemic human and animal epidemics (malaria, trypanosomiasis, yellow fever, schistosomiasis, ...) and to a lesser degree the routes of communication like roads and railway lines. The new ties established between the towns and the fields are accused of damaging the proper development of these countries. Did the colonization not destroy the secular equilibrium between the populations and their framework of life which then led to the degradation of the latter? A contrasting observation comes from the Turan where 20 to 30% of the harvests are lost because of the poor state of repair of the untarred roads.

To this perspective of the inhabitants of the former colonial countries comes the neo-colonialist point of view of the former colonial powers for whom the present post-colonial degradation of the environment results from the inability of these countries and their forces in power to give to the sustainable management of the environment the importance it deserves.

4.2.5
The Judicial and Socio-Cultural Perspective

This results from the fact that in most of the developing countries which are victims of degradation of the environment in tropical ecosystems, in contrast to property in animals and herds, there is no individual ownership of land: the soil belongs to each and everybody. At the same time there is no longer any responsibility for its degradation nor an obligation to pass it on to future generations in a good state. This legal gap is a source of conflict between the owners of livestock and the farmers, a rivalry dating back to the one between the farmer Cain and his brother Abel, the breeder of small livestock, whose offering is preferred by God. Whatever the outcome, the environment will pay for this wrangling between the two groups.

In order to respond to this legal gap, these states have started privatization campaigns for land which, however, do not appear to represent the optimum solutions. In southern Tunisia, e.g. along the southern limit of the semi-arid region the ancient grazing lands were privatized in 1962 and turned into olive orchards, not for production, which is miserable (there is a harvest every five years), but for prestige and to a lesser degree for forage.

4.2.6
The Global Perspective

This relates the living conditions of the Third World people to external factors: the drop of the world market prices for minerals and basic tropical commodities like sugar, coffee, rice, peanuts, etc.; the external debts of these countries in capital and interest; and the creation within these populations of a demand for imported technology, like transistor radios, televisions, bicycles, motor cars, computers, etc., which have contributed to their indebtedness (Fig. 4.1).

4.2.7
The Perspective of Non-Adapted Strategies and Technologies

This attributes the ills of the Third World and especially the degradation of its land to the transfer of strategies from the developed countries, where such strategies have been conceived for developed countries and turn out to be non-adapted to the developing countries. Some examples are:

- "Modern" agricultural technologies damaging the vulnerable ecosystems with fragile soils (multiple disc ploughs so dangerous for e.g. sandy soils with an open vegetation cover).
- Advice for opening up areas without taking into account local peculiarities: large dams, mega-irrigation.
- The introduction of fertilizers, pesticides, and herbicides with unforeseen repercussions on the local ecology.
- The introduction of petrol and the resulting dependence on it.
- Human and veterinary medicine initially intended for developed countries in temperate regions, which after importation into the developing countries have led to demographic pressure by man and his animals. This in turn destabilizes the equilibrium in these countries and contributes to the degradation of their environment. We again encounter the Malthusian perspective here.

4.2.8
The Perspective of Underdevelopment as the Cause for the Degradation of the Environment

For any person the urge to survive will take precedence over the conservation of the environment, which is seen as a luxury. The exploitation of the environment takes priority even at the expense of a future degradation; only the immediate result counts and making provisions for the future is neglected.

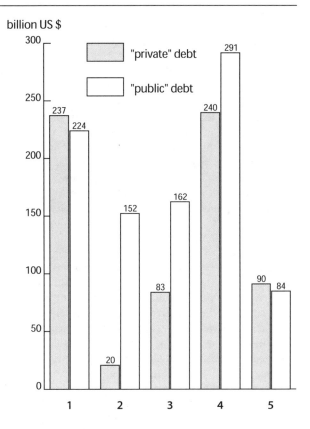

Fig. 4.1. Indebtedness in 1992 in different regions (after UNICEF letter 38 1994). *1* Latin America and the Caribbean; *2* sub-Saharan Africa; *3* North Africa and Middle East; *4* Asia; *5* Central and Eastern Europe and the former USSR

4.2.9
The Moral Perspective

This perspective entails that the immediate maximum benefit takes precedence of over a concern for conserving the natural capital. It results in cash crops, extensive breeding of exceedingly abundant herds, which become increasingly damaging to the environment, and the absence of a reliable banking system as an alternative.

4.2.10
The Perspective of Worldwide Ignorance

This for some is an easy way to forget the other perspectives and causes of degradation of the environment and to vindicate causes that present world ignorance cannot yet invalidate or confirm:

- The thinning of the ozone layer under the influence of CFCs (chlorofluorocarbons) which still has to be proven and is held responsible for an increase in ultraviolet radiation;

- A Strengthening of the greenhouse effect following the increase in surplus CO_2 ascribed to exhaust gases from industry, motor cars, aeroplanes and volcanic emissions;
- Any negative effects of the thermonuclear plants of the developed countries.

In conclusion, it appears that this analysis is quite arbitrary and rather schematic in nature, and that any single explanation of the environmental degradation that does not entail a whole array of causes will only be a caricature. The reality is too complex for such schemes to be satisfactory and it is quite clear that all causes discussed and arbitrarily classified do play a role.

All these principles are by no means unique to drylands and apply to all countries referred to as "developing", but in drylands they assume a particular importance, like the demographic explosion which will be outlined in the chapter devoted to desertification.

The question which should be raised honestly is for the level of development reached in the drylands resulting from a fragile equilibrium between the environment and its occupants qualified as underdevelopment, thereafter by other parts of the world that since one century has known the progress of the so-called developed countries. Its limitations, in parallel to these of the water resources are just beginning to be recognized elsewhere. In other words, does not what we understand under the general term development, especially if we include the term *sustainable development*, carry in itself the germs of a collapse of this equilibrium, which will in future prove to be fatal for those particular zones, unless they have achieved a high level of technological advance which, however, would be expensive?

When the drought crisis, which started in 1968 in the Sahel, appeared to level off in 1985, a consensus appeared which was summed up in the World Bank Report for 1985: "Degradation and destruction of ecosystems and natural resources have taken on severe proportions in certain developing countries and endanger a sustainable development." Myers (1987) added: "It is increasingly recognized that ... in the developing countries the natural production systems are more and more deteriorating: the soils are eroded, the forests cleared, pastures overgrazed and this to a degree which will compromise any sustainable development".

A less pessimistic point of view has been voiced in which the concepts of vulnerability and fragility are replaced by resilience and persistence or resistance, the ability of a system or one of its components to survive. The term elasticity, the rapidity of rehabilitation after a disturbance, is also taken into account. In this context we shall examine the oases and the value of their mutation.

4.3
The Oases – Mutation, Decline or Resurgence?

Considering its historic value and its recent resurgence the phenomenon *oasis* (Photo 4.1) constitutes a wide field of instruction:

- The drylands have actually given rise to a specific type of occupation of space by enclaves or islands making up a discontinuous cover, because of the limited nature of water and soil resources for settled human activities. However, are not some of

4.3 · The Oases – Mutation, Decline or Resurgence?

Photo 4.1. Oasis development and architecture in semi-arid Tunisia. Collective buildings of stamped mud in the oasis of Médénine in southern Tunisia. Man tries to protect himself by domed structures from the nycthermal amplitudes, i.e. hot days and cold nights, characteristic of the drylands. The buildings have become the handicraft market of Médénine (Photo M. Mainguet)

them an – albeit utopian – afterthought of a development and of continued human occupation of the deserts?
- The oasis system might be one of those minimizing most of the ecological risks induced by the climatic variability.
- The couplet oasis agriculture and pastoralism has become the turning-point of an agricultural revolution at a time when rain-fed agriculture becomes increasingly haphazard.

4.3.1
Description of the Oasis Space: Definition of the Term

The simplest and most general definition of the term describes it as an agricultural enclave in or on the edge of a desert. An oasis could be defined by the contrast between an island of dense greenery represented by it and the arid or semi-arid expanses surrounding it, or as stated by George "as the antithesis to the desert in a mainly mineral context". As an island close to a spring or to a point of supply of deep undergroundwaters (wells, boreholes, drainage galleries) or to a strip along a fluvial valley (the Fergana in central Asia, the Draa in Morocco, the Gila with Phoenix, Arizona), an oasis is a spot of sedentary life in an area which because of its weak precipitation is predisposed for nomadism or transhumance. It may also be a contact area between two ways of life, that of the settled populations and that of the herdsmen or breeders. Its three-levelled vegetation results from thousands of years of cultivation, and gives it a paradisiacal appearance which frequently only serves to camouflage great difficulties.

An oasis is a rather unhealthy environment as the anomaly represented by the regular supply of water to the surface leads to a proliferation of parasites. Its management is thus controlled not only by agricultural and social necessities which impose strict disciplines for the utilization of the water, but also by sanitary precautions because of the stagnant nature of these waters. It represents an extremely precarious environment. Furthermore, spacially limited by the area covered by irrigation the oasis is threatened by a double danger: overpopulation and, because of this, a limitation of the resources in soils and water.

The oases are an interruption to the aridity of their surrounding lands and are accompanied by bioclimatic anomalies usually referred to as the "oasis effect". The vegetal cluster is subjected to stronger evapotranspiration than if it were integrated in a continuous vegetation cover, the insolation balance is higher and the heat exchange with the vicinity is disturbed. Although emitted in larger quantities, the water vapour is rapidly carried away by the wind. The droughts of the last 20 years have increased the salinity of the soils and lowered the water tables which are also subjected to the consequences of wind dynamics which themselves are causing deflation, transport and deposition of sand particles. The oases are exposed to invasion by sand because of the obstacles which they form in the path of the sand like, e.g. the Kaouar oases which were invaded by trains of barchans (Bilma).

Whereas most oases are human creations in which man has assured that the necessary conditions like rise of water to the surface, working of the soils and introduction of vegetation are met, natural oases are by no means impossible. A plausible explanation was advanced by Simaïki (1952): "An oasis is formed when the wind reaches the underground water level and then blows away the grains in the ensuing depression. It dies when the water reserves cannot sustain the evaporation or when a sudden drop of the water level brings the water out of the reach of the wind that originally formed the oasis." Collin-Delavaud (1968) described a natural oasis of *garoua* (night dew) from a locality 20 km north of Trujillo on the coast of central Peru, where precipitation is as low as 10–35 mm/year; it was an "oasis with epiphytes, but also grasses, perennial plants and bushes."

The boundaries of the oasis domain coincide with that of rain-fed agriculture. As it subsists on water derived from outside its surrounding ecosystem, its limits cannot be tied to any defined isohyet, mainly because of the interannual variability of the precipitation. Zonality differentiates between oases in dry tropical environments and those in dry temperate ones (Photo 4.2).

4.3.1.1
Oases in Tropical Deserts

These are the oases with date palms (*Phoenix dactylifera*) which are not trees in the proper sense even if they attain heights of some 30 m.

> As the decoration of the desert, the symbol of its permanent occupation by man, the main source of his traditional nourishment and object of his hardest toils, it imparts to the oases the impression of colour and fame. But the landscape so created is ... bordered by hot deserts, i.e. the Sahara,

4.3 · The Oases – Mutation, Decline or Resurgence?

Photo 4.2. Teheran. The Iranian capital is a large urban oasis receiving only 230 mm/year in precipitation. Its development benefits from a water supply derived from the 5268 m high Mt. Elburz, which separates the town from the southern shores of the Caspian Sea. Below the level of permanent snow and glaciers representing precious water reserves, the barren slopes, cut by a dense network of gullies, run out into the town with its rich vegetation of trees (Photo M. Mainguet)

the Arabian Peninsula up to Amman, Palmyra and Abu Kemal, the foothills of the Zagros, southern Iran, Baluchistan, the Punjab (Dresch 1982).

They are also encountered in southern California. Dresch (1982) defined them further:

"... a mean annual temperature of 18 °C and an isotherm of 7 °C in January and 18 °C in July actually coincide with a good maturation of the dates and with the northern limit of the Saharan oases."

A variety of date palms is cultivated in the southern part of France for decorative purposes. They do not occur in the oases of Damascus, Aleppo and Mosul.

As a tree-like plant the date palm possesses a shallow root system and an aerial stalk or stem (false trunk), the cylindrical base of which is formed by imbricated sheath-like leaves. It reproduces by shoots or by seeds. There are three levels of vegetation in the palm groves:

- The upper level, the date palms, is traditionally planted in an irregular pattern using shoots from stumps or *djebars*, whereas recent plantations are laid out in lines with regular densities. Even if the date palms are cultivated mainly for dates,

they also supply seeds for livestock, building timber (carpentry, beams for ridges and rafters), fibre for ropes, material for basketry and eventually fire wood.
- The mid level with fruit trees and creepers: almonds, grenadilla, apricots, figs, and vines.
- The lower level with cereals (wheat, barley, maize), vegetables (broad beans, carrots, tomatoes, gombo and egg-plant) and lucerne.

4.3.1.2
Oases in Deserts of Temperate Latitudes

These are the oases of middle and central Asia where the desert basins receive water from water-courses originating in the surrounding mountain ranges. These courses are accompanied by gallery forests called *tougai* in which poplars (*Populus diversifolia* and *P. pruinosa*) stand above a lower layer rich in tamarind, and willows and a level with grasses and impenetrable reeds along the lower reaches of the Amu and Syr Daria valleys. These tougai have been cut and replaced by industrial irrigated cultures of cotton and rice following a programme launched by the Supreme Soviet of the USSR at the beginning of the 1960s. The oases with cold winters are at a disadvantage against those of warm deserts where several harvests are possible.

The concept of the *oasis space*, introduced by Marouf (1980), describes the complexity of the system into which the oases have turned. This concept was developed during an analysis of the western oases of the Algerian Sahara and its characteristics were defined by the addition of the ecological pressures to the cultural and economic constraints. Marouf wrote: "The term oasis is ... borrowed from the Egyptian. Whereas it has taken on the connotation *Saharan* fertile piece of land, its synonym in autochthonous toponymy only covers it in part. In contrast to this the Arab terminology defines it by setting it against its negation, the desert: *Khla, Qifâr, Frafi*".

According to Martin (1908), the Arab word *touat* is used in the sense of oasis, but Marouf (1980) clarified this definition by stating that *touat* is only a deformed (or arabized) plural of the Zénète[1] word *tît* meaning spring and "... touat only constitutes a reference to water in the oasis complex, the other two subaspects being the land ... and man." An oasis is actually the result of a trilogy:

- The water resources and the hydraulic society
- The land and its unstable equilibrium
- The social actors, the rural community and the possibilities for its sedentary nature

> As an artificial ecosystem because of the importation of plants from the desert margins, of cultivation of cereal, forage, gardening or fruits foreign to the drylands, an oasis is nothing but an artefact: man here selected desert plants, the most characteristic one being the date palm ... Even if one could define an oasis civilization and stress its common characteristics, the oases differ quite considerably between one part of the arid world and the next (Dresch 1982).

The north-Saharan oases of the Maghreb – in Morocco those along the Draa, the Tafilalt group along the Ziz, the Rhéris and in Wadi Guir, those of the high Algerian

---

[1] *Zénète*: faction of the Sahara descended from Getulis who settled in southern Tunisia and then in the Hodna prior to regaining their original lands of the western oases.

4.3 · The Oases – Mutation, Decline or Resurgence?

Photo 4.3. Typical architecture of El Oued, Algeria. Among the particularly beautiful oases of the northern Sahara, El Oued is one of the oldest. Its architecture with cupolas and vaulted roofs, its gleaming whiteness and the shade of the trees are quite representative of the oasis spaces of the Maghreb (Photo M. Mainguet)

steppe, and those surrounding the sebkhas in Tunisia – exhibit all transitions between arid and semi-arid, between the largest desert on our planet and the Mediterranean zone (Photo 4.3). Despite all their differences they possess certain common traits:

- They traditionally represent agrosystems based on irrigation by gravity: dams made of accumulations of rocks in wadis diverting the flood waters towards the séguias[2] and the *foggaras*[3] which lead the undergroundwater towards the soil surface and into wells by gravity, and exploiting the aquifers also (Boubekraoui and Carcemac 1986).
- At subtropical latitudes the oases are based on date palms in contrast to those of the deserts in temperate latitudes as in central Asia or the Chinese part of Asia where poplars dominate.
- They are characterized by the small size of the plots of which they are composed, a feature that is most critical in the oases along the Draa and inhibits their modernization.

---

[2] *Séguia*: an Arab term for earthen irrigation canals.
[3] A *foggara* is an underground gallery tapping an aquifer in the upstream area and bringing its water towards the area to be irrigated. It is built by digging wells which are then connected by a single gallery.

Although traditionally viewed as an opposite to the desert, the oasis lives from it. As a point of rest and new supplies rendered obligatory by the extent of the desert, the Touat oases of Algeria avoid the Chech sand sea, the *Grand Erg Occidental* and the Tadémaït, regions which are only traversed during times of insecurity when the normal routes are interrupted. The southern branch of the *Silk Road* passed the Tarim Basin and having left the Yan 'Kuan and the Yumen Kuan to the east, it turned south, skirting the slopes of the Kuen Lun along which there is a line of oases. In the Himalayas, the oases of Lo Mustang (29° 10' N and 83° 55' E, at an elevation of 3800 m), Phari (at 4200 m) and Leh before the end of the trans-Himalayan trade were obligatory points of passage as crossroads of the tracks in the valley leading to the cols.

The economy of the oases is based on agriculture: dates in the palm groves of the northern Sahara, fruit and early vegetables in the oases of Argentina, and cotton in those of Central Asia. Traditional oases have also introduced handicraft activities. Martin (1908) described the handicraft markets of the Touat-Gourara-Tidikelt oases, which produced woollen or cotton clothing (burnus, haik), sacks and straps. The woollen covers of Gourara were held in high repute. Drinngrass and the fibres of palms were used to make esparto products and basketry (trays, baskets, mats). Blacksmiths and jewellers worked in Timmimoun, Tamentit, Aoulef and In Salah, pottery was produced in Djerba (Tunisia), and the carpets of the Turfan oasis and of Khotan in China are famous.

Oases also may act as refuges for religious minorities like the Jews in Tafilalt (Morocco), Ghardaia (Algeria) and Libya until the 1950s. Their modern functions entail tourism with sunbathing, the dunes and adventures being the main driving forces. In the 1970s tourism became the second most important activity in Djerid (Tunisia) and Phoenix (Arizona). The latter region, also known as Sun Valley, attracts an increasing number of tourists and many retired people become permanent residents here.

Having outlined these general features we shall now abandon the old notion of an oasis based on the topographic situation and its relation to water (Haenel, 1885, in Moulias 1927), and introduce a new one based on the agrotechnical level achieved. Because they were tied to surface water, not enjoying the perennial water of large allochthonous rivers, the oases were initially in a precarious situation, but man has looked for underground or artesian water, either directly in the case of freely flowing artesian water, or indirectly using complicated hydraulic techniques.

4.3.2
Chronological Framework of Oases According to Their Agrotechniques

Marouf (1980) proposed a chronological framework for Algerian oases which proved its value for other oases also, as their origin is always subject to ecological constraints. Man who creates them subjects himself to the demands of the environment, being by no means a docile actor, and because of his ingenuity he tries to free himself of them. His efforts may be retraced in seven stages.

4.3.2.1
Lacustrine or Swamp Sites

The surface hydrography of the Sahara bears witness to more humid periods when the paleogeography of the Quaternary turned into that of the Holocene some

10 000 years ago, with the many lakes described by classical Greek authors. The task is eased by the discovery of numerous wadis which are clearly noticeable on aerial photographs thanks to a higher reflectance resulting from the sand fill highlighting them.

Aumassip (1986) observed the first traces of occupation of the drylands and the oldest forms of human reaction to droughts in the Melrir Basin. With an area of 700 000 km² it encompasses the catchment basins of the Mya, Igharghar and Djeddi wadis and makes up the Lower Sahara between the *Grand Erg Occidental* and *Grand Erg Oriental*. It was tropical warm and humid with a pronounced dry season during the Mio–Pliocene and again so during the lower to middle Quaternary, but it dried up during the middle upper and upper Quaternary with a flora close to the present one, i.e.

> a steppe environment ... During the fifth millennium B.C. the desert became noticeable in certain parts of the depression .... In wadi Mya a flora of Compositae and grasses, identified as coming from the sixth millennium B.C., gave way to a vegetation in which Chenopodiaceae (*Cornulaca monacanta*) predominated,

which may be used as an indicator for the 150 mm/year isohyet.

It is tempting to relate the birth of the Saharan oases to the lacustrine period starting some 10 000 years ago and to view it as the basis of the oasis space. The degradation of the lacustrine system led to the second stage, during which techniques for concentration of rain-wash were developed.

4.3.2.2
The System of Collecting Surface Rain-Wash by Open Drainage

This phase is most clearly illustrated by the system of the Nabatean oases (300 to 200 B.C.) described by Issar (1990) from the Negev. According to Marouf it reappeared in the Algerian Sahara during the fifth century A.D. at the start of the Byzantine empire, when fresh Jewish immigration took place in the Touat. The system imported there is probably that of the Nabateans of the Negev.

Gaullic Celts and Nabateans invaded Asia Minor at the same time. Around 300 B.C. – at a time when according to Issar (1990) there was a short cold spell – the Celts moved from the plains of northern Europe towards the warmer, Mediterranean climates. The Nabateans, consisting of groups of Semitic nomads of Arabic descent, settled in the Negev and later around Petra in southern Jordan. At the end of the first century B.C. they founded an empire based on commerce and irrigated agriculture, as the desert climate had become less arid. As new overlords of the Negev they established trade routes between Arabia to the south and Syria to the north, between the Orient, including India, and the Mediterranean World, transporting spices, silk, ivory, incense, myrrh and medicinal plants by caravan (Hillel 1991).

In the Negev six abandoned towns (Rehovot, Haluza, Nizzona, Shifta, Mamshit and Avdat) date from Nabatean times to the Byzantine epoch (300 B.C. to 600 A.D.). They are surrounded by terraced valleys and yielded wine and oil presses, cisterns and a sophisticated agrarian system. This system gathered water in canals dug on the slopes and led it to terraces of the valley in which the farms were located. The canal network subdivided the slopes into plots of 1–1.5 ha. For each hectare cultivated in the farms 20 ha were required for gathering water. Issar (1990) estimated that precipitation then was 50% higher than today and rain-wash amounted to 15%, leading to a water yield of 4500 m^3 for each hectare of cultivated land. The Nabateans gathered water from their roofs and the paved streets of their towns into cisterns, for drinking water for man and his animals.

According to pollen records, the Nabatean civilization flourished at a time when the climate had become less dry and collapsed when it returned to drier and warmer conditions. The refined techniques for gathering water and the terraces in the valleys proved to be insufficient and the towns were deserted. The reason for this desertion, which occurred around 600 A.D., was an aridification of the climate leading to a proved desertification (Issar 1990). The disappearance of surface waters frequently reveals a natural increase of the aridity up to a certain level at which the loss of autonomy destroys the stability of the settlement.

4.3.2.3
Artesian Wells, Induced "Ain" or "Tît"

Aumassip (1986) has shown that the running waters of the ancient authors were substituted by underground waters. It is the position of the water table which controls its appearance on the surface and consequently – prior to man digging wells – the human population. Aumassip (1986) wrote:

> Nearly all waters of the Lower Sahara are derived from artesian aquifers. The water table is situated a few metres below the surface in the northern part. As the waters are highly saline they are only ... utilized in the Souf. Several overlapping aquifers occur down to a depth of 2000 m in the Mio–Pliocene sands, the deepest ones in limestones, which are Eocene in the north and Senonian in the south. Having been exploited empirically for centuries, the uppermost water table, referred to as "Arab wells", makes itself felt in the wadi Rhir in the form of crater-shaped springs called "chriât" or "behour". They appear to be on the way to exhaustion.

Aumassip (1986) remarked furthermore:

> There is also a deep aquifer enclosed in Albian sandstones, which has been tapped since 1948 to supply the oases. *These recent boreholes reached depths of 1167 m at Zelfana and 1450 m at Ourgla.* Part of these waters is reinfiltrated to feed the upper water level. The Ouargla chott is in actual fact at the present time an artificial chott.

We shall now witness the change from stage 3, the outcrops of the water table to the stages of the oases with wells.

4.3.2.4
The Foggara System

This was introduced to Algeria during the 15th–16th centuries by the Barmaka, who knew the secrets of the Persian *ganât*. Issar (1990) described the invention in Persia and Afghanistan of a hydraulic system referred to as *ghanat* or *ganat* in Persia, *feledj* in Arabia, *karez* in central Asia, *kheriz* in Turkish, *kanayet* in Syria, *foggara* in the Sahara, and *rhettara* in Morocco. It consists of a trail of wells connected by a common drainage gallery excavated in the gravels of the alluvial cones in the foothills of mountain ranges (Fig. 4.2). Issar mentioned that the foggara system was also referred to by the Greek historian Polybios (204–122 B.C.) on his visit to Ecbatana in the Median empire one century after the conquest of Persia by Alexander the Great.

4.3 · The Oases – Mutation, Decline or Resurgence?

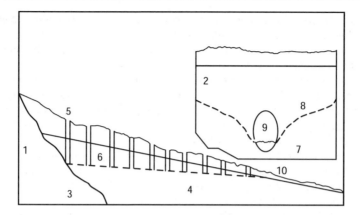

Fig. 4.2. Section through a ghanat (after Issar 1994). *1* Longitudinal section; *2* transverse section; *3* bedrock; *4* alluvial cone; *5* wells; *6* water table prior to construction of the ghanat; *7* underground flow; *8* aquifer; *9* tunnel; *10* emergence of water table. Rain falling onto the land filters down along the contact between the bedrock (*3*) and the alluvial cone (*4*). This water gathers in the aquifer and is recovered by man with the aid of canal constructed with a shallow slope (*7*) and extraction wells (*5*). The water appears on surface at position (*10*)

The Persian foggara system exploits the waters of an endorheic desert basin surrounded by mountains 3000–4000 m high, which form an obstacle for the air flows bringing rain. This applies to the Zagros with respect to winds coming from the Mediterranean in the west, to the Elbrus to the north cutting off the air from the Caspian Sea and to the lower ranges holding back the air coming from the Indian Ocean to the south. This topography leads to an interesting geographical situation: the mountain ranges intercepting the humid air in the north, south and west are themselves reservoirs for precipitation and snow. They supply the desert basins in their foothills with torrential runoff, which is absorbed by the vast alluvial cones, consisting of heterogeneous material, and which is led to the closed depressions where it forms saline lakes or swamps.

The ingenuity of this hydraulic system is twofold:

1. It utilizes the winter and spring floods by subdividing the beds of the wadis at the foot of the mountains into a system of strips opened up on terraces. The terraces are irrigated from the canals deriving their water from the principal feeders.
2. During the dry and warm seasons of summer and autumn it exploits the large quantity of infiltrated water.

Aerial photographs of Iran reveal an impressive density of the ghanat system: thousands of wells arranged in lines at the lower ends of which one finds villages with their fields and orchards. The wells, surrounded by the material excavated from them, are channels connecting the canal and the surface, assuring access to the surface to bring out the rubble, for ventilation and for maintenance of the galleries.

Construction and maintenance of the ghanats in Iran is carried out by specialists, the so-called *mughani*. The frequency of evaporites and of saline layers in the geological formations of the plains and basins, explains why the water percolating through these formations is so highly charged with salts. It thus has to be captured on its exit from the mountains and not utilized until it emerges in the sebkhas. Because of this the ghanats were invented: a gallery with a slope below that of the topography but above that of the piezometric level of the water table. One may suppose that the *ghanats* originated in the valleys where the rivers debouched from the mountains, and that with the growth of the population they progressively spread into the neighbouring plains until they reached a tolerance threshold referred to as level "0" by Marouf, viz. the area of the sebkha.

The ghanats enabled Persia to pass from a system of dispersed oases to a feudal society with large landowners. Cyrus unified Persians and Medians and could forge his empire because of a rural economy which, despite its position in a dryland, was prosperous because of its system of gathering water.

The knowledge of the Ghanats spread towards the east and southeast in China (the Turfan basin), the Pakistani part of Baluchistan, and India; towards the west to Dakhla, and to Kharga in the Egyptian Sahara in the sixth century B.C. When the Persians conquered Egypt, Ghanats were built close to the oases; however, the water was not derived from rain infiltrated into the alluvial cones on the foothills, but instead came from the Nubian Sandstone, which is an excellent reservoir rock. After the Islamic conquest of Persia the ghanat method was exported to the Maghreb in the west, where it was widely applied: in Algeria, in Morocco in the oases of the Draa, Ziz and Rheris wadis, then to the Canary Islands and Spain in the sixth century A.D. and to Libya towards the southwest. Eventually it reached central America and Peru in the 16th century.

4.3.2.5
Processes for Raising Water from Streams: The Archimedean Helix, the Large Water Wheels of Hama on the Orontes

In the eastern Mediterranean several methods for tapping water tables have existed since antiquity: wells with pivoting poles – the *chadouf* in Egypt, vertical manual traction, pulley systems and *norias* driven by camels walking in endless circles.

When the water table is deep, boreholes are necessary. The size of an oasis and of the cultivation surrounding it depends on the volume of water available through pumping. New methods economising on the traction force exerted by man or his animals, such as wind power, motor pumps or large diverting structures, have led to profound transformations of the rural landscape and way of life in regions such as California, the Negev and central Asia. This also raises the problem of recharging the aquifers. In Australia, where the capacity of the Great Artesian Basin had been overestimated, there has been a reduction in its discharge of 15% in a few years. At Denver (Colorado) the water table has dropped by 150 m in only a few years after the start of its exploitation.

4.3.2.6
Artesian Wells of Great Depth

These were introduced by the French army at Gourara and Tichkelt at the start of the 20th century.

4.3.2.7
The Era of the Motor Pump

In this chronological overview the motor pump occupies a separate position because of its economic importance and impact on the environment.

> The eastern Mediterranean and the Middle East were one of the cradles of irrigated agriculture, and since its onset the region has witnessed a coexistence between individual forms of raising and distributing water and collectively organized systems, ranging in scale from the confines of a village to entire fluvial basins. Whereas governments now gladly turn to a policy of large dams, less attention is paid to a phenomenon in the everyday life of the fields which is also relevant, i.e. the multiplication of individually operated pumps (Bazin 1993).

The motor pump is the symbol of modern methods of raising water. Its advantages have been recognized by its rural owners:

- The pumps replace human or animal muscle by mechanical power and permit a higher production per time unit.
- The small pumps are mobile and do not require any permanent installation.
- Drawing water from relatively deep water levels which may be located more than 100 km away has become possible.
- The pumps also may be used for raising water directly from water-courses located below the fields to be irrigated and require neither diversion structures nor canals.
- The pumps have also facilitated irrigated agriculture all year-round. Kosl (1984, in Bazin 1993) recorded how the village of Mushâ, south of the Egyptian town of Assiut, had introduced perennial irrigation with the aid of motor pumps even before the construction of the Aswan Dam.

The disadvantages of this individual pumping affect the environment in particular. This modern means of irrigation has proliferated in an anarchic fashion and without public control. Thus the aquifers of the Damascus oasis have dried up because of overexploitation by individual pumping. It is also because of the wells with individual pumps that numerous *ganâts* of central Iran have dried up and that moribund oases here have degraded (Ehlers 1980).

This chronological framework based on the factor water reveals that an oasis is an arrangement between man and the environment, originally because of the very nature of the desert which demands adaptation. Man determines the modalities of this relation. Where technical capacities are limited, the natural factor has the greatest influence on managing an area and one may talk about an environmental determinism. With the growing capacity of the means employed, man increasingly turns into an active factor with regard to the environment, which then behaves like a passive factor. One has moved from the exclusive dependence on natural factors in a closed site, to an opening into space and resources on a national and international scale.

In this sense a typology might be based on the degrees with which the problems are mastered and in which the factor water is superimposed by exogenic factors. Modern routes and means of transport are integrated into a national and international network, the absence of water from a site is controlled and the hazards of the climatic conditions reduced by modern hydraulic structures, soils are amended, transformed, improved, even completely newly created; the mastering of the climate in greenhouse cultivation gives only a foretaste of this. The management of an oasis is less and less a function of the local site conditions and increasingly marked by the technologies employed and the desired objectives.

The example of the development of the lower Columbia River basin, which has been turned into a range of dams which raise the water levels, regulate the discharge and constitute a navigable route opening the region to the national markets, represents the type area of the most complex modern oasis space. The river water is pumped to the plateau whose surface has been levelled and the soil improved by urban sludge and manure, and which is then irrigated by a system of sprinkler sets that distribute water enriched with liquid waste by means of 400-m-long mobile arms with a full rotation in 15 to 17 h. Working the soil, sowing and harvesting are guaranteed by these powerful machines. The result is an extremely productive agriculture whose products are exported throughout the United States.

At Phoenix, Arizona, the inhabitants proudly talk about a "de-desertification". This oasis comprises residential areas built around artificial lakes, private swimming pools and parks and is a centre of luxury tourism. It supplies agricultural produce for the entire US market like the oases in the southern Californian Imperial Valley. These do not experience winter frosts and supply 85% of the winter lettuce in the USA, attracting at the same time large investments. Phoenix is also a mining centre, this industry employing more than 200 000 persons and offering services for a region 500 km away from next metropolitan area. This urban development of high standing entails a wastage of water and competition between agricultural and urban activities for land.

It is clear that despite a growing independence from the environment the typology proposed here remains valid: A map of water is always also a map of life, and the progress of an oasis logically runs directly parallel to chronological evolution. We shall see that development does not necessarily lead to a mastering of the environment, but sometimes also to its degradation. In conclusion it can be said that faced with growing demographic pressures the oases have become a space which is increasingly sought after but also endangered.

4.3.3
The Oases of the 20th Century: Decadence of a Strategy of Development

After 9000 years of the "civilization of the oases" – Jericho is 9000 years old – the question arises: isn't the end of the 20th century also the end of this civilization so characteristic of the drylands? Are the oases not a system of development on the road to disappearance, their opening to the outside exposing them to another socio-economic framework and requiring more space if they are to become able to furnish a production which permits an autonomous life?

Since the 1950s the exploitation of the deserts has experienced changes because of oil, mining, industrial and urban development. Kuwait is an example in this context. Nevertheless an extractive or industrial activity does not exhibit any specific relationship to the desert, and always leads to the same landscapes, without a transformation of the environment by an enrichment of its biocenoses or a softening of the constraints by the desert. The changes that occurred in the utilization of the oases during the 20th century have transformed their age-old characteristics and their relationship with their geographical context and the rest of the world. Can this evolution, which was considered as a decline by Capot-Rey (1953), Weis (1964), Nesson (1972), Rouvillois Brigol (1972), Despois (1973), Meckelein (1980) and Boubekraoui and Carcemac (1986), not be modified by the positive impact of economic results and by the disquieting consequences for the environment?

4.3.3.1
Exogenic Causes for the Mutation of the Oasis World in the Northern Sahara

Exogenic and intricate endogenic causes which will be distinguished here for the clarity of our argument, will be analysed in order to outline the difference between short-term economic profit and sustainable development, it being understood that no sustainable development is possible without conservation of the environment.

Sedentary Life in the Oases as a Seduction for Herdsmen
The oasis space was described in its twofold nature: an area of sedentary agricultural activity, and, on its periphery, a nomadic grazing area of variable extent, with all complementary cultural and economic ties between them. The former sold agricultural produce to the latter: sugar: barley and dates. It bought wool, skins and meat, and salt from the continental salt deposits of the Sahara. Furthermore, the nomads were the protecting warriors of the oases against heavy taxes.

At the end of the 19th and the onset of the 20th century well-defined territories were assigned to the various tribes in Morocco and Tunisia. With the demographic pressure growing during the second half of the 20th century, the collectively owned grazing lands – as a land reserve the object of many desires – were opened to cultivation. In the semi-arid Maghreb regions, areas of dry forest or woody steppe, with their relatively abundant arboreal vegetation, have disappeared. We have to recall that in the *Rhantorium suavolens* steppe of Tunisia, olive plantations have taken over, and a system of individual exploitation, with a redistribution of the space between grazing areas and agriculture, has replaced the former system of collective pastoralism.

This structural dualism was accompanied by a reduction of the mobility of the herds, a privatization and then the cultivation of the pasture land requiring the food supply system to turn increasingly less pastoral, the grazing land only assuring part of the forage requirements of the livestock. The capacity of the steppic grazing areas in Algeria dropped by 50% within 15 years. The steppe with a carrying capacity of 1 million sheep now has to support 10 million head and the pasture land will thus only satisfy 10% of the demand (Bourbouze 1993). The breeding of sheep in the Algerian steppe becomes increasingly dependent on agriculture (straw and stubble) and on buying cereals and agroindustrial concentrates. Alternative forage materials soften the impact of the variability of the resources of the natural pastures. Bourbouze (1993) estimated that the ratio of free-range pasturage to barley fed to sheep is about 1 : 12 in Morocco. Only small herds remain dependent on the open range land, but in a dry year straw and barley account for 40% of the requirements.

In the Maghreb breeding is in the hands of a few large owners and the purchase of supplementary feed is becaming commonplace. However, with the spread of cultivation the number of animals per hectare grows and overgrazing becomes a general trend. The drought of the 1980s, which affected the three countries to different extents, was overcome with little animal mortality, but the sale of animals increased in order to create funds for the purchase of concentrates. This proved, however, to be insufficient to stem the degradation of the pastures, which goes to shows quite clearly that among the effects of such changes one has to distinguish between those beneficial for the economy on the one hand and detrimental to the environment on the other.

In conclusion it can be said that the transformation of the breeding system and the disappearance of nomadism depend on three mechanisms:

1. A reduction of the mobility of the herds and motorization
2. The transformation of the collectively owned land into private ownership
3. The feeding of the initially only pastoral herds has become agricultural with water and concentrates being brought to the herds by small trucks

The Opening of the North-Saharan Oases Towards the North

Having been open traditionally to the arid and hyperarid Sahara to the south, the oases have now opened up to the north, under the pressure of modern means of communication, of the establishment of administrative services, the promotion of urban centres, the diversification of the economy open to distant markets. All these changes are suppliers of permanent employment and regular salaries, and eventually the advent of tourism. From the cul-de-sac open to the south, they have turned into a pathway and a stage between north and south.

The exodus from the oases, a corollary of this complex evolution, has turned into a veritable desertion. However, the landowners of the oases, themselves not resident there, do not give up the land and some of them even leave it to fall into disrepair: the very beautiful old ksours and the lands are not maintained, the palm groves do not produce any longer, all of which indicates a slow death shattering the structure of the oases. In contrast, others who have a regular income from stable employment in the towns, maintain their lands and build comfortable houses outside the ksours.

> The great facilities of circulation which have grown from ... development of the major road axes have ... stimulated sales to markets farther afield and thus local cultivation of produce native of the Sahara like dates or of produce grown so far only sparingly or not known at all like tomatoes, green peppers, egg plant or zucchini. (Bisson 1993)

However, this opening is not only favourable, but also entails disadvantages:

- The modernization of date cultivation has substituted the monoculture of the *deglet nour* for the hundreds of genotypes grown previously. The *deglet nour* is juicier and of a better market value, but also more sensitive to epidemics: the *bayoud* (palm fusariosis) has destroyed many palm groves in the Moroccan oases and led to the loss of food resources in the face of the demographic explosion, requiring the extension of the areas under wheat at the expense of forage, and leading to soil deterioration. At the same time the death of the palms leads to a loss of firewood and building timber. Drawing more on the surrounding natural environment in searching for fodder and wood, the inhabitants of the oasis will cause a severe degradation of the surrounding area, and will look for paid activities outside the oases, which are the cause for the breakup of the social network.

- Tunisia and Algeria grow early vegetables in plastic foil greenhouses for the urban markets of the north or of Europe, at times when they are not in season there. In Algeria this agriculture, thanks to the road network, compensates for the degradation of the market gardening belts around its urban centres to the north. However, this greenhouse cultivation is not without risk: the greenhouses are actually simplified warm and humid tropical systems (in Tunisia their temperature is maintained by warm water from the Albian aquifer) and they are thus favourable for parasites, bacteria and viruses, requiring the increased application of pesticides.
- Another example is that of the milk from the Tunisian oases. To maintain a supply of indispensable fertilizer, palm grove owners keep small herds, but the subdivision of the land is more favourable for goats. The production of milk is a method of utilizing the forage cultures of lucerne or date seeds, which must be irrigated and are thus expensive. However, the oases can obtain milk from the north more cheaply than the milk from the local goats, so that goat milk can hardly be sold. The only alternative would be the production of cheese which, however, is not a traditional activity here.

The breakup of the social network and of the structure of the oases is just one facet of the degradation of the system. The exogenic causes for the changes in the oases combine with the endogenic ones in causing an exacerbated degradation of the environment, as expressed by desiccation, invasion of sand, salinization and even desertification.

4.3.3.2
Endogenic Causes Leading to the Mutation of the Oases

Desiccation
In the oases of the Maghreb the water demand is dominated by agricultural requirements, which amount to 70% of the total demand in Algeria, 83% in Tunisia and more than 90% in Morocco (Mamou 1993). Yet the phenomenon *oasis* further increases this demand. The oases of the Maghreb possess a potential of deep groundwater which is larger than that of the surface water or shallow groundwater. At the start of the 20th century the French started with drilling boreholes which by now have reached depths of more than 2500 m. These wells operate either artesically or with pumps, and have multiplied rapidly.

The dry areas of Algeria and Tunisia are underlain by sedimentary basins rich in aquifers. In the semi-arid to arid parts of Morocco, however, the crystalline or metamorphic lithology is not suitable for larger aquifers. The deep waters of Tunisia come from two aquifers: (1) the *Continental Intercalaire* which is exploited in the region of the Chott Fejej, where only 10% of a potential 4000 l/s had been utilized in 1993, and (2) the shallower *Continental Terminal* around the Jerid and Nefzuoua oases, which has a S-N flow from the *Grand Erg Oriental* towards the Chotts Djérid and Gharsa, where some 80% of potential discharge of 4500 l/s has been utilized (Mamou 1993). The extracted water has a low salinity (3–3.5 g/l from the *Continental Intercalaire* aquifer at a depth of 1500–2000 m) but it is hot, with a temperature of 60–70 °C, and has to be cooled to be usable. Tozeur and Kilili in Tunisia are oases established after mobilization of springs on outcrops of the water tables or from deeper aquifers along faults. Difficulties in maintaining the aquifers and their exhaustion, frequently resulting from the use of motor pumps, have led to a drying-up of the springs, to progressive abandoning of the foggeras and to the exploitation of the aquifers with boreholes.

The exploitation of the groundwaters has transformed the steppe into a region of modern oases by a growth of the irrigated surfaces, but mobilization of the groundwater resources has led to a number of difficulties:

- Drying-up of the springs as seen above
- Lowering and even disappearance of the water tables
- Diminution of artesian waters, particularly aggravated by overexploitation
- Mobilization of probably non-renewable water resources. "The present state of mobilization of the water resources in the Maghreb countries … reveals that in Tunisia and Morocco more than 50% of the water resources have already been mobilized" (Mamou 1993)
- Failure to take the real cost of water into account after extraction (this amounts to US \$2–3/m^3 in southern Tunisia), a cost to which the future lack of this fossil water must be added. The water costs borne by those irrigating their lands are below the costs of exploitation (labour, energy, repairs). This low price of water is no incentive for reducing wastage or for increasing productivity.

To summarize, the village system of collective utilization of water by diversion of wadis and the use of foggaras and springs has been replaced by individual wells and motor pumps and, recently, by the tapping of deeper aquifers which no longer reach the people of the oases. This mobilization of the water resources has led to a regional desiccation, the least reversible form of degradation.

The decline of the Algerian oases (Kobori *et al.* 1980) accelerated in 1962 when Algeria became independent. Priorities and efforts were indeed set in the north of the country. The agrarian reform of 1976 then hit the Sahara in the catastrophic form of the nationalization of water-rights, which turned the collectively owned free water into a national property to be paid for, a source of revenue for the state and the cause of the further impoverishment of the already most deprived. In most oases the decline was accompanied by a desertion fanned by the attraction of the urban centres or the production of oil. The causes for this decline were numerous:

- *Partly natural*, as several droughts have taken place between 1968 and the 1990s, resulting in a rainfall of effectively zero, whereas prior to 1968 the means were between 10–20 mm/year. In contrast, the Algerian Sahara experienced torrential rains in February 1969 with 24 mm over 2 days, which initiated a degradation by reactivating fixed sand structures.
- *Human*, as because of the nationalization of the water rights the foggara system, which was better adapted to the Saharan oases, became dilapidated because of lack of maintenance, sand invasion and the shrinkage of the supply of water as a result of drought and overexploitation; the authorities undertook to clear the wells but the annual clearing was not sufficient to combat the intensity of the sand invasion. Furthermore, the increased water demand competed with the supply to the foggaras, the most stable and best adapted systems for this environment, and found itself out of equilibrium because of socio-economic changes.

Furthermore, the oasis became overpopulated by the arrival of numerous Touaregs from Mali and Niger fleeing from the droughts there, before they were chased away by the Algerian authorities. At Aoulef in the Tidikelt of the Algerian Sahara the degradation showed itself by a lowering of the water table resulting from overexploitation of the wells and by the population growth which according to Kobori (1980) raised the population from 2856 inhabitants in 1961 to 6800 by 1980.

In a still highly relevant paper Capot-Rey (1962) described the evolution of the Tamentit oasis. As an oasis of the Taouat (Algeria) Tamentit was established on the Colomb–Bechar–Reggane route some 12 km SE of Adrar close to a spring originating at the foot of an outcrop of the *Continental Intercalaire* sandstones on which a ksar was built some 15 m above a depression occupied by a sebkha. The oasis, which was well known during the Middle Ages as a centre of a community of judaized Zénètes, disappeared in 1491 and became re-established between the ksar and the sebkha.

Like the other oases of the Taouat, Gourara and Tidikelt, it belongs to a group of oases with foggaras exploiting the water of the *Continental Intercalaire*. Each foggara possesses a comb-shaped partition (kesria) on its downstream side with intervals exactly calculated so as to allow only a precise volume of water to pass. All these foggaras exhibit the same method of capturing and drawing water, but the distribution system at Tamentit is rather complex: in addition to the combs (14 of the 18 functioning in 1962) which distribute the water, there is a time distribution system, with each landowner receiving all the discharge over a certain time span which is measured in *thmen* (one eighth of a day) or in *kirat* (1/24th of a day). Each receives the entire amount of water and uses it immediately without prior storage (*majen*). With the classical foggaras the water is gathered in a reservoir and watering takes place in the early morning, the most favourable time.

The time-controlled system is the one most widely used in the Maghreb and is also practised in Syria and Iran. The use of both systems concurrently is observed only at Tamentit. Since the early 1950s three negative facts have characterized the evolution of this small oasis:

- Disappearance of a number of crops with date palms and cereals (wheat) being replaced by vegetables and lucerne.
- Introduction of motor pumps for the wells. Capot-Rey (1962) counted only 25 in 1959, but already 110 in 1961, when the development was just gathering momentum. These pumps are utilized for filling the reservoirs (*mejan*) and entail a twofold, also negative consequence: abandoning the foggara system which requires too much labour for maintenance; and the desiccation of the aquifers leading to the exploitation of increasingly deeper and saltier waters. As in all oases of the Maghreb, the increasing subdivision of the land has led to difficulties in its exploitation.

The *Damascus oasis* in the south-west of the Syrian desert was studied by Masanori-Naitou (unpubl. 1984), and is an example of such decay. Fifty years ago, the capital of Syria was an urban oasis surrounded by a ring of small rural oases furnishing fruit, vegetables, cereals and accommodation. The villagers found the administrative centre and the services they required in the town.

The oasis was supplied with water from the Barada River, the two sources of which are located in the eastern foothills of the Anti-Lebanon, a limestone range representing an excellent water reservoir. The strongly flowing sources supplied the Damascus basin and gave rise to a wadi which terminated in the Ataiba sebkha some 35 km below the town. From 1970 onwards the river no longer reached the depression, which consequently dried up.

The Damascus basin, with its rainfall of 200–250 mm/year, requires an irrigated agriculture: The water of the Barada River was subdivided into seven canals which themselves are split up into smaller ones for the irrigation of the fields of the oasis which are divided into two areas: *Ghuta*, the western part on the slopes and *Marj*, the eastern steppe part, used in the past by cattle breeders. Around the periphery local springs flow on the downstream side of the alluvial cone making up the Damascus basin, giving rise to small water-courses supplying village oases to the east. For 30 years this network was sufficient for traditional irrigation based on complex and diverse rules for the distribution of the water.

The deregulation of the water followed the "*agrarian reform*" of 1958 in which land was redistributed. The farmers then dug wells on the *Ghuta*. With the growth of the city, buildings and factories were erected on the better agricultural land and the small streams were built over. Exploitation of groundwater thus became a necessity. During the 1960s the richer owners on the Ghuta bought modern equipment, in particular motor pumps. The number of wells with such equipment grew and in the middle of the 1980s reached a total of about 12 000 throughout the entire oasis. The wastewaters were directly discharged into the irrigation canals, thereby leading to sanitary problems and making cholera endemic. In the boundary area between the Ghuta and the Marj, the springs dried up shortly after the spread of the motor pumps. Because of excess irrigation waterlogging of the soils occurred, followed by salinization.

Prior to the agrarian reform of 1958 the agriculture of the Marj was rain-fed. The subsequent two years were dry and the landowners, the herdsmen and the nomads to whom the land had been given, profited from the irrigation of most of the area. The cereal and cotton harvests were good at the beginning, but irrigation led to the usual drop of the water level. Nowadays it is not possible to find water even at a depth of 100 m and the eastern part of the Marj has been transformed into a desert. The collapse of the irrigation forced the farmers to seek employment in the armed forces and the factories, so that many hectares of land which had been turned into desert were abandoned. This collapse occurred despite a law which has forbidden the digging of wells since 1959, as the farmers do not adhere to it:

- Most of the canals on the Ghuta have been destroyed and the Marj farmers on the downhill side cannot rely on them any more. In the Marj the land belongs to a few large rich owners who have been charged with the maintenance of the canals. With this maintenance no longer being carried out, the irrigation co-operatives collapsed and the individual owners have no other means of survival at their disposal than their own wells.
- The farmers themselves were not able to maintain the canals or to preserve the traditional system. In the Ghuta the few rich landlords which had a co-operative water-management could not resist the motor pumps and the possibilities of making some quick profits.

In conclusion, the rural oases which have been invaded by urban expansion especially to the west of the town and modern pumping technology have superimposed their own new rules onto the traditional rules for irrigation, reducing the aquifers to a critical degree. To the east of the city the degradation of the environment has reached a critical, if not irreversible degree. The thirst for profit has overridden the concern for the conservation of the environment. Under Islamic customs the individual rights and interests have to be respected and as a Muslim cannot be ordered to abandon a profitable activity, the lack of any control in the utilization of the undergroundwater has become the main reason for the degradation of the lands of the Damascus oasis.

Sand Invasion, the Curse of the Oases
The advance of the northern limit of the Sahara to the north, a theoretical concept which led Algeria to construct its green barrier, is contradicted by field observations. In Morocco as well as in the Algerian and Tunisian steppes, it is not the desert that advances northwards, but the steppe which is degraded under human influence. Active dunes are generated in situ from the coarse fraction of local or distal sandy material, after the fine essentially loamy matrix has been winnowed out. Such growing spots of degraded areas may eventually, when extending farther, join up with the desert. It is this mechanism of coalescence which, especially in the oases, has given rise to the impression of superficial observers that the desert is advancing.

The term sand invasion covers several aspects in the Maghreb oases. Along Wadi Draa of Morocco we are dealing with sand tongues or barchanic edifices which invade the séguias (such as Merzouga close to Zagora) and the palm groves to the south, notably the six palm groves of Fezouata on the middle Draa (Photo 4.4).

The effects of sand invasion could already be recognized on aerial photographs taken in 1952, but the situation has deteriorated since, as indicated by satellite imagery, for two reasons:

1. The degradation of the vegetation cover, which opens the alluvial soils to the deflating effects of the *chergui*, a north-easterly wind, and of the *saheli*, a south-westerly one.
2. The construction of the Mansour Eddahbi dam in 1972 at the entrance of the Draa *tarhia* or canyon, some 2 km downstream the confluence of the Dadès and Ouarzazate wadis, in order to avoid low waters and floods and to improve irrigation. This dam actually resulted in a more regular flow of water to the séguia systems and held back the floods, which in addition to their various disadvantages had the one advantage of washing the sand which on the contrary had invaded the beds of the wadis.

In the Tafilalt of eastern Morocco, the Hannabou, El Krair and Sifa areas west of the palm grove are most intensely invaded by sand from barchans or sifs which grow in length and are responsible for the invasion of sand into the khettaras, séguias and palm groves of the village. Because of this, Hannabou has been moved three times already (Mainguet and Chemin 1979; Mainguet *et al.* 1981).

Sand invasion is a widespread process among the Tunisian oases, the Aoudia region in the Tozeur governorship and the new oases of Nefzaoua being good examples

Photo 4.4. Sand invading a séguia on the Wadi Draa. In the Draa valley sand invasion represents a major problem. Here, at Isougha, a crossing of two traditional séguias is invaded by sand, requiring meticulous daily clearing, as indicated by the quantity of rubble (Photo M. Mainguet)

for this. The NNE winds displaced the sand towards the SSW. In the olive orchards on the sandy soils of Ben Gardane the sands have been built into small barchanic structures after some 30 years. The entire surface becomes mobilized because of damaging ploughing of the soil, in particular with multi-disc ploughs. To maintain the humidity necessary for the olives 5–7 ploughings are carried out annually. These activities, and the distance of 25 m between individual trees necessary because of the competition between the root systems, leave bare ground exposed to aeolian erosion. Prior to the formation of these dunes there was an equilibrium between the resources of the steppe and their essentially pastoral utilization. The inherited soils which are slightly reddened and isohumic are slightly aggregated and at present are not replenished when they are eroded by the wind.

Salinization of the Oases
Salinization is the most difficult damage to contain in drylands.

In the oases of the Maghreb salinization does not generally result from the salinity of the underground water, but from the excess of water used and its evaporation

4.3 · The Oases – Mutation, Decline or Resurgence?

Photo 4.5. Irrigation as the cause of secondary salinization (detail of Photo 4.6) (Photo M. Mainguet)

(Photos 4.5 and 4.6). Most of the oases were originally located, as we have seen, along the wadis. The general policy of building large dams allowed an extension of the cultivable lands downstream from the dam by regulating the volume of water in the basin. In the Tafilalt of Morocco the construction of the Hassan Addakhil dam on the Ziz bears a great part of the responsibility for the salinization of the soil, because of poor water distribution. The decline of the southernmost oases of the Fezouata palm groves close to the confluence of the wadis Feya and Draa is also caused by salinization.

The death of the oases on the coastal plain of Batinah (Oman): Scholz (1982) described how after the revolution of 1970 the Sultan of Oman undertook a land reform involving redistribution of land, modern agricultural equipment, fertilizers and reorganization of irrigation. At Batinah the new methods of exploitation and the motor pumps led to a degradation of the water equilibrium, a higher salinity of the irrigation waters, a lowering of the water table, and an intrusion of a network of seawater under the continent. The several centuries old palm groves of the littoral oases died off and the oases themselves declined.

Photo 4.6. Irrigation as the cause of secondary salinization in Tunisia. In the area of Kébili, a continental palm grove in southern Tunisia, excessive irrigation of the palms led to waterlogging of the soils and secondary salinization, because of poor management of the drainage waters. This salinization is a degradational mechanism in oasis regions difficult to reverse (Photo M. Mainguet)

4.3.3.3
The Decline of the Asian Oases

The Evolution of Khoulm

Gentelle (1969) described this Afghan oasis which had been destroyed by Gengis Khan at the start of the 13th century and reconstructed several times thereafter. It forms part of a line of oases 450 km long on the piedmont, at the contact between the Amu Daria and the argillaceous–calcareous plateau bordering the Hindukush and Koh-i-baba ranges to the north. Khoulm lies on the alluvial cone of a tributary of the Amu Daria, which connects Afghanistan with central Asia. Precipitation here is 120 mm/year and the oasis obtains its water from the tributary of the Amu Daria.

In 1956 the male population amounted to 20 000 and some 22 750 ha were irrigated by flooding. Each year 11 450 ha are cultivated, 5320 ha of this in the form of gardens. "The dearth of water is such that only one-third of the fields may be irrigated every year" (Gentelle 1969). The irrigated lands are subdivided into:

- *Baghi* (gardens), which are irrigated throughout the year and yield reliable cultures, orchards of almonds and pomegranates, and grapes.
- *Koram* or *sekoti*, land situated between the gardens and the fields, which benefits from the water not used by the gardens.

4.3 · The Oases – Mutation, Decline or Resurgence?

- The *paikali*, agricultural lands on which crops of open ground like cereal (wheat) and melons are planted for 1 year, lying fallow for two further years.

The water is distributed in canals of tamped earth, leading to high water losses. The distribution system is controlled by the mirab-en-chahr, and each of the 18 canals by the tchakbachi, but both of these have seen their authority whither away in favour of that of the landowners, who have modernized their agriculture with tractors, Russian ploughs and fertilizers, sacrificing quality to quantity in the process. Gentelle (1969) wrote also that the "yields and the massive sales of poor quality products are sufficient for them … although the smaller owners become aware with bitterness that looking for good quality does not lead to much. The sharecroppers start to plant regard-less and one sells even salty melons … as the buyer has no means of tracing the seller".

The decline thus has affected the distribution of the water as well as the quality of the products. A rift has therefore appeared between the owners adhering to older production methods and those who are modernizing, buying tractors and embarking in commerce, speculation and transport. The bazaar or market which was the centre of exchange and point of sale of handicrafts is losing its importance, and has to compete with new trades settled outside the walls and with co-operatives selling karakul sheep, cotton and pomegranates. These changes started in the early 1940s when the camel caravans were substituted by trucks, transforming Khoulm to a mere stop on the poor road connecting Afghanistan and the former USSR. Eventually the tar road of the 1950s turned this beautiful isolated oasis into a gathering point for a few agricultural products. The rural exodus and growing underdevelopment replaced the traditional equilibrium based on self-sufficiency, the sale of surplus products and an almost just distribution of land and water resources.

The Oases of the Taklamakan

Spread out since Antiquity around the desert, the present villages (nine in the south, three in the west and four in the north), which were built on alluvial cones on the foothills of the surrounding mountain ranges, have undergone the same evolution as other oases. Gentelle (1992) described how the meltwater of glaciers enabled the settlement of human groups farther into the desert with progressing development, and their retreat farther uphill when water became scarce because of overconsumption or drought. After centuries of minor vicissitudes, they experienced a demographic explosion since the middle of the 20th century. "Their population more than doubled" between 1950 and 1960 and the demands increased still further because of progress tied to "the production ideology, and modernization", leading to a reshaping since 1958. The land area in Hotan grew from 15 800 ha to 30 500 ha, irrigation was extended and the water consumption doubled from 300×10^6 m^3 in 1943 to over 600×10^6 m^3 in 1980. Degradation caused by overgrazing, salinization and wind erosion, which caused the tugai forests and the reeds to disappear, combined with the socio-political dynamics of the human societies. Since 1987 the desert has been invaded by cars and oil drilling derricks.

The Tamdi Oasis in the Heart of the Kyzylkum

The Kyzylkum is a desert of fragile fixed dunes, which may easily be reactivated by overgrazing in combination with wind erosion. Water occurs at depths of 15–35 m below the surface under the takyrs and in the sands. Artesian aquifers are found at depths between 70–250 m. Such conditions are favourable for nomadism.

Prior to the October Revolution these drylands were frequented by Kazakhs, Kirghizs and Karakalpaks, who are pastoral nomads, and by farmers in the oases. After the Revolution the governments of the republics were of the opinion that the nomads should be included in what they thought was a modern economy, by breaking their isolation and integrating them into a economy of exchange by the creation of co-operatives for buying and selling and of a credit organization. Thus co-operatives were established for the production of fodder and for various crops. The first major difficulties appeared when the first five-year plan attempted by force to nationalize the livestock herds. This plan was intended to be carried out after the cattle breeders had become settled and after transformation of the grazing and breeding techniques.

Instead of a change to real sedentarism, nomadism was actually replaced by subtle forms of transhumance. Dresch (1956) observed these processes: "The mountain nomads were settled in the plains in the *kichlak* but they continued to drive their herds in summer into the mountains towards the *gaylak* … in the custody of shepherds. The kichlak in which the families reside is transformed into a village". The village tends to combine cattle raising with rainfed agriculture and irrigated farming, possesses a hospital and a small school, cultivates fodder plants and sometimes possesses small agronomical and zootechnical experimental stations and a meteorological station. All breeding then was before independence controlled by the kolkhoze organization, the production co-operative.

Tamdi, in the heart of the Kyzylkum, is the centre of a district which in 1956 encompassed ten kolkozes and two sovkhoses or state farms breeding karakul sheep. In each kolkhoze the grazing land is subdivided into sectors receiving 600 head each around an encampment close to a well. In summer the animals are driven into the mountains. In addition to these extensive supervised grazing lands the sedentary breeders of Tamdi also possess plots irrigated with pumped water in dry valleys 100 to 200 km from Tamdi (Dresch 1956).

The kolkhoze members, all producers of karakul skins, meat and milk depend on the market and thereby on the communication axes for the remainder of their food requirements. A mediocre road (the road was, in fact, good in 1997 when we visited Tamdi) connects them to Bukhara, their main market centre 250 km to the south. The older nomads remember their past with nostalgia and yurts are pitched close to the houses. Tamdi has not managed to become a real village, its architecture being ugly and of poor quality. Around the settlements signs of desertification by reactivation of the tops of dunes and by salinization of the cultivated lands are visible on satellite images of the early 1990s. These conditions and the reduced interest in karakul skins and wool made this forced settlement project undeniably a failure that the present government is trying to solve.

In conclusion, all development policies in drylands entail special consequences for the environment because of the particular vulnerability of their main components soil, vegetation cover and dearth of water.

Urbanization, axes of communication, rural development and the expansion of the oases as a common advantage are able to absorb the population surplus. "In the last few years Algeria and Morocco posted a growth rate of 2.4%. Even Tunisia which because of an old sustained birth control policy exhibits a lower birth rate of 1.9% over the period 1980–1990, still exceeds the worldwide average of 1.7%" (Bedrani 1993). In the oases the rate is still higher. Bisson (1991) mentioned the new oasis El Fawar in Tunisia with its 3000 inhabitants where "the growth rate between 1981–1984 amounted to 3.9% and thus was considerably above the Tunisian average rate of 2.3%". The difference in the figures of the two authors is noteworthy.

The former nomads, turned into new citizens of the urbanized oases, retained their taste for breeding and possess abundant livestock. They cannot resist gathering plants growing spontaneously outside the oases to complement the lucerne fed to their animals. The effects of this gathering are aggravated by the use of motorcars for this purpose. Increasingly larger aureoles of degradation have thus been created, and these are clearly visible on satellite images, because of their higher soil reflectance. Methods for rehabilitating such degraded areas are known, but the costs render them difficult if not impossible to apply.

Unsuccessful urbanization and poor management of drainage water are also notable in the form of sewage water, which infiltrate towards the aquifers and waterlog the underlying soils. At El Oued and Ouargla in Algeria this phenomenon has reached such a level that it now endangers the survival of the oases themselves.

For how long will the water resources in the different oasis regions be able to satisfy requirements? What will the demographic evolution be like over the next decades? Is there hope for the necessary slowdown? The answer depends on the mental evolution and on a progress which nobody can predict. The balance of the mutation of the oasis is twofold: the economic effects being beneficial on the short and perhaps medium term which will qualify the human victory and the more hazardous long-term effects which stand against a sustainable development. Are the development of the oases and the degradation of the environment invariably tied to each other?

In face of the dark picture offered by agriculturists and livestock breeders of the arid regions in Asia (Mongolia, Thar Desert) and the peri-Saharan and Sahel regions in Africa, as well as the changing of oases into urban sites, urbanization ex nihilo in arid ecosystems is frequently considered a successful modern example of human ingenuity.

Phoenix (Arizona, USA), which has experienced an exceptional growth in population, i.e. from 100 000 in 1950 to 2.5 million inhabitants in 1995, has a high standard of living despite the initial, unfavourable natural conditions. Las Vegas has 1 million inhabitants although situated in a desert where the average annual precipitation amounts to only 100 mm. Los Angeles, the largest city within the arid areas of the USA, is equally successful with its 15 million inhabitants, a population which outnumbers that of most Sahel countries despite the low annual precipitation of merely 300 mm.

Admittedly, all three cities are situated in the USA, which is the leader among the developed countries. However, in what respect are they, in fact, exemplary? Although they may exemplify brilliant human achievements in arid environments, this is undoubtedly only valid in short and middle terms, as it would be dishonest or thoughtless not to mention here the key notion of the importance of a time scale, as opposed to a spatial scale.

Who can say what Phoenix, Los Angeles and Las Vegas will look like in 50 or 100 years? Phoenix is, in the short term, the city most generously equipped with swimming pools and recreational areas which consume 1000 l water day^{-1}inhabitant^{-1} as compared to 200 to 300 l day^{-1} inhabitant^{-1} in a European city. However, this is achieved at a high price, such as excessive pumping of water from precious aquifers and urbanization at the cost of equally precious agricultural regions. A comparison would be found in Egypt, where the suburbs of Cairo are exploiting – to a pharaonic extent – the even more precious agricultural soil of the Nile valley for mud brick production. With regard to Los Angeles, should one really admire the bold, or unwise, but not unknowing extension of the city into the very active and well-known seismic range of the San Andreas fault?

4.4
Soil Degradation, Irrigation, Salinization and Decadence

Just as rain-fed agriculture in drylands is threatened by water and wind erosion, irrigated agriculture is faced with two other dangers, viz. waterlogging and salinization. At present salinization is a menace for the economy of countries like Egypt, Iraq and Pakistan, dry countries for which irrigation represents the pivot of all agriculture.

The drylands are the place of the most intense salinization as a result of irrigation. Soil degradation is a corollary to all forms of degradation and in particular to that of the vegetation cover. When the soils become degraded production plummets to virtually zero and as a direct consequence the herds are decimated and famine appears. In view of this fact we have to consider that severe soil degradation is one of the expressions of decadence and we have therefore chosen to deal here with salinization, the most acute aspect in the degradation of drylands, the ultimate stage difficult to reverse.

4.4.1
The Degradation of Dryland Soils by Salinization

A direct connection between irrigation and salinization has been demonstrated (Barrow 1991; Rhoades 1991). According to Kayassah and Schenk (1989), some 25×10^6 ha of the entire 92×10^6 ha of irrigated land on our planet are affected by this process, a value which seems to low to Barrow (1991) for whom 30–50% of the irrigated soils or 30–46×10^6 ha are sterile. Not one country in dryland regions is exempt from this. The extreme values are encountered in Pakistan where 67% of the arable land is irrigated and 30% of this affected by salinization (Barrow 1991). According to Bokhari (1980, in Barrow 1991), of 8.49×10^6 ha in the Indus Valley 18% are slightly, 11% moderately and 16% strongly salinized. The situation was better in 1978 with 6, 3 and 6% being affected to the above degrees, respectively.

Gupta (1979) estimated that 7×10^6 ha of cultivated lands in India have become saline. In western Australia the production loss has been estimated as US $94 million annually (UNEP and Commonwealth of Australia 1987). In the Murray River basin this is ascribed to irrigation and defective drainage, and the land has been salinized. Gorbachev (1984) remarked that in the former USSR 30×10^6 ha were threatened by salinization or have fallen victim to it, and that for every newly irrigated hectare of land since 1975 at least 1 ha has been lost due to salinization. Kovda (1983) estimated that the area of salinized land on the entire earth is growing by 1–1.5×10^6 ha annually. Barrow (1991) considered these figures pessimistic and suggested a rate of

160 000 ha/year. He stated that Pakistan alone would have to spend US $317 million for the fight against salinization (Barrow 1987). According to Rhoades (1988), about 10% of the entire area of our planet affected by salinization may no longer be used for agriculture.

Salinization or the accumulation of soluble salts and sodication, the accumulation of sodium ions, are forms of chemical soil degradation. Salinized soils possess an unstable structure, their permeability and porosity decrease, their biological activity and content of organic matter is reduced, their pH drops or rises to above 9–10 and the natural or cultivated vegetation cover is reduced and eventually disappears. The main reasons why irrigation causes salinization or alkalinization are:

- Losses through the unsealed sides of the canals
- Oversupply of water
- Poor drainage

Soil salinization implies the excessive enrichment in it of soluble salts like chlorides, sulphates and carbonates etc. of sodium, magnesium or calcium. When the cations of the salts are essentially sodium one talks about alkalinization or sodication (WMO 1984). A soil is considered as salinized when its content in soluble salts is above 1–2% in the topmost 20 cm (WMO 1983).

When a soil has a texture favouring capillary action, and when an aquifer occurs at rather shallow depths, salts, bases and sodic compounds are easily deposited in the root and soil layer between the water table and the surface. This process has been referred to as "secondary salinization" when caused by human activities. "Soils rich in colloids (clays) may ... lead to a capillary rise of the water in the range of 50 m, but in most instances it takes two years to register a rise by 1 m. As a result in a regularly irrigated soil this phenomenon is not be feared in general" (Durand 1988).

In Turkmenistan salinization appears down to 20 cm 2 years after the start of irrigation and reaches 1 m after 30 years. It modifies the chemical properties of a soil and also its structure and thus its physical properties. In particular salinization modifies by means of the osmotic pressure the tripartite equilibrium between soil, water and vegetation. The solutions circulating in the soil, which are enriched in soluble salts, increase the difficulties of the plants in drawing water (Photo 4.7).

An examination of the extent of salinized soils reveals their clear preponderance in drylands on every continent.

> The majority of the naturally salinized regions in the world are found in poorly drained areas, in topographically low-lying areas with semi-arid to arid climates where the salts leached out from the surrounding higher reliefs are accumulated in groundwaters and in basins where the aquifers come to the surface or are found at shallow depths and where the salts rise in the soil under the effect of a higher rate of evapotranspiration (Rhoades 1991).

Any climatic mechanism supporting aridification or drought, which tends to increase evaporation of surface or undergroundwaters, will lead to a concentration of soluble salts and their precipitation in the form of a pseudo-mycelium, in nodules or as crusts. The surface of the soil and thus its albedo is modified by the appearance of salt efflorescences or the subsoil crusts. The aridity of the climate also increases the salinity of the water in reservoirs behind dams as well as that of the water transported over longer distances in the canals built for irrigation purposes and in the nat-

Photo 4.7. Irrigated cotton cultures in the Aral basin. North of Nukus (Karakalpakstan), with a precipitation of 100 mm/year, in 1990, cultivation of cotton irrigated from the Amu Daria, one of the two tributaries of Lake Aral. The irrigation canals are arranged at right angles to each other. The agglomeration shown is a modern village with corrugated iron roofs. The white field in the background is affected by secondary salinization from the waterlogging of the area which inhibits a proper drainage (Photo M. Mainguet)

ural water-courses. In the Nile below Aswan the dissolved solids increase from 110–180 to 120–230 mg/l (White 1988).

The salts are derived from a variety of sources: oceanic, thermal, volcanic, meteoric, aeolian, or from irrigation:

- *Oceanic origin*: aeolian entrainment of coastal spray. Seawater usually has a salt content of 35‰: which may rise to 40‰: in the Red Sea because of the strong evaporation taking place there. In contrast to this, the Baltic Sea is fed by a large number of water-courses and thus has a salinity of only 10‰.
- *Intrusion of seawater*: the cultivated lands of Bangladesh suffer from increasing salinization because of the intrusion of a wedge of seawater into the delta system and the drop in the river discharge because of irrigation along the Ganges River (Barrow 1991). Egypt and Libya suffer from the same problem.
- *Thermal springs and volcanic eruptions.*
- *Weathering of igneous and metamorphic rocks* liberates sodium ions and also magnesium in some rare cases of salinization.

- *Aerosols blown out of salt depressions*: the Maringouins sebkha is deflated by the trade winds, leading to the deposition of salt-rich clay on the soils cultivated in the delta of the Senegal and its surroundings (Tricart 1994).
- *Evaporites deposited during an earlier lacustrine phase.*
- *Irrigation* is the most frequent cause for the type of salinization referred to as "secondary", especially in drylands as a result of evaporation. Water supplied to the soil in excess of the requirements of the plants percolates beyond the root system, dissolving salts on the way and contributing to waterlogging. When these mechanisms take place the salts dissolved from below the soil are drawn to the surface and precipitated either at the root level or at the surface of the soil. Soils above shallow water tables become salinized in this way.

Sprinkler irrigation mobilizes more salts than flooding, but facilitates the use of more saline water than the other methods, except for drop irrigation where water is used in reasonable quantities, i.e. just above the amount of evapotranspiration. In California areas of irrigated land are underlain by salt-bearing marine shales. Irrigation here leads to a rise of the water table and the salts go into solution.

Heathcote (1980) described a rather paradoxical case of salinization from the far northern plains of the USA where agriculture has been replaced by grazing. Because of the lower evapotranspiration of the new pastures the water levels are rising and this leads to salinization.

The rehabilitation of salinized soils is difficult and costly, and prevention should rather be attempted. For all cases there are adequate solutions but the problems are not of a technical but an economic, political and social nature. Salinized soils may be rehabilitated by improved drainage and the sodium excess adsorbed should be replaced by calcium or magnesium. The principle is to inhibit precipitation of salts by limiting the amount of irrigation water which infiltrates and is not consumed by the vegetation. This may be done by avoiding waterlogging and by keeping the water table at a level where the capillary rise does not reach the root system. Methods for the rehabilitation of salinized soils have been described by the US Salinity Laboratory Staff (1954), Rhoades (1982) and Rhoades and Loveday (1990). It is not a simple problem, as leaching of soils may lead to subsidence (*suffusion* in French) and compaction, with loess areas being most vulnerable in this respect.

4.4.2
Irrigation and Decadence: Four Case Studies

After a phase of enthusiasm at many places irrigated agriculture has come under reconsideration without, however, being questioned altogether. In addition to the effects of reservoir dams previously described in several examples, there are disturbing effects of irrigation itself. Experience has shown that far from representing a miracle solution, large-scale irrigation also entails risks:

- It is responsible for temporary anaerobic conditions in a soil. The alternation between flooding and desiccation leads to disturbances in the life of the aerobic bacteria upon which pedogenesis depends. The slow-down in the formation of nitrates and their degradation by denitrifying bacteria require a supplementary

supply of these salts (calcium nitrates). In an oxygen-deficient environment the absorption of phosphorus stops, to start again after the water has been resorbed and aerobic conditions reappear. However, this occurs only when the water saturation of the soil has not lasted for a prolonged period.
- According to Durand (1988), the content in organic matter grows with the salinity of the irrigation water, but it is not humified and the nitrate content drops.
- In addition to the disturbances of the chemical situation in the soil, there are also effects on its physical state: the scalding of the soil or the formation of crusting in the case of sprinkler irrigation have to be mentioned first. On becoming moist, the soil clays form an impermeable layer. The glazing effect increases with the salinity of the water and its sodium content. Water proportionately higher in calcium than in sodium can, however, remedy the problem. The formation of desiccation cracks after the disappearance of the layer of irrigation water represents another problem for plant roots.
- Irrigation is responsible for an oversupply to the aquifers and for waterlogging the soils, as it is difficult to supply only just the amount of water necessary for plant growth. The excess water percolates under the root system, raise the water table so that it comes close to the surface and waterlogs the soils. On coming to the surface the water evaporates leading to salinization of the soils. All these processes may be observed on the Amu Daria delta and in Khoresm south of Lake Aral.
- The soluble salts are mobilized in the soils according to their degree of solubility, the most soluble ones also being the most mobile ones. In mixture they exhibit solubilities which differ from those of the individual compounds. Massive supply of irrigation water leads the dissolution of nutrients and their removal beyond the reach of the plants. Irrigation thus has to be supported by the supply of fertilizers adapted to the crop and soil type. Gypsum ($CaSO_4 \times 1.5H_2O$) is the most prominent salt in the soils of the drylands, its solubility depending on the concentration of NaCl and on the temperature.

This said, there are still questions: Is irrigation a solution for the drought problem? Are the developing countries in a position to shoulder the astronomic costs of developing major irrigation projects and to preserve the equilibrium in a combination of agriculture and livestock breeding? What can be said about the large dams which engulf the lands of the small farmers only to hand the edges of the water-courses to large companies or major landowners as happened along the Sao Francisco River in the Brazilian Nordeste, where the small farmers were driven into the caatinga?

To illustrate the negative aspects of irrigation we shall have a look at the problems of waterlogging caused by irrigation in Pakistan, at salinization in Mesopotamia, then on a wider scale at the degradation of the environment resulting from the irrigation in the Aral basin, and finally in the Tarim plain.

4.4.2.1
Irrigation and Waterlogging of Soils in Western Pakistan

During the first half of the 20th century the desert parts of the Indus plain were transformed into an irrigated area with the aid of low barrages built on the river and diversions leading water into a complex system of canals irrigating some 10 million ha of land. Waterlogging appeared and was difficult to control, with the excess water resulting from the rise of the water table. Under pronounced evaporation this leads to a salinization of the soils which then become unproductive.

At the start of the 1960s the production of food, which had been above the demand until then, showed a deficit balance because of the population growth. With about 20 million ha under irrigation in the early 1980s the annual losses of land amounted to 40 000 ha. Efforts to stop leakage through the sides of the canals and to start proper drainage have been undertaken since, but the storage of these drainage waters, polluted by herbicides, pesticides and other chemical products, still has to be solved.

4.4.2.2
The Salinization of Mesopotamia

On the plains of Euphrates and Tigris the dense network of old canals and ruins shows that agriculture once prospered in what is now a desert. Jacobsen and Adams (1958) demonstrated that soil salinization and alluvial sedimentation led to the disappearance of the ancient civilizations. They also pointed out that modern irrigation and other agrarian innovations in semi-arid tropical environments entail disconcerting effects not anticipated on the basis of investigations carried out in the temperate countries of Europe and America.

The following phases of salinization and degradation have been observed.

From 4400 to at Least 3700 B.P. Middle Bronze Age, dry period of the XVth–XVIth dynasties of Egypt with the Hyksos invading and dominating a Middle Kingdom weakened by the droughts. During these droughts the XVIIIth and XIXth dynasties authorized the biblical *Exodus* from Egypt. Weiss *et al.* (1993) rejected the concept of salinization having played a role in the downfall of the northern Mesopotamian cultures around 4000 B.P., ascribing this only to the drought. Competition for water and the first phase of salinization nevertheless constitute the most probable cause for the shifting of the centres of political power of the south towards the centres of the countries. Salinized fields are described in a document dating from 4100 B.P. Wheat which is less tolerant of salts was replaced by the more tolerant barley. Whereas both were grown in approximately equal amounts around 5000 B.P. (Helback, in Jacobsen and Adams 1958), the ratio between wheat and barley changed to 2 and 98% around 4100 B.P. in the Girzou area. Around 3700 B.P. wheat was abandoned in the south and at the same time the yield of cereals shrank from 2537 l/ha to 1460 l/ha in the area of Larsa and fell to 897 l/ha. This represents the warmest period of transition between the upper Bronze Age and the start of the Iron Age in the Middle East (Issar 1994). Excessive sedimentation accompanied this salinization: About 10 m of loamy-silty sediment were deposited over 5000 years, the great Sumerian urban centres declined, turned into mere villages and then fell into ruins.

A Less Severe Phase Touched the Centre Between 3300–2900 B.P. Decadence was interrupted by the Achemenid era which lasted from 2539–2331 B.P. and possibly aided by the conquest of Mesopotamia by Alexander the Great. Irrigation in combination with a considerable population increase reached its maximum during the

Parthian period (2100–1700 B.P.) after the toppling of the empire founded by Alexander, a period that was cool and moist in the Middle East (Issar 1994).

During this second phase of major development, water and land were exploited to the maximum and the urban centres spread. A sophisticated network of canals with the 300-km-long *Nahrvan canal* in the south, closed off by lock-gates, was constructed according to a new irrigation concept. This concept reshaped the topography but required technical know-how and an organized labour force.

Later, from 1200 B.P. to the Present. The Nahrvan region east of Baghdad declined because of salinization and excessive sedimentation. This is the second Islamic period in the Middle East, during which the peak of heat and drought lasted for 150 years (Issar 1994).

The Mesopotamian hydraulic system bore in itself the roots of decadence. Because of their low gradients, the canals easily became clogged, requiring much labour and a strong authority to organize it. From the start of the social and political troubles local authorities began to replace the strong central power. Maintenance of the system broke down during the Sassanid period and the region was abandoned. During the Islamic period a resurgence occurred but the decline began again in the 11th–12th centuries and lasted into our own times.

Whereas the phases of decline proceed from political and social problems, as has been demonstrated for the dry areas of China by Zhu Zhenda and Liu Shu (1983), sedimentation of 1 m in 500 years, which reduced the amount of water available for irrigation, may possibly be ascribed to an aridification or to droughts. During the time of Abassid Caliph around the 11th–12th centuries sedimentation was the main cause of decadence, having both natural roots like the drought and social ones stemming from a lower degree of organization and from the reduced availability of labour. The Mongol invasion of a century later was possibly unjustly accused as the reason for this decadence.

4.4.2.3
Lake Aral and Its Multi-Faceted Decadence

The desiccation of Lake Aral in the central south of the drylands of the former USSR within one generation represents a major ecological drama (Fig. 4.3). Five states – Kazakhstan, Kyrgyzstan, Uzbekistan with its Karakalpakstan region, Tadjikistan and Turkmenistan – share the 1 815 000 km^2 of the basin and its inhabitants. Two features are particularly notable, viz. the wide extend of the environment and the high population.

Since the waters of the Syr Daria and Amu Daria, the main tributaries of the lake, started to become used for irrigated cultivation of cotton on an industrial scale, the lake has lost 60% of its volume and 45% of its surface area, shrinking from 66 000 km^2 in 1960 to 40 780 km^2 in 1987 and 34 800 km^2 in 1990. Its shoreline receded by 65 km in areas of the east and the south, its water level fell by 12–13 m or about 0.8–1.1 m/year, resulting in some 2 million ha of the lake floor falling dry.

However, the drying out of this body of water is but one aspect of the multi-faceted degradation, the result of an ignorance of the natural data and of an inability to take into account the regional scale, its topography and its hydrology. It represents the collapse of a politico-economic mechanism of forced utilization of a dryland area.

4.4 · Soil Degradation, Irrigation, Salinization and Decadence

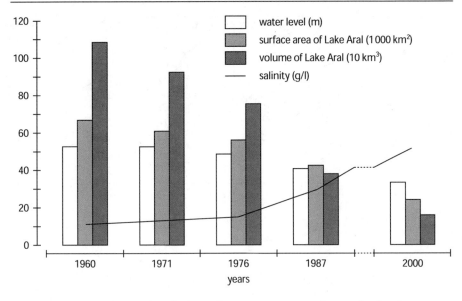

Fig. 4.3. Diminution of the surface of Lake Aral between 1960–1987 and extrapolated to the year 2000 (after Raskin et al. 1990)

The Natural Roots of the Problem

Endorheism and allochthony are the natural roots of the crisis, as the water resources of the lake depend essentially on its two tributaries the Amu Daria and the Syr Daria which supply a mean of 56 km^3/year from the melting of snow and glaciers and an additional 5 km^3 from precipitation. The maximum discharge was 91.5 km^3 in 1969 and the minimum 41.9 km^3 in 1986.

Continentality and aridity are two other factors that have been neglected. Precipitation amounts to 90–120 mm/year in the subtropical central and southern plains, in the range of 100 mm/year in the vicinity of the lake itself and below 30 mm/year in the hunger Steppe to the southeast of Tashkent. None of these conditions is sufficient to assure survival and to permit any but irrigated agriculture (Fig. 4.4).

The flat nature of the area with mean gradients of 0.2–0.4 m/km, although beneficial for irrigation, has turned out to be a handicap. Because of endorheism wastewaters have only three ways to go: into the water-courses, the depressions or the groundwater, all three of which they pollute. Neither treatment, nor evacuation or neutralization of the pollutants have ever been considered.

Incoherence in Management

The Tsar had started to develop the cultivation of cotton in the 1870s, and at the start of the 20th century Turkmenistan supplied some 40% of the cotton cloth used by the factories along the Volga. At the beginning of the 1960s, as we have seen already, the Supreme Soviet started the *Conquest of the Virgin Lands*, (deserts), with the aid of unrestrained expansion of irrigation, to produce rice and especially cotton in a

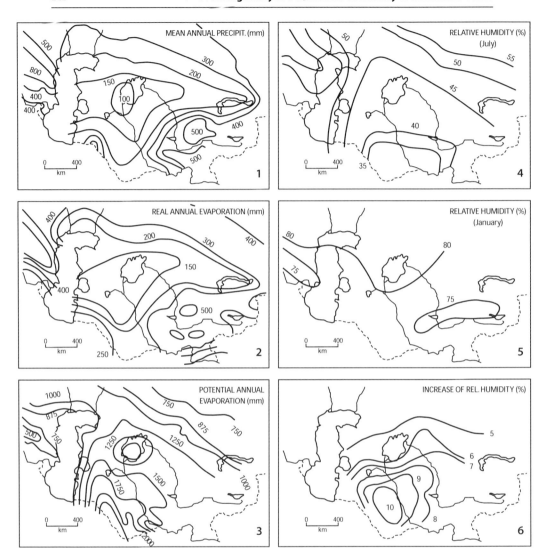

Fig. 4.4. Aridity parameters of the Aral basin (after Létolle and Mainguet 1993). *1* Mean annual precipitation (mm/year); *2* real evaporation (mm/year); *3* potential evaporation (mm/year); *4–5* relative air humidity in summer and winter; *6* increase in relative humidity under the vegetation cover from north to south of the Aral basin

quasi-monoculture. The largest hydraulic engineering projects in the world were undertaken and among these the 1300 km long Karakum Canal. Kotlyakov (1990) estimated that by 1965 some 4.5 million ha were already irrigated, consuming water at a rate of 50–55 km³/year. During the following 25 years some 2.6 million ha more were irrigated, using more than 50 km³/year. In Uzbekistan the overall length of the canals reached 180 000 km or 4.5 times the length of the Equator. Only some 12% of this canal length has been sealed, explaining the widespread infiltration, the rise of the water table and the waterlogging which have affected soils and vegetation (Zaletaev and Novikova 1990). Since the early 1960s the surface area of the irrigated lands in Uzbekistan and Tadjikistan has increased by 50%, in Kazakhstan by 70% and in Turkmenistan by 140%.

The Disastrous Consequences

Climatic changes: endorheic interior water bodies are valuable climatic regulators. The indifference of the central powers of the former USSR with respect to the disappearance of Lake Aral results from ignorance. Was it known that its loss would entail reduced summer evaporation, the loss of the water surface as a heat reservoir in winter, and a bulge of warm air blocking off the intrusion of the north winds? The reduction of the surface area of Lake Aral led to a continentalization of the climate with an increase in the seasonal temperature amplitudes by 4 °C, much later night frosts in spring and much earlier ones in autumn. The season without night frosts on the Amu Daria delta was shortened to 170 days and is now well below the minimum of 200 days required for growing cotton.

The salinity of the lake grew from 10–24‰: in 1942 to 28–30‰, mostly because of the growing salt content of its tributaries, as e.g. in the lower reaches of the Syr Daria, where it increased from 0.8 g/l in 1960 to 20.8 g/l in 1985. The total salt content of the water remained relatively stable with $10-11 \times 10^9$ t since 1960 save for a drop in carbonates and gypsum due to sedimentation. In 1960 the lake was rich in calcium sulphate and carbonate ions, but it has become salinized in the last 20 years, predominantly with sodium sulphate and sodium chloride. In the north and especially in the east, where the retreat of the waters has lead to the uncovering of more than 3 million ha, some 3–6 t of mineral and saline compounds, of which 250–800 kg are soluble, have been deposited per hectare. In the water of the canal built in the palaeochannel of Akcha Daria, we measured a salinity of 17 g/l in September 1997.

The salinization of the soils by rain, the mineral content of which has trebled by the 1980s in comparison to the 1960s to 1970s, has been supplemented by aeolian supply and by irrigation, which introduced salts at a rate of 10–20 t/ha annually to the surface of the fields and 15–25 t/ha into the soils. This accumulation results from the evaporation of excess water which causes waterlogging, but as proper drainage would permit the elimination of only 30–40 t/ha annually, the danger of waterlogging represents a limiting factor.

According to Khakimov (1989), the percentage of irrigated fields degraded by moderate to severe salinization reaches 80% in Turkmenistan, 60–70% in Kazakhstan, 60% in Uzbekistan, 40% in Kyrgyzstan, and 35% in Tadjikistan. The total harvests in Turkmenistan dropped by 1986 to the level of 1980. In Kazakhstan and Kyrgyzstan the harvests of raw cotton have not grown since 1980. In Uzbekistan, where the harvest had reached 3.9 million t in 1973, it had grown to 5.3 million t by 1975, but fell to 5.1 million t by 1985.

The part of the lake floor that became dry made salts available to the winds, which carry them mainly from north to south over tens or hundreds of kilometres (according to the different sources of information), polluting the air, surface water, soil and through the latter also the groundwater. Estimates of the quantity of salts and dust carried by the winds range from $30-150 \times 10^6$ t/year (Photo 4.8).

Photo 4.8. The desiccation of Lake Aral. The overexploitation of the waters of the two main tributaries of Lake Aral, the Syr Daria and the Amu Daria, for the development of irrigated cotton since the early 1960s, explains the retreat of the shore line shown. At Mouinak on the southern edge of the Amu Daria delta, a hotel once built on the edge of the lake now lies isolated in a sandy expanse after the retreat of the shore line (Photo M. Mainguet)

To make this expensive irrigated agriculture economically viable, increasing amounts of pesticides, herbicides and fertilizers had to be applied. The fertilizers are required by the increasingly impoverished soils and the pesticides because in this virtual monoculture resistant strains of parasites have developed, and without these compounds the intended range of products could not be achieved. Their impact on human health has also been neglected. The rates of morbidity and infant mortality are the highest in the world.

- In conclusion, it can be said that the crisis of Lake Aral bears witness of a number of errors:
- Bureaucratic dirigism from far away
- A poor choice for placing the production forces in a dry ecosystem
- A development strategy by mega-irrigation which is faulty because of the investment and maintenance costs, the difficulty in managing the wastewater, the damage to the environment and especially because of the salinization inherent in this method
- The uncontrolled massive application of fertilizers, pesticides and herbicides for the protection of the plants grown, and of defoliants which strongly pollute the drainage and groundwater

- The orientation of the economy towards the production of raw materials in combination with poor facilities for storage, communication and commercialization
- The neglect of any control of population growth as the birth rate was 0.95% for the entire former USSR in 1990, but 2.75% around the Aral Basin and 3.2%, the highest value, in Tadjikistan. This was aggravated by massive forced immigration between 1928–1940, leading to underemployment and unemployment, in particular among the young. In Kyrgyzstan more than 50% of the population had migrated into the area. In Tadjikistan "where the fertility is the highest in the former USSR, 52% of the population are below 18 years in age" and "unemployment exceeds 600 000 persons, or 24% of the active population" (Sablier 1992).

Recalling these mistakes teaches us lessons for the entire drylands which, as we remember, represent 37% of the continents of our planet, and for the decision makers in the field of the so-called development programmes. The preliminary study of expectable impacts should be entrusted to competent persons who are not dependent on the political powers. The decision should consider the existing situation and the natural and human resources; all this should be investigated more rigorously, as such projects are of wide extent and with a monitoring without concessions.

This situation is reminiscent of Lake Chad, also a basin extending into several states. Its surface has shrunk by 90% within 30 years, as a result of droughts, withdrawal of water and the unco-ordinated construction of dams on all tributaries. This has endangered the survival of 9 million people by the destruction of traditional agricultural systems, the reduction of pastures, the disappearance of wildlife, and the decline of fishing, from 140 000 t in 1960 to 70 000 t in 1990.

4.4.2.4
"The Water of Life" or the Tarim River in the Xinjiang Desert

China is divided by the 100 mm/year isohyet into hyperarid, arid, and semi-arid drylands, with grass steppe in the west and a humid ecosystem in the east. The distribution of precipitation and runoff in China contrasts with the population densities: 81% of the water resources, but only 19% of the cultivated lands, are located south of the Yangtze-Kiang. The Yellow River, the Huaihé, Haihé and the Liaohé basins encompass 9% of the water resources, but 42% of the agricultural land. The seasonal concentration of the precipitation into 4 months per year and the annual contrasts between floods and low waters are additional disadvantages.

The 58 500 km^2 of glaciers contain 5.1×10^{18} m^3 of water and furnish 56×10^9 m^3 of meltwaters (Li Lierong, pers. comm. 1993). The water resources of China are substantial, and with 2.8×10^{18} m^3 they place this country in fifth position behind Brazil, Canada, the United States and Indonesia. However, because of its immense population compared to the rest of the world, it has only 25% of the worldwide average of water per inhabitant and only 75% of the average quantity of arable land per inhabitant (Li Lierong 1988).

In the dry parts of China development is based on the exploitation of allochthonous water-courses (the Tarim being one of glacial origin), on groundwater and consequently on irrigation. Using 80.7% of the total water consumption, irrigation supports 72% of the annual grain harvests of 215×10^6 t. With 730×10^6 mu (15 mu = 1 ha)

China possesses the largest area of irrigated lands on our planet. The water demand will represent one of the major problems faced by China in the 21st century, and it will be felt most acutely in its drylands.

The Desert Framework of the Tarim Plain

In the arid north-west of China, the Tarim is the main river of the autonomous Uigur republic of Xinjiang, referred to as "the water of life" by the Chinese. In the last three decades severe degradation has become apparent as a result of poor exploitation: a drop in water quality, deterioration of the hydrology, vegetation and soils of the plain and a piecemeal implementation of changes, without understanding the desert environment as a whole. Cheng Qichou (1992) has retraced the chain of processes leading to this result.

Topographic data: the Tarim River, before it brings water to the largest sand desert of central Asia, the Taklamakan, receives water both from the glaciers and eternal snow on the north slopes of the over 7800-m-high Karakorum, via the Hotan and Yarkant rivers, and from the Tienshan via the Aksu. It extends over 1280 km between the confluence of the Hotan, the Yarkant and the Aksu and Lake Taitmar (Fig. 4.5), and is classed as an endorheic water-course. Including the Yarkant it is 2200 km long, and its basin covers some 198 000 km^2.

The climate of this arid, continental and windy desert is rather unfavourable:

- Mean precipitation amounts to 55 mm/year on the upper reaches of the watercourses, dropping to less than 2 mm/year on the lower reaches.
- The potential evaporation varies from 2100-3000 mm/year.
- Relative humidity does not exceed 40–50%.
- Insolation is high with 2811–3133 h of sunshine annually.
- The mean annual temperature is 10 °C.
- There are 185–214 days of frost annually.

Aridity, poor soils of mostly sandy texture and acquired salinity explain the low species variety of the vegetation cover of the Tarim depression, with the exception of

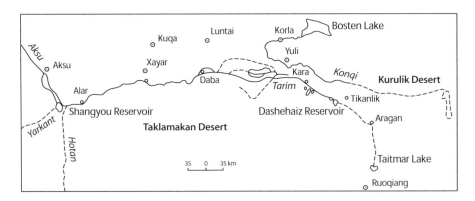

Fig. 4.5. Geographical reference points in the Tarim valley in China (after Cheng Qichou 1992)

the valleys, where gallery forests contrast with the gravel plains (*Gobi*) of the foothills of the Tienshan and the Karakorum and with the vast active dune fields.

Degradation of the Environment

Hydrological Degradation. Three large dams were built under a development programme started in 1958: at Xiaohaiz, the largest of the Chinese reservoir in the plains; at Shangyou below the Yarkant River; and at Dashehaiz. These structures transformed the basic hydraulic conditions of the Tarim.

The salt content of the Tarim water grew continuously with irrigation, with the discharge of drainage water from the irrigated soils, and with the strong evaporation and concentration of the water in the large but shallow reservoirs. In addition to an increase in the salt content of the water, symptoms of soil salinization were observed in the Aksu plain, where there was also a considerable increase in the salinity of the groundwater. Waterlogging of the soils is ascribed to a rise of the water table from a depth of 4–8 m to only 1.5–3 m. In the low-lying, poorly drained lands the waters of the aquifers, which had salt contents of 10–30 g/l prior to the start of irrigation, now have 10–40 g/l.

Degradation of the Vegetation. Prior to modern management, dense tugai forests of the poplars, *Populus diversifolia* Schrenk and *P. pruinosa*, grew on the lower terrace due to a higher runoff (Fig. 4.6), whereas only *Populus diversifolia* Schrenk was present on the middle terrace. Later, the total forest area fell from 400 000 ha in 1958 to 175 000 ha in 1978, or by 57%. The quantity of timber was estimated as 3.8×10^6 m^3 in 1958 but only 2.4×10^6 m^3 20 years later in 1978, this drop by 37.6% resulting directly from the clearing of the forests for agricultural purposes and from the withdrawal of the waters for irrigation (Cheng Quichou 1972).

Degradation and Wind Activity. During the Han and Tang dynasties (608–907 A.D.), the Tarim plain was the main Chinese axis of the *Silk Road*. The Korla–Ruoquing road and the Qinhai–Xinjiang railway line were built here during the 20th century. At present, the route is highly damaged by the invasion of sand, its maintenance being rather costly, as e.g. in 1992 continuous repairs were required at 168 points. This damage results from the degradation of the Tarim plain.

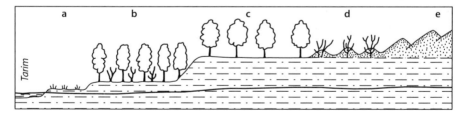

Fig. 4.6. Schematic geographical section of the Tarim terraces in the Goskhose section (after Petrov 1962). *a* Floodable terrace with mesophilic vegetation; *b* terrace with young *tograk* (*Populus diversifolia* and *P. pruinosa*); *c* terrace with old tograk of *P. diversifolia*; *d* hummocky sand with *Tamarix ramosissima*; *e* mobile sands and barchans of the Taklamakan

According to Chen Qijou (1972), Chinese scientists estimated that some 840 000 ha or 59.8% of the land on the middle reaches of the Tarim was degraded, with the rate at 66.1% for the lower reaches being still higher. Between Kara and Tikanlik 7000 ha of private irrigated land had to be abandoned; moreover, dunes are forming again.

The 10–20 km wide sector between Kara and Taitmar on the lower Tarim with its forest of poplars, reeds and willows was a most valuable corridor as it represented not only a green barrier against the Taklamakan and Kurulik deserts, but also an economic and strategic axis connecting Xinjiang to the rest of China. With the reduced water discharge from the upper reaches and the remaining water being held back in the Dashehaiz reservoir and by local demand, the groundwater levels fell and the water quality decreased. As a consequence, the vegetation of this corridor deteriorated and the countryside became sandy and covered by dunes, obliterating some 200 km of the green barrier near Aragan.

The degradation of the Tarim plain is just another example of the ill effects of an erroneous development in a dryland with large-scale irrigation as the only proposed method, without any rational utilization of the limited water resources and without co-ordination of activities on the upper middle and lower reaches of the watercourses. The difficulty of maintaining the discharge of the water-courses is aggravated by political boundaries, because the sources of the Aksu, which supply 70% of the waters of the Tarim, and those of the rivers Kunmalic and Toxhu, are actually situated in the former USSR.

A More Reasonable Future?
The World Bank and the FAO decided to supply US $120 million for an irrigation project in the Yarkant and Ogan valleys under one condition, viz. the protection of the ecosystem of the Tarim basin. An administrative office for the Tarim river was established in 1991, a group of scientists was selected to study the course of the river between Alar and Lake Taitmar and a number of sites for studies of the ecology were set up; these should have become operational by 1995.

In 1985 some 10 million mu or 660 000 ha were under irrigation, equivalent to 3.6 mu per inhabitant. Cotton and fruits like pears, peaches, apricots, figs, cherries and pomegranates as well as raisins and melons and even silk are the chief products of these regions. Additionally, medicinal herbs and roots are produced. In this oasis region salinization of the soils appears to be under control and in order to control erosion by wind the government had a number of windbreaks built. A water conservation programme was approved in 1991 on the basis of the investigations of the Tarim ecosystem. The total investment amounted to 1.1 billion yuan of which 58% is in the form of a loan from the World Bank.

4.5
Desertification – An Expression of Decadence?

In 1992 the UNEP stressed the impact of land degradation by the following evaluation:

> About 75% of the pastures, 47% of the cultivated and 30% of the irrigated land in drylands (arid, semi-arid, subhumid dry) of our planet are affected, or a total of more than 3.6×10^9 ha or 25% of the emerged land, supporting about 300 million people, equivalent to one-sixth of the world population, the loss of revenue amounting to US $42.3 billion annually.

4.5 · Desertification – An Expression of Decadence?

Any attempt to define the term desertification will end in ambiguity, and in 1997 the sense in which this term is used was still fluctuating. When speaking of desertification, do we consider it a true, i.e. irreversible desertification or only as a degradation of the environment which has reached a most severe extent but is not irreversible? The first question to be asked is of an epistemological nature and thereafter we shall try to describe what has really happened, and to establish a relationship between the levels of development, famine and drought crises in drylands.

4.5.1
Difficulties with the Sense of the Word Desertification

The word is derived from Latin etymology:

- *fication*, the act of doing, derived from *fieri*, the passive form of the active word *facere* (to have done, to produce, to do, to arrive)
- *Desert*, with a twofold Latin origin:
  - The adjective *desertus*: uninhabited
  - The noun *desertum*: a deserted area

According to Budge (in El-Baz 1988), the word *desert* is derived from a hieroglyphic sign pronounced *tesert* which signifies an abandoned place or one that has been left aside. From the verb *deserere* is derived *desertum*, a barren or wild place, and *desertus*, abandoned, which in English leads to desertified or deserted, words used erroneously without distinction as for certain industrial or rural areas which are not yet desertified but only deserted.

In Gilbert's *Dictionnaire des mots nouveaux* (Hachette-Tchou, p. 158) the term desertification has the meaning of "more or less entire disappearance of any human activity from a region gradually abandoned by its inhabitants." A number of other terms were created concurrently:

- *Sahelization or steppization*, transformation of a Sudanian region into a Sahelian one, i.e. the replacement of a closed vegetal formation by an open one.
- *Sudanization*, transformation of a Guinean region into a Sudanian one with the virtually complete disappearance of the tree layer.
- *Aridification, aridization*, a climatic evolution towards an increasingly drier and eventually even arid climate.
- *Desertization*, proposed by Le Houérou for emphasizing the human contribution to degradation; however, this term has not found acceptance in the scientific community.

In his article entitled *Les forêts du Sahara* (The Saharan forests), Lavauden was in 1927 the first to use the term *desertification* in a scientific sense to describe the extreme impoverishment of the plant cover in this desert at the beginning of the 20th century and to claim that man bears the responsibility for it. "The fact that the desertification, if I dare to say so, is purely artificial applies to the whole zone about which we have been talking. It is solely man-made. Incidentally, it is relatively recent and still could be combated and held in check by very simple human means."

Aubréville (1949) observed in the north of the dry subhumid tropics of the former Oubangui Chari (now the Central African Republic): "These are real deserts which are coming into being under our very eyes in countries with annual rainfall reaching 700 to at least 1500 mm." As made by this forest expert, this statement was neither premonition nor intuition, but the fruit of accurate observation of the terrain in a dry forest at the northern border of the grass savannah.

The interest in the tropical world led the UNESCO to initiate its research program on the arid zones in the 1950s; 10 years later, 30 volumes of high technical and scientific quality had been produced, supported by 200 research institutes for arid zones which had been established, especially for this purpose, in all arid areas of the Earth.

With the appearance of L. Carson's (1962) book, *The Silent Spring,* the problems of pollution and environmental degradation went beyond the domain of science alone and reached a wide public, showing the connections between agrochemical pollution and degradation of nature under the effect of human activities.

Ten years later, in 1972, the Club of Rome published its reflections of the *Limits of Growth* and demonstrated the destructive effects of agriculture on the environment when economic efficiency is given priority.

Scientific examination of the interrelations between human activities and environmental degradation, on the one hand, and the explosion of a severe drought between 1968 and 1972 afflicting the seven countries of the Sahel–Sudan zone from the Atlantic coast to Somaliland and famine, on the other, caused the United Nations to organize the "United Nations Conference on Human Environment" in Stockholm in 1972. The governments and the International Community formed a Committee for Drought Control in the Sahel: the CILSS and the Sahel Club were created by the Donators' Community, the OCDE, to mobilize donors and to coordinate aid to drought-stricken countries. The United Nations Saharo-Sahelian Office (UNSO) was placed under the supervision of the UN which was made responsible for overall coordination.

In 1975, when the Sahel was again afflicted by a recurrent drought, the General Assembly of the United Nations (UNGA) planned a conference in Nairobi (Kenya) in August and September of 1977, the UN Conference on Desertification (UNCOD), which suggested a Program of Activities to Combat Desertification (PACD) including 28 recommendations detailing the activities to be undertaken, and entrusted the UNEP (United Nations Environment Program), a new organization founded in Nairobi, to realize and carry through the PACD task. This conference officially used the term *desertification* in its resolution No. 7: "Desertification is the diminution or destruction of the biological potential of the land, and can lead ultimately to desert-like conditions. It is an aspect of the widespread deterioration of ecosystems, and has diminished or destroyed the biological potential, i.e. plant and animal production, for multiple use purposes at a time when increased productivity is needed to support growing populations in quest of development." (Fig. 4.7)

Despite this official definition scientists working in different field have proposed, depending on their own observations, more than 130 definitions (according to Glantz) which may be subdivided into five major categories:

1. Appearance of desert landscapes outside the deserts themselves, along their edges; this definition led to the unhappy concept of the advance of the deserts.

4.5 · Desertification – An Expression of Decadence?

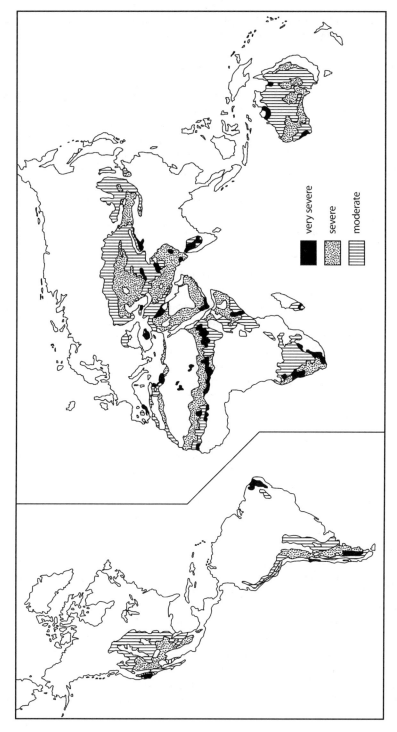

Fig. 4.7. Map of areas prone to desertification (after United Nations Conference on Desertification, Nairobi, 1977)

2. An explosion, in the semi-arid to subhumid dry ecosystems, of the physical mechanisms of land degradation (erosion by wind and water) which are characteristic of arid regions.
3. Decline of the biological productivity of the land with deterioration of the vegetation cover and the soils.
4. Degradation of the socio-economic system because of degradation of the environment. However, would it not be logical to reverse these two concepts, environmental degradation being the consequence and not the cause of the deterioration of the socio-economic systems? "Nature" is not really responsible for the process of desertification. It cannot withstand human occupation, which does not take into account this inability and unleashes the process of degradation, once sustainable levels are exceeded. Within the concept of an ecological equilibrium between man and his environment it is man himself who is responsible for the desertification.
5. When the ultimate stage of environmental degradation has reached an irreversible level, the concept of irreversibility has to be discussed.

After 1977 and the UNCOD, several national projects to combat desertification were proposed. Cardy (1994) noted that the major activities since 1977 have been predominantly of theoretical and methodological nature, aiming at the definition and nature of desertification, with one major difficulty, the communication of scientific and technical information.

Parallel to this, the WMO (World Meteorological Organisation) focused after 1977 on interdependencies between climatic changes and soil degradation, while in 1985 the FAO (Food and Agricultural Organisation) adopted the "Behaviour Project in Tropical Forests" a theme which became the central topic at the "Sommet de l'Arbre" in Paris in 1989.

In February 1990, the UNEP (United Nations Environmental Program), aware of the danger of inflationary definitions of the term *desertification*, formed an ad hoc committee for the "Global Evaluation of the Desertification: State and Methods of Investigation" (UNEP 1991). It suggested the following definition: "Desertification is land degradation in arid, semi-arid and dry subhumid areas resulting from adverse human impact." Land degradation includes decline in harvest yields, thinned plant covers, exacerbation of physical mechanisms on the soil surface, qualitative and quantitative decrease in the water resources, soil degradation, and air pollution.

This definition was only partially accepted at the Soil Conference in Rio de Janeiro in 1992, which suggested the following revision of the definition: "Desertification is land degradation in arid, semi-arid and dry subhumid areas resulting from various factors including climatic variations and human activities."

Those two latter definitions merit discussion. In admitting that land degradation represents a continuous problem and causes a reduction in the potential of resources, they overlook two essential concepts: the heterogeneity of desertification both in its causes and mechanisms and in its consequences. Man's responsibility is considerably diluted in the definition given by the Rio conference, which fails to differentiate between degrees in severity of desertification, in particular the last irreversible state, which must be considered as the basis of the term *desertification*. Irreversibility is defined on the human scale of one generation, if the scale generation causing destruction is not capable of restoring the original state for economic and/or technical and/or political reasons.

If one accepts desertification as being equivalent to irreversible landscape degradation, then it represents today's most dramatic environmental problem. This conceptual idea of irreversibility as the ultimate state of the sequence of processes leading to a sterile environment, however, is nearly inexistent in the present technical and economic context. According to Dregne (1983), only 0.2% of our planet is affected, an example of a statement which needs to be realized!

The consensus admitted by the UN definition holds that the term environmental *desertification* must be restricted to irreversible degradation within the time span of only one generation and can refer exclusively to dry ecological systems. The best example is that of Gibuès in southern Piaui (Brazil): situated in the middle of the humid tropics with a mean annual precipitation of 1200 mm, this region offers the devastating picture of a completely barren landscape where the forest has vanished, to be replaced by badlands due to intense hydrologic erosion following a brutal search for diamonds. This sort of degradation, although irreversible by our definition, cannot be called desertification, since tropical precipitation rates permit completely different and less complicated rehabilitation procedures than can be employed in dry ecosystems.

A revival of interest in desertification led to a meeting summoned by the UNESCO in Paris in June 1994, which concluded with the signing of a convention to combat desertification by the INCD (Intergovernmental Negotiation Committee for a Convention to Combat Desertification). In the course of the discussions, the following main statements became apparent:

- only a very small percentage of the global financing destined to combat desertification has actually attained its goal, i.e. 1% of the credits granted was put to good use;
- the developing countries consider that too much money has been used to fund foreign experts and that, in future, local investigators will be sufficient;
- continents such as Latin America, which are not to any extent involved in desertification, have begun to consider that they should play a more important role;
- the Western countries rejected the creation of new structures, demanding more efficiency from those already established. This rejection also applied to any financial increase.

This chronological analysis clearly demonstrates that the term *creation of an uninhabitable region* was made more precise by field observations, which have identified regions becoming uninhabitable as a result of human and animal overpopulation that exceeds the resource capacities and thus causes an environmental degradation, which in turn prevents humans from living in the affected region. The idea of environmental desertification is preferred to that of abandonment of a terrain, as the worst consequences of desertification are famine and ecological refugees. Altogether the fact needs to be emphasized that mankind transforms a habitable area into one where survival is no longer possible. Hence, mankind is both the creator and the main victim of this phenomenon.

If, as it has been shown, two fundamental reservations are included in the term *desertification* as a global, environmental and socio-economic concept, namely temporal limitation through irreversibility on the scale of one generation and spatial limitation through the term's exclusive application to dry ecosystems, then according to Dregne (1984), the key criterion, in practice, should be the increasing difficulty in seed germination. We suggest including in the definition the difficulties inherent in degraded soil, such as the growing of young seedlings and the continually declining crop yield potential, which all originate in textural and structural changes in the soil consistency: reduction in aggregativity, in infiltration, in water retention and in loss of erosion resistance to wind and water by a lowered threshold of gully formation. According to Le Houérou, the key criterion for desertification should be the disappearance of perennial varieties of plants, a state exclusively caused by mankind.

When we admit that the concept of desertification is synonymous with land degradation it corresponds to the most dramatic environmental problem at present. However, if it contains the idea of irreversibility in the present technical and economic context because it terminates a series of processes leading to a definitely sterile environment, it is well-nigh non-existent as according to Dregne (1983) only 0.2% of our planet may be classed as such.

We have proposed the following definition (Mainguet 1991) after three decades of observations in the drylands of Africa and Asia:

> Desertification, revealed by drought, results from human activities when the carrying capacity of the land is surpassed. It proceeds by natural mechanisms which are exacerbated or induced by man, showing itself in the form of a deterioration of the vegetation and the soils and on a human timescale leads to a diminution or irreversible destruction of the biological potential of the lands or of their capacity to support the population living on them.

Human activities lead in a few decades to situations which on a geological timescale would take several millennia.

This definition stresses the human causes and considers climatic parameters only as factors modifying the process. Has not UNEP (1990) proposed, in order to limit the imprecision, to substitute the terms land degradation or environmental degradation for the term desertification? Degradation of the environment because of human activities takes place in all ecozones, but it is largely in the driest ones that man contributes most to giving them the appearance of a desert. Because of this it is justified to restrict the term desertification to the arid, semi-arid and subhumid dry zones.

The term desertification frequently has been given the meaning of an irreversible extension of the desert. This introduces a certain ambiguity as the meaning of the word irreversible is by no means clear. In our definition irreversible is limited on a human timescale to one generation or 25 years, implying that one human generation is not able to restore the destroyed environment on which its survival depends. Beyond these 25 years the degradation is considered as irreversible by this generation (Photo 4.9).

However, this irreversibility is also relative in comparison to the means available to the states in which desertification takes place. In the USA and Australia it is of limited extent but it will lead to famine in the poor countries of Africa. Irreversibility may also have other reasons as in ancient Mesopotamia between the Euphrates and the Tigris, where the lands irrigated for centuries are encrusted with salts and in front of our very eyes shine like snow, because nothing is done about draining the soils.

4.5 · Desertification – An Expression of Decadence?

Photo 4.9. Nomadism and desertification in Somalia. Red aggregated dunes from an earlier climatic period, which are otherwise fixed by a tree steppe with *Acacia tortilis* as shown in the background, were completely eradicated here by 1982 by overgrazing and cutting of trees. This grass hut of the last nomads along the road between Raas Daay and Raas Filfile in Somalia represents the miserable vestiges of a possibility for human occupation (Photo M. Mainguet)

Because of its common root with the word desert, desertification entails the risk of creating the impression that it generates ipso facto the desert, and that because of this one talks erroneously about the advance of the deserts, as we have already seen in the analysis of droughts on a geological timescale in the first part of the book. Two authors have advanced convincing arguments based on a human time scale:

1. Mabbut (1985) rejected the idea of a desertification which one might call centrifugal, i.e. extending outwards from the deserts, and proposed an extension which could be qualified as centripetal, starting at a distance from a real desert by a mechanism that is comparable to regressive erosion observed in water-courses. Degradation affects the peripheral areas in particular, their vegetation cover and then their soils, until the physical mechanisms have become more efficient so that they evolve towards a desertification and create *desert-like* areas, which by coalescing and extending end up joining the neighbouring deserts. In conclusion it can be said that it is not a desert that advances, but it is generated in pockets sometimes extending outwards from its edges by mechanisms characteristic of oases.
2. Mortimore (1982) excluded from the processes of ecological degradation leading to desertification: deforestation which he sees as a normal prelude to the agricultural exploitation of land and considers as reversible; salinization of irrigated land because of poor drainage; and erosion of the soils by water.

For Mortimore, the processes of degradation are: erosion of the soils by wind; formation or reactivation of dunes; disappearance or degradation of the vegetation; desiccation of the soil profile and the lowering of the water table.

The deterioration of the environment finds its expression in a degradation of the vegetation cover and, especially in the semi-arid regions, in the replacement of the perennial species by annuals, resulting in a seasonal opening of the vegetation cover favouring the degradation of the underlying soils. Instead of the advance of the deserts we should rather talk about the spreading of the degrading mechanisms characteristic of the drylands, consisting (in the semi-arid areas up to the 600–650 mm/year isohyets) of an exacerbation of the aeolian mechanisms and (below the 400 mm/year isohyet in the Sahelian to the Sudanian–Sahelian domain) of an increase in the influence of the water erosion.

After the peak of the 1984 drought in the semi-arid ecosystem of Kordofan in the Sudan, Olson (1989) estimated, on the basis of studies of Landsat images and field observations, that signs of desertification could be noted only during the paroxysmal phases of the drought, and he was not able to prove a real trend of environmental degradation in the area investigated. Ahlcrona (1988) confirmed that not one species of tree appears to have been eradicated entirely from the area and that over the last 80 years no definite shift of the limit of the drylands towards the south had taken place.

In China, scientists of the University of Lanzhou recognized that the problem of land degradation is at least 2000 years old and that it occurs essentially during times of political instability.

Four years after the drought of the early 1980s Toulmin (1988), writing in *Science*, described lusciously green areas from Mali. This serves to show that it is absolutely necessary to differentiate land degradation, which may be resolved in most cases, from what rightfully may be referred to as desertification *s. str*. To be treated, this requires very expensive and sophisticated techniques especially in the developing countries.

Livestock breeding has been considered as one of the most damaging activities for drylands and as one of the major causes of land degradation. Many aspects of even traditional breeding have been considered as inconsistent with and damaging for situations of aridity and drought. During recent years these points of view have come under serious reconsideration by Walker (1987) and Skarpe (1991), under the influence of field observations which have shown a resilience of the steppe and savannah ecosystems under pressure from climatic hazards, fires or overgrazing.

The notion of the carrying capacity is among the ones most hotly contested. Behnke and Scoones (1992) discussed this question and showed that establishing a direct connection between the carrying capacity and natural conditions is too simplistic. Instead they proposed a connection between the carrying capacity and the objectives of the breeders, who do not want maximum livestock densities. Even if this were their aim, they could not achieve it as the pressure of prolonged droughts on the vegetation cover is such that the livestock would die prior to degradation reaching its peak. This point of view is shared by Perkins (1990), Mace (1991) and Warren and Khogali (1992).

In Botswana the number of head of livestock has not stopped growing since the start of the 20th century, with regressive phases during droughts and epidemics (Arntzen and Veenendaal 1986; Thomas and Shaw 1991). In 1986 the country counted close to 3 million head of cattle and small livestock (Perkins and Thomas 1993). This growth trend has been supported since 1966, the year of independence, by the clear policy of the government to widely establish wells and to exploit the groundwater, aided by financial assistance from international bodies. The strategy is still adhered to today despite warnings of scientists such as Abel and Blaiki (1989), Perkins (1991), Campbell et al. (1991) and White (1992).

4.5.2
What Is Really Happening?

The concept of desertification implies a transformation of the affected land by degradation of the vegetation cover, by soil erosion, salinization, crusting, the reduction of the organic matter content of the soil and the degradation of its structure. Even if this phenomenon is of great spatial extent, the controlling human causes are of a local nature.

What do we know today about desertification?

- According to UNEP the total of the desertified land amounted to 950 million ha at the end of the 1980s and 4500 million ha were threatened by desertification.
- For the drylands alone, also according to UNEP, 61% of the 3257 million ha of productive lands were desertified, with 2556 million ha pastures (62% of this type), 570 million ha rain-fed crops (60%) and 131 million ha irrigated land (30%). The worldwide estimate comes to 35% of all land of our planet or 45 million km^2. Ahmad and Kassas (1987) presented different figures: throughout the world 60 000 km^2 or 6 million ha are lost every year, or 3.2 million ha of pastures, 2.5 million ha of land cultivated by rain-fed agriculture and 125 000 ha of irrigated lands. To this the authors added lands the productivity of which is reduced by soil erosion, salinization as well as all other forms of degradation and by a drop in fertility. Along the southern fringe of the Sahara 650 000 km^2 or 65 million ha have been turned into desert during the last 50 years.
- Compilations again by UNEP have shown that annually in Africa alone 3.7 million ha of forests and woodlands disappear and that 20–50 million t of soil are carried away by wind and water erosion annually. More than a quarter of the continent or 740 million ha are thus on the way to becoming sterile under the influence of this process, representing a loss of about US $1.5 billion annually.
- In 1977 the United Nations estimated that the number of persons affected by desertification amounted to one-sixth or one-fifth of the world population. Ahmad and Kassas (1987) gave an estimate of 850 million people. More than 108 million people in Africa alone or more than 61% of its population are seriously affected (Harrison 1987). In June 1994 during the fifth session of the intergovernmental negotiating committee for the International Convention on Desertification (ICOD), it was estimated that 900 million people suffer from the effects of drought and desertification throughout the world, of which 200 million live in rural Africa.
- According to the World Map of Desertification compiled by UNEP in the 1970s (Fig. 4.7) the 38 countries most affected by desertification are spread over all continents, but the area most affected is circum-Saharan Africa. But outside Africa there is still the drama of Lake Aral ...

To recognize what is really happening requires:

- A balance of the causes, which should distinguish between natural ones like rapid climatic changes and human ones.
- The examination of the distribution of the areas in which desertification is most dramatic.
- A reconsideration of the ties between desertification, level of development and famine.

4.5.2.1
The Physical Dimension of Desertification

Let us recall that the Sahel has experienced four droughts since the start of the 20th century: 1900–1903 1911–1920 1939–1944 and 1968–1985 with the greatest rain deficits in 1972–1973 and 1982–1984. In total 47 years with a deficit rainfall have been counted between 1900 and 1990. Concurrently and without any connection, southern Africa between 20–30° S has experienced two dry phases since 1921, during which precipitation fell by 75% against the normal values: 1981–1983 and 1991–1993, and since 1963 a total of 16 years with a deficit rainfall stands against 14 above-average years (Laing 1994).

The dry parts of China experienced similar deficit phases: 1900–1909 1913–1930 and 1965–1990, but without any increasing or decreasing trend in number or severity of the droughts in its drylands during the same period.

When set against the human responsibilities, the droughts, which are seen as revealing mechanisms, force us to take the different scales of the causes into account: the long-term climatic variations on a global scale interfere with short-term fluctuations acting on a regional or local scale, and are in turn superimposed onto human causes always of a short-term and local nature.

4.5.2.2
The Human Dimension of Desertification

Several factors are of importance here.

The stagnation and even the decline of economies, in particular in the African drylands. Of the countries in the Sudanian–Sahelian region, only four were able to post an economic growth rate of 3% between 1980–1985, whereas all others have seen their growth rates falling (Warren and Khogali 1992)

Civil wars, tribal conflicts, archaic nationalistic wars and the resulting disturbances (Mozambique, Angola, Somalia, Ruanda). Invasions by Mongols and Turks have repeatedly destroyed the irrigation systems of the Aral basin.

Excessive establishment of deep wells and boreholes: in the Senegalese Ferlo, the Maastrichian aquifer, for example, had been exploited since the 1950s by only 35 boreholes commissioned between 1950–1957, but by 1958 boreholes between 1963–1969. Barral *et al.* (1983) summed up the consequences:

- Decline of transhumance
- Abandonment of control of the open spaces by the pastoralists

4.5 · Desertification – An Expression of Decadence?

- Severe degradation of the vegetation and even of the soils around the boreholes, because of the establishment of permanent settlements. In 1973 Zebu cattle died from hunger with their feet in the water near a borehole with a discharge of 500 m^3/day.

The most economic strategy for breeding is a multiplication of the watering points fed by rain-wash and not the exploitation of groundwater, which should be used for watering the animals only in cases of exceptional emergencies (Chevallier and Claude 1989) because of the cost of the installations and the necessary technical education for the villages.

According to the French Geological Research Office (BRGM), a borehole itself will cost 3 million CFA-francs in northern Burkina Faso, and the necessary technical education another 1 million CFA-francs. The presence of groundwater aquifers is by no means a long-term feature (meaning several generations), and they should be exploited with great prudence. In total their replenishment appears to be all the more limited the deeper the layers are located.

The lack of proper attention by the states and the selection of inappropriate agricultural choices: at the time when the drought in the Sahel was hardest in 1983–1984 five states – Burkina Faso, Mali, Niger, Senegal and Chad – reaped a harvest of 154 million t in 1984, against 22.7 million t in 1961–1962. However, this was essentially only cotton grown by small farmers instead of food crops. In the same year the Sahel imported 1.77 million t of cereal, compared to only 200 000 t/year at the end of the 1960s, which also represented a record (Timberlake 1985). Unfortunately, this went together with an increase in food prices and a drop in the cotton price.

The impact of the demographic development which like in all drylands, mainly in Africa and Asia, has surpassed the growth rate of agricultural production. During the last half century all drylands experienced a demographic growth rate of 2.5–3%, equivalent to a doubling of the population within one generation. The *World Map of Soil Degradation* (Oldeman et al. 1990) clearly reveals a correlation between soil degradation and demographic development, especially in the drylands of western China, the Sahel, the Maghreb, the Near and Middle East as well as East Africa and there especially Kenya, where the growth rate exceeds 4% annually.

The demographic explosion is responsible for:

- The deterioration of the vegetation cover
- A shortening of the fallow periods with its consequences of impoverished soils and a higher demand for labour with meagre results
- The cultivation of marginal lands
- Soil degradation.

In the sandy steppe of central Mongolia wind erosion annually carries away a layer of 6–20 mm from previously fixed soils, equivalent to a rate of 200–300 t/ha annually (Zhu Zhenda and Liu Shu 1986).

Demography, and this includes the number of animals because of overgrazing especially in sub-Saharan Africa, is a problem also in the central Asian countries Kazakhstan, Kirghistan, Uzbekistan, Tadjikistan and Turkmenistan, in China especially in Xingjiang, in Latin and North America, and less so in Australia. The advance of veterinary medicine and the increase in the number of livestock at a rate above that of the global economy are also causes of desertification.

4.5.2.3
Where Is the Soil Degradation Most Severe?

Eight years after the severe drought of 1984 the pastoral areas of the Sahara–Sahelian region were on the way to rehabilitation. This is least obvious on the agricultural lands of the Sahel, where farmers and pastoral people compete for the same ground, and in the lands where rain-fed agriculture has been established on forests cleared by fire (Table 4.1).

The analysis of two photographic missions – one from the start of the 1950s and the second one in 1987 and Spot images from 1987 – covering a strip from the southern margin of Mauretania (15° 40' N) to the northern edge of Guinea (10° 50' N) combines natural and human data, leading to the following astonishing result (Collective 1989): It is the subhumid dry area between 650–800 mm/year mean precipitation that is the most affected by environmental degradation. It is also in this area that the densest relocations of the breeders fleeing from drought in the north have taken place and here since the start of the 1950s the highest densities of the rural population have developed with 10 and locally even 20 inhabitants/km^2. The zone right at the top of the 800 mm/year isohyet is the one that has experienced, as we have already said, the least pronounced shift of only up to 40 km, during excess or deficit oscillations of the rainfall; this explains the concentration at this latitude of people migrating from north to south during the drought crisis.

In 1987 the northern half of the Sudanian zone exhibited the highest percentage of denuded lands with 7.2%, whereas in 1952 barren ground did not exist in these ecosystems. In 1987 they accounted for less than 1% north of the latitude of Nara (15° 13' N) in the Saharan–Sahelian zone and 0.5% to the south of Yanfolila (10° 10' N) in the Sahelo–Sudanian zone. There was no general rule about the evolution of the vegetation along this strip except a retreat of the tree savannah in favour of the

Table 4.1. Areas affected by degradation in drylands (UNEP 1991)

Continent	Arid zones (10^6 ha)	Arid zones and dry areas (%)	Dry arable land	
			Total (10^6 ha)	Already degraded (%)
Africa	672	61	1433	73
Asia	277	46	1881	70
Australia	0	75	701	54
Europe	0	32	146	65
North America	3	34	578	74
South America	26	31	421	72
World total	*978*		*5160*	

wooded savannah in the Sudanian ecosystem and of the bush savannah between Nara and Mourdiah (14° 35' N). The retreat of the tree savannah could be noted all along the transect, in places where it was present in 1952.

The land degradation observed thus resulted more from human pressures increased by drought, than by drought itself. Although irreversible desertification is difficult to prove in the Sahel, it appears that the northern Sudanian zone with 800 mm/year precipitation and the greatest density of soil utilization is the most severely affected by soil degradation, not the Sahel itself.

4.5.2.4
Financial Implications of Desertification

The financial implications of desertification are difficult to estimate. How many people are affected, and what surface area of the productive arable land? How much has desertification cost already? What amount of money is necessary to get to grips with the problem?

UNEP estimated that the cost of rehabilitation of the degraded lands exceeded US $26–42.3 billion annually between 1980 and 1991, and of this US $6.5 billion were required to combat desertification in developing countries. According to Dregne (1984) donor organizations made available US $10 billion between 1978–1984 for programmes which included the battle against desertification. According to Kassas (1987) the worldwide annual losses resulting from degradation of drylands amounted to US $26 billion, in the form of a drop in agricultural productivity, which would be equivalent to the supply of cereals to some 80 million people. Ahmad and Kassas (1987) estimated that for the fight against degradation annually some US $4.5 billion would be necessary, of which more than half was required in the developing countries.

Barrow (1991) compiled several data sets:

- Between 1975 and 1984 donor organizations made available US $600 million for fighting the desertification of pastures in Africa.
- In 1982 the countries of the Sudanian–Sahelian zone received US $40 per capita annually or about US $7.45 million.
- In 1986 UNSO financed US $47.6 million for projects proposed for the control of desertification.

4.5.2.5
Connection Between Desertification, State of Development and Famine

In 1985 famine broke out in Africa from the Atlantic coast to the Horn of Africa, towards the south to Mozambique and over the entire Bantustans of South Africa (Timberlake 1985). This situation is the consequence of a chain of tragic events which had already started during the drought from 1968–1984. The Horn of Africa which had been less affected by the first phase of the drought, suffered more during the second phase, particularly in 1984. Some 35 million people experienced the famine and 10 million became ecological refugees in the parts of Africa affected by the drought.

Photo 4.10. Desertification in Mali in 1985. On the northern border of the Malian Sahel the village of Boyté has been abandoned because of the drought during the preceding years. In front we note a palisade of spiny branches, at the back a grain store raised on rocks. The round living hut is built with traditional techniques: walls of dried mud bricks and wooden posts supporting a roof made of straw of *Andropogon gayanus* (Photo M. Mainguet)

Droughts are natural mechanisms, but the rain deficit alone is not sufficient to explain the famine. The overexploitation, which is as it were endemic to the Third World because of the demographic growth, turned the droughts into major disasters affecting an increasing number of people. In the 1960s the drought affected 18.5 million people annually, during the 1970s 24.4 million and in Africa alone in 1985 some 80 million (Photos 4.10 and 4.11).

The couplet drought–famine only makes its appearance in a context of erroneous land management, of economic and political disorganization, of underdevelopment and in particular of poverty. Australia and the United States experienced a severe drought crisis in 1983 which, however, did not lead to deaths from famine. The only victims were economic victims, and exactly those who were indebted prior to the famine were affected. Desertification is a phenomenon that varies with the state of development. Where financial resources and techniques are sufficient, when an analysis of the natural and human situation has been carried out, when the political situation is quiet, and when there is an equilibrium between the population density and agricultural production, the degradation of the land is not really a problem.

Desertification and its effects may only be distinguished from the effects of drought crises by the time scales: Drought and its effects are a short-term recurring problem whereas desertification is a chronic long-term evil.

4.5 · Desertification – An Expression of Decadence?

Photo 4.11. Refugee camp in the Sudan. At El-Obeid in March 1985, after the severe drought of 1968 to 1984 which culminated in 1983–1984, we find a makeshift shelter of jute sacks and old rags at the edge of a camp of ecological refugees from the Chad. Note that the *Goz* soil, a layer of red sand once fixed by the wooded steppe, has been completely denuded over a perimeter of several kilometres (Photo M. Mainguet)

Among the solutions an attack on the causes – whether they are natural, due the effects of the drought, or human, resulting from overexploitation or wrong management – will lead to a solution only on the long term and at a very high cost. Medium- to long-term solutions will have to be found for the physical and physico-chemical mechanisms of degradation, viz. erosion by wind and water as well as salinization, and both time frames have to be taken into account simultaneously.

Is desertification really the problem? Droughts have always alternated with humid phases in the climatic history of Africa, but, in the absence of human beings, no desertification as rapid as during the 20th century, has appeared.

The droughts in the Sahel are not synonymous only with poor harvests as there is also evidence for some streaks of optimism. In the Maradi region of Niger with a mean precipitation of 300–600 mm/year and 3–4 rainy months in the south and 2–3 in the north, with a PET of 2380 mm/year, the drought of 1983–1984 imposed certain new constraints on the agricultural systems. The farmers responded to these by replacing traditional varieties with earlier ones and abandoning water-demanding cultures and the compacted soils between hillocky dunes in favour of fixed dunes with less pronounced hydrological constraints (soils on these sandy areas are able to bridge interruptions in precipitation lasting 2–4 weeks). Furthermore, such soils are easier to work. The reduction of the livestock herds due to the drop in pastoral resources represents another adaptation (Koechlin 1989).

Other examples of substitutions or combinations of produce have been reported e.g. Beauvillain (1981) from northern Cameroon for the 1930 drought, when the hardiest plants like peanuts, manioc and *mouskouari* were given preference. Gallais (1989) reported from Chad that during the drought of 1969–1974 rain-fed sorghum with a long vegetal cycle was replaced by *mouskouari*, a type of sorghum of subsidence cultures Seignobos (1984). Around the Koutous in Niger reports that peanuts were pushed back by millet, sorghum and niébé (beans) during the 1980s. In the Sertao the drought was accompanied by a return to subsidence farming with maize, beans and manioc replacing the more commercial plants. In the High Borgou of the Peoples' Republic of Benin in the Sahelian region, because of the variability of the rains and of the seasons for two decades, the farmers have adopted mobile agricultural calendars, starting when the moisture front has reached down 10–15 cm several hours after the end of the rains. The agronomists must have recognized that this depth of the moisture front corresponds to a real evapotranspiration equivalent to half the PET, revealing an excellent knowledge of the environment and an ability to adapt to a new framework of aridity (Boko 1989).

Modern utilization of deserts with the aid of large dams, like those on the Colorado, the Nile, the Euphrates, the Indus and the Yellow River, strengthened by the wish to concentrate the population onto clearly demarcated ground in order to better control productive activity and amortize investments, has so far proved to be a success only in the American West. The failure of the projects on the Aral and Indus as well as in Mali and Gezira in the Sudan necessitates a more opportunistic evolutionary ability in the short term, and in the long term requires research by the states concerned, which takes into consideration the rural world, combinations of agricultural products, terms of exchange and the reconsideration of economic alternatives.

4.6
A Glimmer of Hope

Among the expressions of human intelligence there is the ability for adaptation: in the light of the devastating consequences of many management projects, understanding and prudence are beginning to be valued for such high-risk environments like the drylands. Parsimonious utilization of the resources at the beginning of the 21st century becomes evident and a new agrarian civilization on a high scientific level appears which is based on the multiscale and multitemporal investigation of satellite imagery, its data supplementing the still irreplaceable aerial photographs. Research stations in Saudi Arabia, which has become an exporter of wheat since 1988, in Iran and the former USSR increasingly reveal the behaviour of soils under the influence of wind and water erosion.

Wind is no longer considered as a local phenomenon. We know by now that it is a process on a continental scale, and the description of the large transcontinental currents of the Sahara (Mainguet 1972) forced a review of the strategy for combating the invasion of roads by sand as e.g. in Mauretania the *route de l'espoir* and the sand encroachment in towns Nouakchott or in Morocco the agglomeration of Layyoune ...

Since the early 1980s we have known that Algeria, China, Iran and other countries with a large percentage of desert territory and a high birth rate should "develop" the former in order to respond to the latter. The known mineral resources and the vast groundwater aquifers should enable them to do this, provided that this development takes the fossil nature of these aquifers into account.

That plant growth only requires light, water and nutrients is by no means a truism: in contrast to the general opinion, water is not the only constraint explaining the poor productivity of the drylands, but also the availability of nutrients particularly nitrogen. By introducing nitrogen-fixing Leguminosae Penning de Vries and Van Heemst were able to improve pastures in Australia, and they concluded that appropriate amount of fertilizers would increase the production of dry plant matter from 0–4000 kg/ha to 5000–9000 kg/ha depending on the nature of the soil with precipitation increasing from 230–540 mm/year. The results depend not only on the pluviometric regime but also on the storage of the rain in the soil, the way of protecting the water from evaporation and on making it available to the plants. They furthermore showed that the longer the water stays in the soil, the longer the growth of the plants will last and the higher the biomass will be.

It is generally admitted by now that irrigation is a large consumer of water leading to extreme waste and that a large part of the water lost joined up with the groundwater, causing a rise of the water table and secondary mineralization and waterlogging, as we have seen. This has resulted in a new concern for the economic use of water and has brought about new know-how:

- Sensors placed in the soil initiate spraying when the plants require it.
- One of the aims of this research is the utilization of the precipitation, which is still insufficiently taken up by the water demand of the plants, the key parameter of yield, but lost to runoff and in unproductive drainage.
- The artificial preparation of the structure of the soils – for better water management – by the establishment of "sleeves" around the roots with the aid of emulsifiers and artificial aggregates.
- The selection of plants started during the Neolithic: in the 20th century it led to biotechnology permitting the isolation of proteins, which increase the resistance of the most widely cultivated crops (maize, rice, barley) to drought.
- A reduction of evaporation by the spreading of molecular layers of chemical products.
- An increase of the rain-wash coefficient by the use of synthetic materials for the establishment of artificial drainage areas.
- The direct use of salt water for agriculture and livestock breeding:
  - In greenhouse cultures in the Negev this is done experimentally with success for vegetables, but this salt water is by no means seawater with a salt content of 35 g/l.
  - Fenced-in camels and karakul sheep drink water containing 10–12 g/l and even 15–16 g/l salts and still grow and reproduce normally (Carmont 1967).

However, the use of salty water in agriculture and livestock breeding only shifts the problem as the salts will be replenished and there still is the question of drainage in particular in endorheic areas like the Aral, for example:

- In addition to the parsimonious use of water one has to consider the potential for financing such complicated and expensive hydraulic engineering projects.

In a biography called *Hidden Harvest* the IIED (International Institute for Environment and Development) has shown that in addition to cereals, tubers and domestic animals, the agricultural populations take recourse to wild foodstuffs, which are of particular value as they are available during droughts, and to hunting small animals like rats, mice, crickets etc. However, the range of wild foodstuffs has decreased because of the growth of the planted areas and the modifications of the methods of cultivation. A way out would be the domestication of these foodstuffs. In Israel a new agricultural philosophy intends to multiply the varieties of domesticated plants for the markets of the 21st century and to abandon the plants which require too much water.

From these scientific tests and common sense a renewed view of development has come forward based on the recognition of the virtue of modest programmes: should they eventually fail the impact will be small, but should they succeed there will be hope that they can grow to the largest dimensions, under proper adaptation to each "natural" and human context.

All these approaches are based on a high technological level and more precisely on a combination of techniques, but their efficacy will remain ridiculously low unless they are integrated into the global development system and into the rural environments in what Tricart (1994) refers to as ecological development.

Does not the success of the efforts undertaken in the Taklamakan to make the wealth of the underground mineral resources a means for permanently populating

Photo 4.12. The Great Wall of China. To protect themselves from invasions by the Mongols, the Chinese erected the Great Wall, built from stone in the eastern part and from dried clay bricks in the west. It is shown here south of the Badan Jilin desert and east of the Tengger, a two thousand year old defensive system which, although in clay, still remains recognizable because of the drought in the semi-desert area of Yanshi County (Photo M. Mainguet)

the area depend on the way in which agriculture is developed and the soils, vegetation and water resources are conserved, which in turn themselves depend on the supply of water from the glaciers covering the surrounding mountain ranges? These examples serve to show that mineral exploitation and technical progress are not sufficient to reap the best benefits from the sporadic resources of the drylands. Knowledge of the natural data and of the social and cultural realities remain an indispensable prerequisite for any type of development.

It would be naive to take into account under these notes of hope only the technological aspects. The Cairo world summit on demography in 1994 and the one that opened on March 6th 1995 in Copenhagen on "social development", which focused on the underdevelopment endemic to our planet, were both organized by the UN. They were evidence for another advance in which it has been recognized that the control of demographic growth and progress of the human resources are the two pivotal points for the sustainable development so much desired.

General Conclusion

Definition and geographical positioning of the drylands, of aridity, drought, soils and vegetation were the object of the first part of this book. The second part was dedicated to water, its scarcity and its erosive force and lead to the statement that this basic resource is constraint number 1, the wind being constraint number 2. We furthermore described the high degree of sensibility of the drylands, which are unstable and are exposed to greater ecological risks than most other climatic zones, and at the same time possess a lower agricultural potential. They appear more demanding than the temperate and perhaps also the tropical humid environments, and in the context of sustainable development they are more vulnerable. However, paradoxical as it might appear, the drylands are attractive and because of this they exhibit high population densities. The equilibrium between these low potentials and the population growth remains difficult to preserve. A series of a few dry years will suffice to break the equilibrium between the survival of the society, the exploitation of the natural resources and the conservation of the environment.

A tremendous creativity is thus required to allow man to populate the desert and to live here with a minimum of comfort. His different responses to aridity and droughts, the ingenuity which he exhibits to accommodate these constraints and to live with them, were the object of the third part of this book.

This analysis suggests that climate and its two key parameters discussed, viz. aridity and drought, have been able to govern the evolution from Archaic man to modern man – the steppe man – as well as the history of civilization, invention of agriculture, i.e. the hatching of the Neolithic revolution and its two most important innovations in the history of man: the pastoral life of Abel and the farming life of Cain, who were already then antagonists. In this second way of life the agrarian novelties of water management and irrigation appeared. After thousands of years of tribal dispersal an urban civilization also appeared, with the art of writing and especially the measuring of time by taking into account the seasonal rhythms: the floods bringing the waters and the low waters leading to lack of water.

The balance of this historic co-existence between man and aridity has led in the fourth part of the book to the observation of a decline and to the necessary new perspectives of an equilibrium between man and his climatic environment in the drylands. Man cannot prevent neither aridity nor droughts which are events taking place on different time scales and for the second several times in a century or millennium. Man does not know their reasons or rhythms, but he is able to combat their effects, i.e. he can foresee the consequences and he can adapt by an improved approach to the mechanisms and the functioning of the ecological systems.

Since they are subject to natural constraints tied to climate and also to human constraints, drylands nevertheless possess an ability for rehabilitation after drought crises (resilience), which is quite astonishing and is only now being discovered by science.

It is technically possible to live with aridity in our present time, but the main handicap is "an excessive, sometimes overwhelming birth rate …". In its report on the *State of the World Population* (1992), the United Nations Population Fund estimated that:

- The world population would rise to 6 billion by 1998.
- The growth rate amounts to three births per second or 250 000 every day.
- At the beginning of the 1990s the annual growth amounted to 93 million and it will increase to 100 million annually at the end of this decade.
- Ninety-five percent of this growth takes place in developing countries located in drylands, with the exception of those in Australia and the USA.
- More than half of the population in the developing countries will be under the age of 25 years by the year 2000.
- The time required for an increase by 1 billion people has been shrinking and it took the world population:
  - One century from 1830–1930 to grow from 1 to 2 billion
  - 30 years, from 1930–1960, to reach a total of 3 billion
  - 15 years, from 1960–1975, to reach 4 billion
  - 12 years, from 1975–1987, to reach 5 billion
- The population of the developing countries has more than doubled within 55 years, growing from 1.7 billion in 1930 to 4.1 billion by 1990, and will reach 5 billion from a total world population of 6.21 billion by the year 2001. In contrast to this the population of the so-called developed countries (Europe, the former USSR, North America, Japan, Australia and New Zealand) grew from 832 million in 1950 to 1.5 billion in 1992 with a decline to 1.26 billion predicted for the year 2001.
- Between 2020 and 2025 the population growth of the latter countries will account for not more than 3% of the total world population growth and they will account for less than 5% of the world population. This increase will exert a heavy pressure on resources and will represent a menace for the environment (Fig. 5.1.)
- A corollary to this growth is the migration towards the towns, and the United Nations estimates that around the year 2000 some 40% of the population of Africa and Asia (excluding Japan) and 76% of that in Latin America will live in cities. However, in the drylands where the march to the urban areas is accompanied by an uncontrolled extension of the metropolitan regions, this under-equipment of the infrastructure and the rise in water demand will lead to a situation where there is less water available for each inhabitant, with the usual consequence for sanitation and hygiene. Will rural areas, alleviated by this desertion for the towns, then witness a reduction in their degradation?

The battle against the effects of the couplet aridity-drought and the overexploitation of the environment are the two faces of the same question in view of which Hillel (1991) distinguished an optimistic and a pessimistic position.

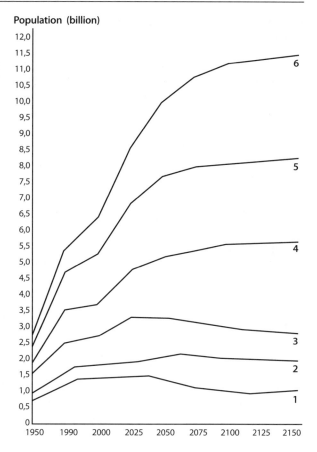

Fig. 5.1. Demographic projections by region (based on mean variables). *1* Developed countries; *2* Latin America; *3* China; *4* India; *5* other Asian countries; *6* Africa (after UNICEF letter 37 1994)

The pessimists would believe that:

- There is a trend towards an aridification of the climate.
- Agriculture in the drylands has reached a level of production beyond which no sustainable development will be possible any longer.
- The production of foodstuffs can only decrease because of the degradation of the soil.
- The major development programmes have failed – they are large consumers of land, water and biomass in already impoverished areas.
- Pollution and environmental degradation are the inevitable consequences of development and demographic expansion.
- The fight against demographic expansion cannot be suppressed, as it is a component inherent in man.
- The small supplies of energy available and their cost are going to put pesticides, fertilisers, underground water and mechanical transport beyond the means of small farmers.
- Mankind eventually will be condemned to live at a level of lower quality and reduced security.

In contrast, the optimists or, as the pessimists would call them, the utopists, are of the opinion that:

- The rapid population growth will create human resources and productive forces; according to a recent trend of thought it will be the factor necessary for the intensification of agriculture, but there is another precondition: proper land management, which is indissolubly tied to the degree of participation of the farmers and pastoralists in the decision-making processes.
- The growth of the population and of exploitation of natural resources have led man to take into account the limited nature of these natural resources, and the limitations for civilization in using them.
- These new approaches – a major advance of the 20th century – have led to new concepts of management, in which the interdependence between development and conservation of the environment takes priority: we know now that we have to preserve resources for our descendants.
- The drylands are subject to interconnected natural constraints, tied to climate, and to human constraints, but they also possess an encouraging ability for rehabilitation after a drought.
- The fertility of the world population has been falling since the 1960s and this trend may be supported by education and economic progress.
- Even if the production potential is not unlimited, technical progress may still increase further, with poverty resulting more from the poor usage of resources and from bad behaviour in relation to them, than from a lack of resources in the true sense.
- The value of farming methods and the capacity of the farmers to adapt to new ecological and economic conditions will be decisive.

Most developing countries in the drylands of Asia and the Middle East already have at their disposal structures of research and technical assistance. Their main problem is the integration of the results of these activities into their production systems, and especially into small-scale exploitation to bring the research to the small producers. In Africa the situation requires "at least to encourage the farming initiatives in the sense of a reproducible food security" (Marty 1989) and to organize, from the conception of the projects, a more permanent co-operation between the users of the land and the managers.

Development in dry ecosystems is not continuous. More than in any other ecosystem it is characterized by progress and regression. The Saharan oases are an example: after centuries of prosperity they are now abandoned more due to changing life habits and communication intersections than as a result of climatic variations.

The same applies to environmental degradation. In climatic crises which reveals rather than cause man's destructive action, environmental degradation seems to develop an acceleration which exceeds our capacity to influence, that is to diminish its progress.

If at present it is commonly admitted that dry ecosystems belong to Earth's most degraded ones, it must also be admitted that arid ecosystems are the most vulnerable of all. This was realized already at the beginning of the 1970s when the first drought crisis spread over the Sahel from 1968 to 1973, proving to be a terrible revelation whose consequences, environmental degradation, famine, ecological refugees, and disappearance of cattle herds, caused worldwide shock and led the UN to organize the UN Conference on Desertification (UNCOD).

A crisis of recurrent droughts in the Sahel between 1983 and 1984 crippled both the Sahel and Saharan countries and, simultaneously, the United Nations. During this stage of panic, the irreversibility of the situation seemed definite.

Now at the end of the 1990s, another stage has been reached: desertification in the arid ecosystems of both Africa and Asia is no longer considered a fatality. A new degree of maturity has become visible through numerous local approaches developed by groups of ingenious cattle farmers and agriculturists to escape climatic adversity, aridity and drought. These solutions no longer follow traditional patterns of land use or programs imported from developing countries, but represent adaptive behaviour developed on site in the face of two major challenges: the rapidly growing population and the environmental degradation. These new approaches try to replace extensive environmental exploitation – still a common practice at the beginning of the 1950s – by intensive land use. It is still too early to predict whether these new strategies will be the beginning of a lasting development.

Nowadays, the participative strategies proclaimed for more than a decade became obsolete; accompanying strategies have become absolutely necessary. The majority of projects are planned from a national or regional point of view. It is, however, probable that higher efficiency, leading to better development, is more in need of individual solutions, or microdecisions made by the affected populations themselves on site – their identification, their following through, and finally the amelioration of their situation could be the fruitful level on which to attain the desired long-lasting development.

To support efficiently the creative efforts that are already being furthered in agriculture in the semi-arid developing countries, it seems to be the right moment to let these countries profit from the scientific acquisitions of these past three decades. However, these scientific acquisitions will remain ineffective if they are not accepted, integrated and assimilated by the rural population into the local socio-economic systems of resource and production management, based on local surroundings that to a greater or lesser degree have been transformed by man, i.e., by droughts and/or attempts at agricultural intensification.

We must be aware that there is a delay both in efficient and practical investigation of multiple risk factors – regardless of their possible hierarchy and without denying their interdependences – and between the very rapidly achieved scientific progress in the developing countries during the past decades and their ability to exploit it. It is not merely a question of wishing to bring knowledge to the developing countries, but also of putting the technical skills into practice for the land users, and foreseeing possible consequences which might interfere with the new dynamics. For instance, how can this knowledge be combined with the new ways of appropriating natural resources (e.g. soil, water, plants)?

The so-called modern techniques which scientists can furnish can alleviate resource shortages, but cannot replace them. For example, agriculture with saline water is feasible, increase in water retentiveness of soils is also practicable using polymers, but agriculture without water is inconceivable.

It is generally claimed that water shortage is the limiting factor to development in arid ecosystems; however, a calculation by R. Letolle (pers. comm.) shows that a Frenchman has 2000 m^3/year, whereas in the arid ecosystems in Central Asia (in Tourane) each inhabitant has 2200 m^3/year. What do these amazing figures signify? Is the water deficiency only apparent?

Although priority is given to water by the construction of huge barrages, canals, and the development of gigantic irrigation systems, apparently not sufficient consideration is given to land protection and above all to the serious drainage of wastewater, especially in closed basins. This negligence will very quickly and inevitably lead to chemical pollution and soil salinization: the beginning of the 21st century will thus become in the drylands the century of soil deficiency rather than, as at present envisaged, of water deficiency.

References

Abrahams AD, Parsons AJ (eds) (1994) Geomorphology of desert environments. Chapman & Hall, London, 674 pp
Adams R McC (1974) Historic patterns of Mesopotamian irrigation agriculture. In: Downing TE, Gibson McG (eds) Irrigation's impact on society. Antropological Papers of the University of Arizona 25, Tucson, Arizona, pp 1-6
Ahlcrona E (1988) The impact of climate and man on land transformation in central Sudan. Applications of remote sensing. Meddelanden fran Lunds Universitets Geografiska Institution, Avhandlingar, Sweden, 103, 140 pp
Ahmad MU (1990) The state of the art hydrology for the development of Sahelian water resources. UNESCO, Proc of the Sahel Forum, The state-of-the-art of hydrology and hydrogeology in the arid and semi-arid areas of Africa. Ouagadougou, Burkina Faso, 18-23 Feb 1989. International Water Resources Association, Urbana, Illinois, pp 813-823
Ahmad YJ, Kassas M (1987) Desertification: financial support for the biosphere. Hodder and Stoughton, London, 187 pp
Albergel J, Casenave A, Ribstein P, Valentin C (1992) Aridité climatique, aridité édaphique: étude des conditions de l'infiltrabilité en Afrique tropicale sèche. In: Le Floc'h E, Grouzis M, Cornet A, Bille JC (éd sci) L'Aridité, une contrainte au développement.ORSTOM, Paris, pp 123-130
Amiran DHK (1966) Man in arid land. In: Hills ES (ed) Arid lands. New York, pp 219-287
Anderson RS, Haff PK (1988) Wind modification and bed response during saltation of sand in air. NATO Advanced Research Workshop on Sand, dust and soil in their relation to aeolian and littoral processes, 14-18 May, Sandjberg, University of Aarhus, Denmark. In: Barndorff-Nielsen OE, Willets BB (eds) Acta Mechanica (Suppl 2). Springer, Berlin Heidelberg New York, 250 pp, 83 figs
Anderson RS, Sorensen M, Willetts BB (1990) A review of recent progress in our understanding of aeolian sediment transport. Department of Theoretical Statistics, University of Aarhus, Res Rep 213, 48 pp
Andrianov BV (1985) A study in the typology of irrigation and farming in central Asia and Kazakhstan (the end of the 19th, the beginning of the 20th centuries). Typology of the main elements of traditional culture. UNESCO, New Delhi, pp 36-113
Andrianov BV (1990) History of development of Aral region economy and its influence on nature. Nukus Int Conf, Karakalpakstan, Chap 2, 26 pp
Armbrust DV, Paulsen GM (1973) Effect of wind and sandblast injury on nitrate accumulation and on nitrate reductase activity in soyabean seedlings. Commun Soil Sci Plant 4(3):197-204
Arntzen JW, Veenendaal EM (1986) A profile of environment and development in Botswana. Institute for Environmental Studies, Free University, Amsterdam and University of Botswana, Gaborone
Ash JE, Wasson RJ (1983) Vegetation and sand mobility in the Australian desert dunefield. Z Geomorphol (Suppl) 45:7-25
Aubreville A (1949) Climat, forêts et désertification de l'Afrique tropicale. Société d'Editions Géographiques et Coloniales, Paris. 352 pp
Aufrère L (1931) Le cycle morphologique des dunes. Ann Géogr (Paris) 40(226):362-385
Aufrère L (1935) Essai sur les dunes du Sahara algérien. Géografiska Aunolei Sven Heolin (Stockholm) XVII:481-500
Aumassip G (1986) Le Bas-Sahara dans la préhistoire. Editions du CNRS, Paris, 612 pp
Aurenche O (1982) Préhistoire des sociétés hydrauliques du Proche-Orient ancien. In: L'homme et l'eau en Méditerranée et au Proche-Orient, Travaux de la maison de l'Orient. Presses Universitaires de Lyon, no 2, vol 1, pp 31-44
Awad M (1954) The assimilation of nomads in Egypt. Geogr Rev 1954:240-252
Bagnold RA (1941) The physics of blown sand and desert dunes. Methuen, London, 265 pp

Bagnold RA (1953) The surface movement of blown sand in relation to meteorology. In: Desert research, 89-96. Research Council of Israel, Spec Publ 2, Jerusalem
Baker VR, Kochel RC, Patton PC (1988) Flood geomorphology. Wiley, New York, 503 pp
Bakre M, Bethemont J, Commère R, Vant A (1980) L'Egypte et le haut barrage d'Assouan de l'impact à la valorisation. Presse de l'université de Saint-Etienne, 191 pp
Balek J (1977) Hydrology and water resources in tropical Africa. Elsevier, New York, 208 pp
Balek J (1983) Hydrology and water ressources in tropical regions. Elsevier, New York
Balek J (1989) Groundwater resources assessment. Elsevier, New York, 249 pp
Banque Mondiale (1985) La désertification dans les zones sahélienne et soudanienne de l'Afrique de l'Ouest. Washington, DC, 71 pp
Barral H et al. (1983) Systèmes de production d'élevage au Sénégal dans la région du Ferlo. ORSTOM, Paris, 172 pp
Barrow CJ (1987) Water resources and agricultural development in the tropics. Longman, Harlow
Barrow CJ (1991) Land degradation - development and breakdown of terrestrial environments. Cambridge University Press, Cambridge, 295 pp
Barth HK (1990) Implications pour l'environnement et l'agriculture de la construction des barrages sur la vallée du Niger au Mali. UNESCO, Proc Sahel Forum,The state-of-the-art of Hydrology and hydrogeology in the arid and semi-arid areas of Africa. Ouagadougou, Burkina Faso, 18-23 Feb 1989. International Water Resources Association, Urbana, Illinois, pp 857-869
Bazin M (1993) Des réseaux collectifs aux pompages individuels. Observations en Turquie et à Chypre. In: L'eau, la terre et les hommes, au fil de l'eau, hommage à René Frécaut. Presses Universitaires de Nancy Nancy, pp 411-418
Beadnell HJL (1934) Libyan desert dunes. Geogr J 84(4):337-340
Beaudet G, Gabert P (1993) Les déserts littoraux brumeux. Aspects et problèmes géomorphologiques. Colloque R Coque, 11-12 fév, Paris
Beaumont P (1989) Environmental management and development in drylands. Routledge, London
Beauvillain A (1981) Un espace de migrations frontalières importantes, le Nord Cameroun. In: Cahiers géographiques de Rouen, no 15, Etudes sahéliennes. Mobilité, réserves d'espace et frontières 1970-1980
Bedrani S (1993) Les politiques maghrebines dans les zones arides et désertiques. Cours sur le développement des zones arides et désertiques, 8 nov-3 déc, IRA, Médénine, Tunisie, 29 pp (inédit)
Behnke RH, Scoones I (1992) Rethinking range ecology: implications for rangeland management in Africa. IIED Drylands Network Programme Paper 33, International Institute for Environment and Development, London
Belal AE (1993) Sustainable development in Wadi Allagi in Egypt. Desertification Control Bull, UNEP, Nairobi, Kenya, no 23, pp 39-43
Bell RH (1987) Conservation with a human face: conflict and reconciliation. African land use planning. In: Anderson D, Grove R (eds) Conservation in Africa: people, policies and practice. Cambridge University Press, Cambridge, pp 79-101
Bendali F, Floret C, le Floc'h E, Pontanier R (1990) The dynamics of vegetation and sand mobility in arid regions of Tunisia. J Arid Environ 18:21-32
Bernus E (1974) Géographie humaine de la zone sahélienne. In: Le Sahel: bases écologiques de l'aménagement. Notes techniques du MAB, UNESCO, Paris, pp 67-73
Bernus E (1989) La sécheresse dans la tradition touarègue. In: Bret B (éd) Les hommes face aux sécheresses.EST-IHEAL, Paris, no 42, pp 251-256
Bernus E, Savonnet G (1973) Les problèmes de la sécheresse dans l'Afrique de l'Ouest. Présence Afr 88(4):113-138
Berry L (1974) The Sahel: climate and soils. In: The Sahel: ecological approaches to land use. MAB Technical Notes, UNESCO Press, Paris, pp 9-17
Bethemont J (1982) Sur les origines de l'agriculture hydraulique. In: L'homme et l'eau en Méditerranée et au Proche-Orient, Travaux de la maison de l'Orient, Presses Universitaires de Lyon, no 2, vol 2, pp 7-30
Bille JC (1971) Etude d'un écosystème subdésertique. Principaux caractères de la végétation herbacée du Sahel sénégalais. ORSTOM, Paris, multigr, 51 pp, 15 figs
Bille JC (1972-1973) Graines et diaspores des plantes herbacées du Sahel. Description, production et dynamique. ORSTOM, Paris, multigr, 2 vols, 105 pp
Bille JC (1973) L'écosystème sahélien de Fété Olé. Essai de bilan au niveau de la production primaire de cette annuelle. ORSTOM, Paris, multigr, 66 pp, 13 figs, 7 tab
Bisson J (1991) Un front pionnier au Sahara tunisien: le Nefzaoua. Bull Assoc Geogr. Fr, Paris 4:99-309

Bisson J (1993) Paysanneries du Sahara maghrebin. Dynamiques locales et politiques de développement. Cours sur le Développement des zones arides et désertiques, 8 nov-3 déc, IRA, Médénine, Tunisie, 23 pp (inédit)
Biswas AK (1987) Water development and management for desertification control: a review ot the past decade. Land Use Policy 4(4):401-411
Biswas AK (1990) Conservation and management of water resources. In: Goudie AS (ed) Techniques for desert reclamation. Wiley, New York, pp 251-265
Blackwelder E (1934) Yardangs. Geol Soc Am Bull 45:159-166, pls 1-7
Blanford WT (1876) On the physical geography of the Great Indian Desert with especial reference to the former existence of the sea in the Indios Valley, and on the origin and mode of formation of the sand-hills. J Asiatic Soc Bengal (Calcutta) 45:86-103
Bokhari SM (1980) Case study on waterlogging and salinity problems in Pakistan. Water Supply Manage 4(2):171-192
Boko M (1990) Bilans d'eau des sols et calendriers agricoles traditionnels dans le Haut Borgou (RP Bénin). UNESCO, Proc Sahel Forum, The state-of-the-art of hydrology and hydrogeology in the arid and semi-arid areas of Africa. Ouagadougou, Burkina Faso, 18-23 Feb 1989. International Water Rseources Association, Urbana, Illinois, pp 958-963
Boubekraoui M, Carcemac C (1986) Le Tafilalt aujourd'hui. Régression écologique et sociale d'une palmeraie sud marocaine. Rev Géogr Pyrénées Sud-Ouest (Toulouse) 57(3):449-463
Boudet G (1972) Désertification de l'Afrique tropicale sèche. Adansonia, Sér 2, 12(4):505-524
Boudet G (1974) Les pâturages et l'élevage au Sahel. In: Le Sahel: bases écologiques de l'aménagement. Notes techniques du MAB, UNESCO, Paris, pp 29-33
Boudet G (1975) Inventaire et cartographie des pâturages en Afrique de l'Ouest. In: Inventaire et cartographie des pâturages tropicaux africains. Act Colloq Bamako 3-8 Mars (Mali):57-77
Boulanger A (1990) La sécheresse hydrologique. Sécheresse 1(4):238-239
Bourbouze A (1993) Transformation des systèmes d'élevage en zones arides. Cours sur le développement des zones arides et désertiques, 8 nov-3 déc, IRA, Médénine, Tunisie (inédit)
Bouwer H (1989) Estimating and enhancing groundwater recharge. In: Sharma ML (ed) Groundwater recharge. Balkema, Rotterdam, pp 1-10
Bovill EW (1929) The encroachment of the Sahara on the Sudan. J R Afr Soc 20:175-259-259
Bowler JM (1976) Aridity in Australia: age, origin and expression in aeolian landforms and sediments. Earth Sci Rev 12:279-310
Braquaval R (1957) Etude d'écoulement en régime désertique. Massif de l'Ennedi et région nord du Mortcha. Commission scientifique du Logone et Tchad, ORSTOM, Paris, polycopié
Breed C, Grow T (1979) Morphology and distribution of dunes in sand seas observed by remote sensing. In: A study of global sand seas. US Govt Print Office, Washington, pp 252-304
Breed CS, Mc Cauley JF, Davis PA (1987) Sand sheets of the eastern Sahara and ripple blankets on Mars. In: Frostick LE, Reid I (eds) Deserts sediments: ancient and modern. Geol Soc Spec Publ 35. Blackwell, Oxford, pp 337-359
Bret B (ed) (1989) Les hommes face aux sécheresses. Nordeste brésilien, Sahel africain. Travaux et Mémoires de l'IHEAL, IHEAL and EST, Paris, 422 pp
Brown GF (1960) Geomorphology of western and central Saudi Arabia. 21e Congr Int Géol, Copenhague, vol 21, pp 150-159
Brown JC (1885) Hydrology of South Africa or details of the former hydrological conditions of the causes of its present aridity. Kircaldy, p 260
Bryson RA, Baerreis DA (1967) Possibilities of major climatic modifications and their implication: North West India, a case for study. Bull Am Meteorol Soc 48
Butzer K (1976) Early hydraulic civilization in Egypt: a study in cultural ecology. University of Chicago Press, Chicago, 134 pp
Campbell, Main and Associates (1991) Western Sandveld remote area dwellers. The Associates, Gaborone
Capot-Rey R (1943) La morphologie de l'Erg Occidental. Inst Rech Sahariennes, Dakar, Senegal, vol 2, pp 1-35, 69-106
Capot-Rey R (1953) Le Sahara français. Presses Univeritaires de France, Paris, 564 pp
Carmont J (1967) Peut-on fertiliser les zones arides? Diagrammes, no 130, ed CAP, Monte-Carlo, 90 pp
Carson L (1962) Silent spring.
Casenave A, Valentin C (1989) Les états de surface de la zone sahélienne. Influence sur l'infiltration. ORSTOM, Paris, 229 pp
Cauvin J (1981) "Le problème de l'eau" au Proche-Orient. De l'homme prédateur aux premières sociétés hydrauliques. In: L'homme et l'eau en Méditerranée et au Proche-Orient, Travaux de la maison de l'Orient, Presses Universitaires de Lyon, no 2, vol 1, pp 23-30

Cavaille B (1989) La cohabitation de l'homme avec la sécheresse dans le Nordeste brésilien. In: Bret B (éd) Les hommes face aux sécheresses. EST-IHEAL, Paris, no 42, pp 303–307
Cavalcanti C (1989) Dimension socio-économique de la sécheresse de 1979–80 dans le Nordeste du Brésil. In: Bret B (éd) Les hommes face aux sécheresses. EST-IHEAL, Paris, no 42, pp 405–408
Charney J (1975) Dynamics of deserts and drought in the Sahel. Q J R Meteorol Soc 101:193–202
Cheng Qichou (1993) The research on the Tarim River. Hohai University Press, Hohai, China, 216 pp (in Chinese and English)
Chepil WS (1945) Dynamics of wind erosion. Soil Sci 60:305–320, 397–411, 475–780
Chepil WS, Woodruff NP (1963) The physics of wind erosion and its control. Adv Agron 15:211–302
Chepil WS, Siddoway FH, Armbrust DV (1964) In the Great Plains: prevailing soil-erosion direction. J Soil Water Conserv 19:67–70
Chernet T (1992) A hydrological map of Ethiopia (scale 1:2 000 000). Hydrology 1–2:29–35
Chevalier A (1950) La progresion de l'aridité, du desséchement et de l'ensablement et la décadence des sols en Afrique occidentale française. CR Acad Sci Paris 230:1530–1533
Chevallier P, Claude J (1990) Exploitation de points d'eau de surface temporaires pour l'amélioration de la gestion des pâturages sahéliens. UNESCO, Proc Sahel Forum, The state-of-the-art of hydrology and hydrogeology in the arid and semi-arid areas of Africa.Ouagadougou, Burkina Faso, 18–23 Feb 1989. International Water Resources Association, Urbana, Illinois, pp 947–956
Chlopin IN (1964) The Goksir group of settlements in the eneolithic epoch, M-L. (in Russian)
Chu KC (1973) A preliminary study on the climatic fluctuations during the last 5000 years in China. Sci Sin 16–2:226–256
Chudeau R (1920) L'étude sur les dunes sahariennes. Ann Géogr 29:334–351
Chudeau R (1921) Le problème du desséchement en Afrique occidentale. Bulletin du Comité d'Etudes Historiques et Scientifiques de l'AOF, 4, pp 352–370
CNUED (Conférence des Nations Unies pour l'Environnement et le Dévelopement), UNCED (United Nations Conference for Environment and Development), 1992, New York
Coelho J (1989) Potentiel agricole de la région semi-aride du Nordeste brésilien. In: Bret B (éd) Les hommes face aux sécheresses. EST-IHEAL, Paris, no 42, p 315
Collectif (1989) Evolution des milieux sahélien et soudanien 1957–1987. Etude des critères accessibles par télédétection. Transect Mauritanie-Mali. Institut Géographique National (IGN France International), ORSTOM, Université de Reims (LGPZ), UNEP, Ministère des Affaires Etrangères, Ministère de la Coopération Française, Paris. 117 pp
Collectif (1992) Supplément au no 137 d'Environnement actualité Mars 1992. Ministère de l'Environnement, Paris, pp 10–11
Collin-Delavaud C (1968) Le piémont côtier du Pérou septentrional. Thèse de Doctorat d'Etat, Univ de Paris, Paris, 591 pp
Cornet A (1992) Relations entre la structure spatiale des peuplements végétaux et le bilan hydrique des sols de quelques phytocénoses en zone aride, pp 245 à 264 In: Le Floc'h E, Grouzis M, Cornet A, Bille JC (éds) L'aridité, une contrainte au développement. Editions ORSTOM, Paris, 597 pp
Cornish V (1897) On the formation of Sand-Dunes. Geogr J 9:278–302
Courel MF (1989) Variations de l'albedo de surface dans le Sahel et sécheresse. In: Bret B (éd) Les hommes face aux sécheresses. EST-IHEAL, Paris, no 42, pp 155–163
Dasman RF (1984) Environmental conservation, 5th edn. Wiley, New York, 486 pp
Demangeot J (1972) Le continent brésilien. Etude géographique. SEDES, Paris, 172 pp
Demangeot J (1981) Les milieux naturels désertiques. SEDES, Paris, 261 pp
Demangeot J (1990) Les milieux "naturels" du globe, 3rd edn. Masson, Paris, 277 pp
Derrick J (1984) West Africa's worst year of famine. African affairs, vol 83. London, pp 281–299
Despois J (1935) Le Djebel Nefousa. Larose, Paris, 349 pp
DeVries JJ (1984) Holocene deflation and active recharge of the Kalahari groundwaters. A review and an indicative model. J Hydrol 70:221–32
De Vries JJ, Von Hoyer M (1988) Groundwater recharge studies in semi-arid Botswana. A review. In: Summers I (ed) Estimates of natural groundwater recharge. NATO ASI Series C, vol 222, pp 339–347
De Wispelaere G, Peyre de Fabrègues B (1991) Evaluation et suivi des ressources pastorales par télédétection spatiale dans la région du Sud-Tamesna (Niger). CIRAD-IEMVT, Paris, vol 1, Synthèse, 93 pp, vol 2, Annexes, 413 pp
Diagne Y, Lo M, Mamaty I (1993) Commerce international, dette, politique d'ajustement structurel et désertification. Conf Int ONG sur la Désertification, Bamako, Mali, 16–20 août, Environnement et Développement du Tiers-Monde, Dakar, Sénégal (inédit)
Dincer T, Mugain A, Zimmerman V (1974) Study of the infiltration and recharge through the sand dunes in arid zones with special reference to stable isotopes and thermonuclear tritium. J Hydrol 23:79–109

Dixon JC (1994) Duricrusts. In: Abrahams AD, Parsons AJ (eds) Geomorphology of desert environments. Chapman & Hall, London, pp 82–105
Dolukhanov PM (1986) Foragers anf farmers in west central Asia. In: Swelebil M (ed) Hunters in transition. Mesolithic societies of temperate Eurasia and their transition to farming. Cambridge, pp 121–132
Dregne HE (1984) Combating desertification: evaluation of progress. Environ Conserv 11(2):115–121
Dregne H, Tucker CJ (1988) Desert encroachment. Desertification Control Bulletin 16, UNEP, Nairobi, Kenya, pp 16–19
Dregne H, Kassas M, Rosanov B (1991) A new assessment of the status of desertification. Desertification Control Bulletin, United Nations Environment Programme (UNEP), no 20, Nairobi, pp 6–18
Dresch J (1956) Le Kyzylkoum et la sédentarisation des nomades, Bull, AGF mars-avril, pp 98–108
Dresch J (1968) Reconnaissance dans le Lut (Iran). Bull AGF, no 362–363, pp 134–153
Dresch J (1982) Géographie des régions arides. Presses Universitaires de France, Paris, 277 pp
Dubief J (1947) Les pluies au Sahara Central. Travaux de l'Institut de Recherches Sahariennes (TIRS), Dakar, Sénégal, vol IV, pp 7–23
Dubief J (1963) Le climat du Sahara, 2 vols. Mémoire hors série de l'Inst de Rech Sahariennes, Alger, vol 1, 312 pp, vol 2, 275 pp
Duchaufour P (1965) Précis de pédologie, 2nd edn. Masson, Paris, 481 pp
Durand GH (1949) Essai de nomenclature des croûtes. Bull Soc Sci Nat Tunisie 3/4:141–142
Durand JH (1983) Les sols irrigables. Etude pédologique. Techniques vivantes. Collection publiée par l'Agence de Coopération Culturelle et Technique avec la collaboration du Conseil International de la Langue Française, PUF, Paris, 339 pp
Durand JH (1988) Arrêter le désert. Techniques vivantes, PUF, Paris, 416 pp
Durand-Dastès F (1977) Systèmes d'utilisation de l'eau dans le monde. SEDES, Paris, 147 pp
Economic Development Institute of The World Bank (1987) Seminar on Land and water resources management. Collected Papers (non paginé)
Einstein HA, El-Samni EA (1949) Hydrodynamic forces on a rough wall. Rev Mod Phys 21:520–524
El-Baz F (1988) Origin and evolution of the desert. Interdiscip Sci Rev J, W. Arrowsmith Ltd, vol. 13, no 4, pp 331–347
El-Baz F (1989) Monitoring Lake Nasser by space photography. Remote sensing and large-scale global processes. Proc IAHS 3rd Int Asembly, Baltimore, Maryland, IAHS Publ no 186
Elhai H (1968) Biogéographie. Les cours de Sorbonne. A Colin, Paris 121 pp
Ellison WD (1944) Studies of raindrop erosion. Agric Eng 25(4):131–136; 25(5):181–182
Elouard P (1976) Oscillations climatiques de l'Holocène à nos jours en Mauritanie et dans la vallée du Sénégal. In: La désertification au sud du Sahara. Colloque de Nouakchott, Nouvelles. Editions Africaines, Dakar
Enquist F (1932) The relation between dune form and wind direction. Geol Fören. I Stockholm Förhandl, vol 54, no 338
FAO (1986) Ressources naturelles et environnement pour l'alimentation et l'agriculture en Afrique. Cahiers FAO Environnement et énergie, Rome, no 6, 87 pp
FAO-UNEP (1978) Report of an expert consultation on methodology for assessing soil degradation. Rome, Nairobi
FAO-UNESCO (1971–1979) Carte des sols du monde, vols 1–10. Paris
Farmer G, Wigley TML (1985) Climatic trends for tropical Africa. A research report for the overseas development administration. Climatic research unit. School of environment sciences. Univ East Anglia, Norwich, 136 pp
Farrington I (1974) Irrigation and settlement pattern: preliminary research results from the north coast of Peru. In: Downing TE, Gibson McG (eds) Irrigation's impact on society. Antropological Papers of the University of Arizona 25, Tucson, pp 83–94
Finkel M (1992) Drought contingency planning. Lessons from Kenya. Splash 8(2):4–6
Flint RF, Bond G (1968) Pleistocene sand ridges and pans in western Rhodesia. Bull Geol Soc Am 79:299–314
Floret C, Pontanier R (1982) L'aridité en Tunisie présaharienne. Trav et doc ORSTOM, Paris, no 150, 544 pp
Folk R (1971) Longitudinal dunes of the northwestern edge of the Simpson Desert, Northern Territory, Australia. I. Geomorphology and grain size relationships. Sedimentology 16(1–2):5–54
Ford J (1971) The role of Trypanosomiasis in African ecology. Oxford University Press, Oxford
Francis CF (1994) Plants on desert hillslopes. In: Abrahams AD, Parsons AJ (eds) Geomorphology of desert environments. Chapman & Hall, London, pp 243–254
Frere HBE (1870) Note on the Runn of Cutch. J R Geogr Soc 40:181–207
Freund J (1984) La décadence. Histoire sociologique et philosophique d'une catégorie de l'expérience humaine. Sirey, Paris, 408 pp

Friedman R (1989) Grazing cattle can change local climate. New Sci 123(1680):30
Fryberger SG (1979) Dunes forms and wind regime. In: McKee E (ed) A study of global sand seas. USGPO, Washington, DC
Furon E (1963) Le problème de l'eau dans le monde. Payot, Paris, 251 pp
Gallais J (1975) Pasteurs et paysans du Gourma. La condition sahélienne. Mém/ CEGET, Bordeaux, 239 pp
Gallais J (1989) Systèmes sociaux et systèmes alimentaires à l'épreuve de la sécheresse. In: Bret B (éd) Les hommes face aux sécheresses. EST-IHEAL, Paris, no 42, pp 229–236
Gautier EF (1935) Sahara, the Great Desert. Columbia Univ Press, New York, 264 pp
Gibson McG (1974) Violation of fallow and engineered disaster in Mesopotamian civilization. In: Downing TE, Gibson McG (eds) Irrigation's impact on society. Antropological Papers of the University of Arizona 25, Tucson, pp 7–19
Gifford GF, Merzougi M, Achouri M (1986) Spatial variability characteristics of infiltration rates on a seeded rangeland site in Utah, USA. In: Joss PJ, Lynch PW, Williams OB (eds) Rangeland: a resource under siege. Australian Academy of Sciences, Canberra, pp 46–47
Gilbert P (1971) Dictionnaire des mots nouveaux, Hachette-Tchou, Paris, p 158
Gillet H (1967) Essai d'évaluation de la biomasse végétale en zone sahélienne (végétation annuelle). J Agron Trop Bot Appl:123–158
Gillet H (1974) Tapis végétal et pâturages du Sahel. In: Le Sahel: bases écologiques de l'aménagement. Notes techniques du MAB, UNESCO, Paris, pp 21–27
Golson J (1977) No room at the top: agricultural intensification in the New Guinea Highlands. In: Allen J, Golson J, Jones R (eds) Sunda and Sahul: prehistoric studies in South-East Asia, Melanesia and Australia. Academic Press, London, pp 601–638
Golson J, Hughes PJ (1976) The appearance of plant and animal domestication in New Guinea. In: Garanger J (ed) La Préhistoire Océanienne, IXe Congr l'Union Int Sci Préhistoriques et Protohistoriques, Nice, pp 88–100
Gonzalez-Hidalgo C (1991) Aspect, vegetation and erosion on slopes in the Violada area, Zaragosa. PhD Thesis, University of Zaragosa
Gorman CF (1974) Modèles a priori et préhistoire de la Thailande. Etudes Rurales 53
Goudie A (1972) The concept of post-glacial progressive desiccation. School of Geography, University of Oxford, Res Pap 4, 48 pp
Goudie A (1983) Calcrete. In: Goudie A, Pye K (eds) Chemical sediments and geomorphology: precipitates and residua in the near surface environment. Academic Press, New York, pp 93–131
Goudie A (1990) Desert degradation. In: Goudie AS(ed) Techniques for desert reclamation. Wiley, New York, 271 pp
Granier P (1975) Note sur les interactions plante/animal en zone sahélienne. In: Inventaire et cartographie des pâturages tropicaux africains. Actes Colloq Bamako, 3–8 mars (Mali), pp 225–228
Gribi A, Sai N, Younsi N (1992) Carte hydrogéologique du Hoggar et des Tassilis à 1:1 000 000 (Algérie). Hydrogéologie, BRGM, Orléans, pp 69–77
Grousset R, Leonard EG (ed) (1956) Histoire Universelle, tome 1. Des origines à l'Islam, Encyclopédie de la Pléiade, Gallimard, Paris
Grouzis M (1987) Structure, productivité et dynamique des systèmes écologiques sahéliens (Mare d'Oursi, Burkina Faso). Thèse Doctorat d'Etat, Univ Paris sud Orsay, 336 pp
Grove AT (1958) The ancien erg of Hausaland and similar formations on the south side of the Sahara. Geogr J 124:526–533
Guillemin C, Roux JC (1992) Pollution des eaux souterraines en France; bilan des connaissances, impacts et moyens de prévention. BRGM, Orléans, 262 pp
Guiraud R (1989) Les barrages d'inféroflux; leur intérêt pour l'Afrique saharienne et sahélienne. In: UNESCO, 1990. The state-of-the-art of hydrology and hydrogeology in the arid and semi-arid areas of Africa. Proc Sahel Forum, Ouagadougou, Burkina Faso, 18–23 Feb 1989. International Water Resources Association, Urbana, Illinois, pp 321–329
Gunn R, Kinzer GD (1949) Terminal velocity of water droplets in stagnant air. J Meteorol 6:243
Gupta IC (1979) Use of saline water in agriculture: in semi-arid zones of India. Oxford and IBH Publishing, New Delhi
Guyot G (1983) Manuel sur l'utilisation des brise-vent dans les zones arides. Les cahiers de conservation des sols. FAO, Rome
Guyot G (1987) Les effets aérodynamiques et microclimatiques des brise-vent et des aménagements régionaux. In: Reifsnyder WS, Darnhofer TO (eds) Meteorology and agroforestry. ICRAF, WMO, UNEP, GTZ, Nairobi, pp 485–520
Gvozdetskii NA, Mikhailov NE (1978) Géographie physique de l'URSS, 3rd edn. MISL, Moscow (in Russian)
Hack JT (1941) Dunes of the western Navajo county. Geogr Rev 31:240–263

Hammer UT (1986) Saline lake ecosystems of the world. Junk, Dordrecht, 616 pp
Hanna SR (1969) Formation of longitudinal sand dunes by large helical eddies in the atmosphere. J Appl Meteorol 8(6):874-883
Harrison P (1987) The greening of Africa: breaking through in the battle for land and food. Paladin Grafton Books, London, 380 pp
Hassan FA, Stucki BR (1987) Nil floods and climatic change. In: Rampino MR, Sanders JE, Newman WS, Konigsson LK (eds) Climate, history, periodicity and predictability. Van Nostrand Reinhold, New York, pp 37-46
Hastings JD (1971) Sand streets. Meteorol Mag 100(1186):155-159
Heathcote RL (1980) Perception of desertification. UN University, Tokyo
Heathcote RL (1983) The arid lands: their use and abuse. Longman, London
Hedin S (1903) Central Asia and Tibet, vols 1, 2. Charles Scribners New York, 608 pp
Hellden U (1984) Drought impact monitoring; a remote sensing study of desertification in Kordofan, Sudan. Lunds Universitets Naturgeografiska Institution, Rapporter och Notiser, 61, Lund, Sweden, 61 pp
Hellden U (1988) Desertification monitoring: is the desert encroaching? Desertification Control Bulletin 17, UNEP, Nairobi, Kenya, pp 8-11
Ho PT (1975) The cradle of the East. Chinese Univ Press and Univ Chicago Press, Hong Kong and Chicago
Hollis GE (1990) Environmental impacts of development on wetlands in arid and semi-arid lands. Hydrol Sci (J Sci Hydrol) 35(4):411-428
Hornburg CD (1987) Desalination for remote areas. In: Developing world water. Grosvenor Press Int, Hong Kong, pp 230-232
Hubert H (1920) Le dessèchement progressif en Afrique ocidentale. Bulletin du Comité d'Etudes Historiques et Scientifiques de l'AOF, oct-déc, pp 401-463
Hubert P, Carbonnel JP, Chaouche A (1989) Segmentation des séries hydrométéorologiques; applications à des séries de précipitations et de débits de l'Afrique de l'Ouest. J Hydrol 110:349-367
Hubl K (1986) The nomadic livestock production system of Somalia. In: Conze P, Labahn T (eds) Agriculture in the winds of change. Epi Dokumentation No 2, Somalia, pp 55-72
Hudson N (1963) Rainfall size distribution in high intensity storms. Rhod J Agric Res 1(1):6-11
Hudson N (1971) Soil conservation. Batsford Ltd, London, 330 pp
Huetz De Lemps A (1970) La végétation de la terre, Masson, Paris, 130 pp
Hunt E, Hunt RC (1974) Irrigation, conflict and politics: a Mexican case. In: Downing TE, Gibson McG (eds) Irrigation's impact on society. Antropological Papers of the University of Arizona 25, Tucson, Arizona, pp 129-158
Hurault J (1975) Surpâturage et transformation du milieu physique. Formations végétales, hydrologie de surface, géomorphologie. L'exemple des hauts plateaux de l'Adamaoua (Cameroun). Etudes de photointerprétation, Inst Geogr Nation 7, Paris, 218 pp
Issar AS (1990) Water shall flow from the rock. Hydrogeology and climate in the lands of the Bible. Springer, Berlin Heidelberg New York, 213 pp
Jacks GV (1939) Soil erosion. The rape of the earth. A survey of soil erosion. In: Jacks GV, Whyte RO (eds) Soil erosion and its control. Faber, London, 313 pp
Jackson RD, Idso SB (1975) Surface albedo and desertification. Science 189:1012-1015
Jacobsen T, Adams RM (1958) Salt and silt in ancient Mesopotamian agriculture; progressive changes in soils salinity and sedimentation contributed to the breakup of past civilizations. Science 128(3334):1251-1257
Janzen J (1986) The process of nomadic sedentarisation - distinguishing features, problems and consequences for Somali development policy. In: Conze P, Labahn T (eds) Agriculture in the winds of change. Epi Dokumentation, Somalia, No 2, pp 73-89
Jarman TRW, Butler KE (1971) Livestock management and production in the Kalahari. In: Proc Conf on Sustained production from semi-arid areas. Botswana Notes and Records, Special Edition 1, pp 132-139
Joly F (1957) Les milieux arides. Définition. Extension. Notes marocaines. Rabat 8:15-30
Jordan WM (1964) Prevalence of sand-dune types in the Sahara desert. Geol Soc Am (Spec Pap) 82:104-105
Joss PJ, Lynch PW, Williams OB (1986) Rangelands: a resource under siege. Cambridge University Press, Cambridge
Kalaora B (no date) Une eau "Moderne" Pour une politique de l'eau, Passage - Sretie info, Paris, pp 10-11
Kalenov GS, Muklanmedov G (1992) Vegetation of the Karakum. In: Kar A, Abichandani RK, Anantharam K, Joshi DC (eds) Perspectives on the Thar and the Karakum. Dept of Science and Technology, Min of Sci and Technol, Govt of India, New Delhi, pp 110-117

Kanthack FE (1930) The alledged desiccation of South Africa. Geogr J 76:516-521
Kassas M (1987) Drought and desertification. Land Use Policy 4(4):389-400
Kassas M (1995) Desertification: a general review. J Arid Environ 30:115-128
Kayasseh M, Schenk C (1989) Reclamation of saline soils using calcium sulphate from the titanium industry. Ambio XVIII(2):124-127
Khan MA (1989) Development of surface water resources. In: Kolarkar AS, Joshi DC, Sharma KD (eds) Rehabilitation of arid ecosystem. Scientific Publishers, Jodhpur, India, pp 136-143
King D (1956) The Quaternary stratigraphic record of Lake Eyre North and the evolution of existing topographic forms. R Soc South Aust Trans 79:93-103
Kobori I, Kubo S, Takahas Y (1980) Foggara in the Algerian Sahara. Université de Tokyo, Tokyo, 20 pp
Koechlin J (1989) Adaptation des systèmes agro-pastoraux aux milieux du Niger et dans la Paraiba. In: Bret B (éd) Les hommes face aux sécheresses. EST-IHEAL, Paris, no 42, pp 317-321
Kononova MM (1975) In: Giesenking JE (ed) Soils components, vol 1. Springer, Berlin Heidelberg New York, pp 475-526
Koster EA (1988) Ancient and modern cold-climate aeolian sand deposition: a review. J Quat Sci 3:69-83
Kovda VA (1983) Loss of productive land due to salinization. Ambio XII(2):91-93
Kundzewick ZW, Gottschalk L, Webb B (1991) Hydrology 2000. IAHS, Paris, Publ 171, 100 pp
Labahn T (1982) Nomadenansedlungen in Somalia. In: Scholz F, Janzen J (eds) Nomadismus - Ein Enwicklungsproblem? Berlin, pp 81-95
Lageat Y (1994) Le désert duNamib central. Un ancien désert toujours vivant. Ann Géogr (Paris)103:339-360
Laing MV (1994) Drought monitoring and advisory services in South Africa. Drought Network News 6(1):12-15
Lamachère JM, Serpantié G (1990) Valorisation agricole des eaux de ruissellement en zone soudano-sahélienne (Burkina-Faso), province de Yatenga, région de Bidi. UNESCO. The state-of-the-art of hydrology and hydrogeology in the arid and semi-arid areas of Africa. Proc Sahel Forum, Ouagadougou, Burkina Faso, 18-23 Feb 1989, International Water Resources Association, Urbana, Illinois, pp 934-944
Lamb P(198). Rainfall in Subsaharan West Africa during 1941-83. Z Gletscherkd Glazialgeol 21:131-139
Lampert JR (1967) Horticulture in the New Guinea highlands - C_{14} dating. Antiquity 41:307-309
Lamprey HF (1975) Report on the desert encroachment reconnaissance in northern Sudan. 21 Oct to 10 Nov, UNESCO, Paris/UNEP, Nairobi, mimeo, 16 pp
Lancaster N (1981) Grain size characteristics of Namib Desert linear dunes. Sedimentology 28:115-122
Lancaster N (1982) Linear dunes. Progr Physical Geogr 6:475-504
Lancaster N (1989) Star dunes. Progr Physical Geogr13(1):67-91
Lavauden L (1927) Les forêts du Sahara. Rev Eaux For (Paris) 6(LXV):265-277; 7(LXV):329-341
Laws JO (1941) Measurements of fall-velocity of water-drops and rain-drops. Trans Am Geophy Union 22, 709 pp
Le Houérou HN (1989) La variabilité de la pluviosité annuelle dans quelques régions arides du monde; ses conséquences écologiques. In: Bret B (éd) Les hommes face aux sécheresses. EST-IHEAL, Paris, no 42, pp 127-138
Leprun JC (1989) Etude comparée des facteurs et des effets de l'érosion dans le Nordeste du Brésil et en Afrique de l'Ouest. In: Bret B (éd) Les hommes face aux sécheresses. EST-IHEAL, Paris, no 42, pp 139-154
Leroi-Gourhan A (1956) La Préhistoire, Histoire Universelle 1. Des origines à l'Islam, Encyclopédie de la Pléiade, NRF, Paris, 1863 pp
Létolle R, Mainguet M (1993) Aral. Springer, Berlin Heidelberg New York, 357 pp
Létolle R, Bendjoudi H (1997) Histoires d'une mer au Sahara. L'Harmattan, Paris, 221 pp
Loup J (1974) Les eaux terrestres; hydrologie continentale. Masson, Paris, 174 pp
Lovenstein H, Zohar RY, Aronson J (1987) Water-harvesting based agroforestry in the arid regions of Israel. Meteorology and agroforestry. Int Worksh on the Application of meteorology to agroforestry systems planning and managment, Beer Sheva, Israel, pp 241-244
Lowdermilk WC (1935) Man-made deserts. Pacific affairs. Univ British Columbia. In: Morgan RPC (ed) Soil erosion and its control. Hutchinson Ross Publ, Van Nostrand Reinhold, New York, pp 409-419
Lozet J, Mathieu C (1986) Dictionnaire de science du sol. Technique et Documentation, Paris, 269 pp
Mabbutt JA (1985) Desertification of the world's rangelands. Desertification Control Bull 12:1-11
Mabbutt JA, Sullivan ME (1968) The formation of longitudinal dunes. Evidence from the Simpson Desert. Aust Geogr 10:483-487
Mace R (1991) Overgrazing overstated. Nature 349:280-281
Madigan CT (1936) The Australian sand ridge deserts. Geogr Rev, New York, vol XXV, pp 205-227

References

Madigan CT (1946) The sand formations. Simpson Desert Expedition, 1939, Sci Rep 6. Geol R Soc South Aust Trans 70(1):43–63

Mainguet M (1968) Le Borkou, aspects d'un modelé éolien. Ann Géogr 77:296–322

Mainguet M (1972) Le Modelé des Grès. Problèmes généraux. Institut Géographique National (in Etudes de Photo-Interprétation), 2 vols, Paris, 637 pp

Mainguet M (1984) A classification of dunes based on aeolian dynamics and the sand budget. In: El-Baz F(ed) Deserts and arid lands. Martinus Nijhoff, The Hague, pp 31–58

Mainguet M (1992) A global open wind action system: the Sahara and the Sahel. In: Sadek A (ed) 1st Int Conf on Geology of the Arab World, vol. 2. Remote Sensing Session, 19–23 Jan 1992, pp 33–42

Mainguet M (1994a) Desertification. Natural background and human mismanagement, 2nd edn. Springer Study Edition. Springer, Berlin Heidelberg New York, 314 pp

Mainguet M (1994b) Désertification: quels sont les vrais problèmes? L'information Géographique 58. A Colin, Paris, pp 58–62

Mainguet M (1995) L'homme et la sécheresse. Masson, Paris, 335 pp

Mainguet M, Callot Y (1978) L'erg de Fachi Bilma (Tchad-Niger). Contribution à la connaissance de la dynamique des ergs et des dunes des zones arides chaudes, vol 19. Mém Docum, CNRS, Paris, 184 pp

Mainguet M, Chemin M-C (1979) Lutte contre l'ensablement des palmeraies et des oasis dans le Sud marocain. Rapport technique: étude préliminaire de l'avancement du sable. FAO, Rome, 32 pp

Mainguet M, Chemin M-C (1987) Nappes vives et nappes vêtues du Niger. Quantification de leur vulnérabilité vis-à-vis des actions éoliennes. Processus et mesures de l'érosion. 25ème Congr Int Géographie (UGI), Paris, 1984, CNRS, Paris, 1 vol, pp 101–112

Mainguet M, Chemin M-C (1988) La notion de budget sédimentaire sablo-éolien appliquée aux dépôts de sable du Sahara et du Sahel dans l'analyse de l'état de surface des Grands Ergs Oriental et Occidental. In honour of Fernand Joly. CERCG du CNRS, Paris, pp 113–130

Mainguet M, Chemin M-C (1991) Wind degradation on the sandy soils of the Sahel of Mali and Niger: its part in desertification. NATO advanced research workshop on sand, dust and soil in their relation to aeolian and littoral processes, 14–18 May, Sandjberg, University of Aarhus, Denmark. In: Barndorff-Nielsen OE, Willets BB (ed) Acta Machanica/Suppl 2, 250 pp

Mainguet M, Létolle R (1994) L'Aral est-il un lac hydro-éolien? Rev Géomorphol Dynam XLIII(1):27–34

Mainguet M, Chemin MC, Mozet MP (1981) Lutte contre l'ensablement des palmeraies et des oasis dans le Sud marocain. Rapport technique no 2. FAO, Rome, 196 pp

Maley J (1981) Etudes palynologiques dans le bassin du lac Tchad et paléoclimatologie de l'Afrique nord-tropicale de 30 000 ans à l'époque actuelle. ORSTOM, Paris, mém, 593 pp

Mamou A (1993) Ressources en eau des régions arides et désertiques du Maghreb. Cours sur le Développement des zones arides et désertiques, 8 nov–3 déc, IRA, Médénine, Tunisie, 36 pp (inédit)

Margat J (1992) Suggestions pour une cartographie des ressources en eau du continent africain à petite échelle. Hydrologie BRGM 1–2, Orléans, pp 119–122

Marouf N (1980) Lecture de l'espace oasien. Sinbad, Paris, 281 pp

Marques-Pereira J (1989) Economie et politique de la faim dans le nordeste brésilien. In: Bret B (éd) Les hommes face aux sécheresses. EST-IHEAL, Paris, no 42, pp 257–263

Martin AGP (1908) Les oasis sahariennes. Imprimerie algérienne, Alger, 339 pp

Martonne de E, Aufrère L (1928) L'extension des régions privées d'écoulement vers l'océan. Union Géographque Internationale, publ no 3, Paris, 200 pp

Marty A (1989) Stratégies pastorales et logiques d'intervention face à la sécheresse au Mali. In: Bret B (éd) Les hommes face aux sécheresses. EST-IHEAL, Paris, no 42, pp 289–294

Masson VM (1970) Djeitun settlement (the problem of farming of producing economy). Papers of the Archaelogical Institute, Moscow, 180 pp

Mazor E, Verhagen B, Sellschop J, Jones M, Robins N, Hutton N, Jennings C (1977) Northern Kalahari groundwaters: hydrologic, isotopic and chemical studies at Orapa, Botswana. J Hydrol 34:203–33

McCauley JF, Grolier MJ, Breeds CS (1977) Yardangs of Peru and other desert regions. US Dept of the Interior, Geological Survey, Flagstaff, Arizona, 177 pp

McCauley JF, Breeds CS, Schaber GG, McHugh WP et al. (1986) Paleodrainages of the eastern Sahara – the radar rivers revisited. Institute of Electrical and Electronics Engineers, Transactions of Geoscience and Remote Sensing, Flagstaff, Arizona, GE-24, pp 624–648

Meckelein W (1980) Saharan oases in crises, vol 95. Stuttgarter Geographische Studien, Stuttgart, pp 173–203

Meigs P (1953) World distribution of arid and semi-arid homoclimates. Arid zone hydrology. UNESCO arid zone research series 1, Paris, pp 203–209

Middleton NJ (1990a) Wind erosion and dust-storm control. In: Goudie A (ed) Techniques for desert reclamation. Wiley, Chichester, pp 86–108

Mietton M (1989) Tentatives de maîtrise de l'eau de surface au Burkina-Faso. In: Bret B (éd) Les hommes face aux sécheresses. EST-IHEAL, Paris, no 42, pp 377–383

Molinier M, Cadier E, Gusmao A (1989) Les sécheresses du Nordeste brésilien. In: Bret B (ed) Les hommes face aux sécheresses. EST-IHEAL, Paris, pp 85–92

Monod T (1958) Majâbat al-Koubrâ. Contribution à l'étude de "l'Empty Quater" ouest-saharien. Mém IFAN 53, Dakar, 406 pp

Montesquieu de C (1748) L'Esprit des lois, Livre XIVe, Des lois dans le rapport qu'elles ont avec la nature du climat, Chapitre II. Combien les hommes sont différents dans les divers climats. Bibliothèque Française, Plon, 1912

Mortimore M (1987) Shifting sands and human sorrow: social response to drought and desertification. Desertification Control Bull 14:1–14

Moseley ME (1974) Organizational preadaptation to irrigation: the evolution of early water-management systems in coastal Peru. In: Downing TE, Gibson McG (eds) Irrigation's impact on society. Antropological Papers of the University of Arizona 25, Tucson, Arizona, pp 77–82

Murray GW (1955) Water from the desert. Geogr J CXXI:171–181

Myers N (1987) Natural resource systems for sustainable development. In: Economic Development Institute of The World Bank, 1987. Seminar 1986 on Land and water resources management, Wahington, DC, pp 2–22

Navar J, Bryan R (1990) Interception loss and rainfall redistribution by three semi-arid growing shrubs in northeastern Mexico. J Hydrol 115:51–63

Neely JA (1974) Sassanian and early Islamic water-control and irrigation systems on the Deh Luran Plain, Iran. In: Downing TE, Gibson McG (eds) Irrigation's impact on society. Antropological Papers of the University of Arizona 25, Tucson, Arizona, pp 21–42

Nesson C (1972) Densité des puits piézométriques dans les palmeraies de l'Oued Righ. In: Les problèmes de développement du Sahara septentrional. Coll de Ouargla, 25–26 Sept 1971, Alger, pp 181–203

NGO (Non-Governmental Organization, France) (1993) Contribution à la Conférence préparatoire de Nairobi – Vers une convention sur la désertification. Collectif de liaison des organisations de solidarité internationale Paris, 25 pp

Nieman WA et al. (1978) A note on precipitation at Swakopmund. Madoqua 11(1):69–73

Noj A, Blaikie PM (1989) Land degradation, stocking rates and conservation policies in the communal rangeland of Botswana and Zimbabwe. Land Degrad Rehabil 1:101–123

Oberlander TM (1994) Global desert: a geomorphic comparison. In: Abrahams AD, Parsons AJ (eds) Geomorphology of desert environments. Chapman & Hall, London, pp 13–35

Oldeman LR, Hakkeling RTA, Sombroek WG (1990) World map of human-induced soil degradation. ISRIC, Wageningen and UNEP, Nairobi, colour map, 3 sheets 1/13 000

Olivry JC, Chastenet M (1989) Evolution de l'hydraulicité du fleuve Sénégal et des précipitations dans son cours inférieur, depuis le milieu du XXe siècle. In: Bret B (ed) Les hommes face aux sécheresses, Paris. IHEAL-EST, Samuel Tasted éd, coll Trav et Mém de l'IHEAL no 42, pp 115–124

Ollier CD (1977) Early landform evolution. In: Jeans DN (ed) Australia: a geography. St Martin's Press, London

Ollier CD, Tuddenham WG (1961) Inselbergs of central Australia. Z Geomorphol 5:257–276

Olsson L (1985) An integrated study of desertification: applications of remote sensing, GIS and spatial models in semi-arid Sudan. Meddelanden fran Lunds Universitets Geografiska Institution, Avhandlingar, Suède, 98, 170 pp

O'Sullivan R (1992) Irrigation in the USSR. World Bank Technical-Pap 178, Washington, DC, pp 77–88

Otte M (1994) Origine de l'homme moderne: approche comportementale. CR Acad Sci Paris 318(II):267–272

Oulheri T (1990) Etude aérologique et calcul des déplacements potentiels des sables dans la province de Laayoune. Association marocaine permanente des Congrès de la route, 3e Congr National de la route, Fez, pp 135–150

Pagney P (1976) Les climats de la terre. Masson, Paris, 151 pp

Pélissier P (1980) L'arbre dans les paysages agraires de l'Afrique noire. L'arbre en Afrique tropicale, la fonction et le signe. Cahiers ORSTOM, Série Sciences Humaines, XVI (3–4):131–136

Pélissier P (1989) Rapport général. Sécheresses, Sociétés et Développement. In: Bret B (éd) Les hommes face aux sécheresses. EST-IHEAL 42, Paris, pp 19–26

Perkins JS (1990) Drought, cattle-keeping and range degradation in the Kalahari, Botswana. In: Stone GJ (ed) Pastoralists responses to drought. Aberdeen University Press, Aberdeen

Pessoa D (1989) Sécheresses du Nordeste. Variations des interprétations et des politiques publiques. In: Bret B (éd) Les hommes face aux sécheresses, EST-IHEAL, Paris, no 42, pp 399–403

Perkins JS, Thomas DSG (1993) Spreading deserts or spatial confined environmental impacts? Land degradation and cattle ranching in the Kalahari desert of Botswana. Land Degrad Rehabil 4(3):178–194

Petrov M (1962)Types de déserts de l'Asie Centrale. Ann Géogr 381(LXXI):131–155

Pettersen S (1940) Weather analysis and forecasting. A textbook on synaptic meteorology. McGraw-Hill, New York, 505 pp

Planhol de X (1958) De la plaine pamphylienne aux lacs pisidiens: nomadisme et vie paysanne. Biblioth histor et archéol de l'Instit franç d'archéol d'Istanbul, III, Paris, 495 pp

Planhol de X, Rognon P (1970) Les zones tropicales arides et subtropicales. A Colin, Paris, 487 pp

PNUD (United Nations Development Programme) (1994) Update, vol 7, no 5 du 14-03-1994

Poissonet J, Chambris F, Touré I (1992) Equilibre et déséquilibre des phytocénoses herbacées sahéliennes. Influence de la pluviosité annuelle et de la proximité des points d'eau. In: Le Floc'h E, Grouzis M, Cornet A, Billé JC (éds) L'aridité, une contrainte au développement. Editions ORSTOM, Paris, pp 283–296

Poncet Y (1974) La sécheresse en Afrique sahélienne; une étude micro-régionale en République du Niger. La région des Dallols. OCDE, Paris, 50 pp

Poursin G (1974) A propos des oscillations climatiques: la sécheresse au Sahel. Annales (Economies, Sociétés, Civilisations), no 3, Paris

Pouyaud B (1985) Contribution à l'évaluation de l'évaporation de nappes d'eau libre en climat tropical sec. Exemple du Lac de Bam et de la Mare d'Oursi (Burkina Faso), du Lac Tchad et d'açudes du Nordeste brésilien. Thèse Université Paris sud, Paris

Pye K (1982) Morphological development of coastal dunes in a humid tropical environment, Cape Bedford and Cape Flattery, North Queensand. Geogr Ann A 64:213–227

Pye K, Tsoar H (1990) Aeolian sand and sand dunes. Unwin Hyman, London, 396 pp

Querroum J (1990) Lutte contre l'ensablement des routes en régions sahariennes. Association marocaine permanente des Congrès de la route, 3e Congr Natl de la route, Fez, pp 125–134

Raskin P, Hansen E, Zhu Z, Stavisky D (1992) Simulation of water supply and demand in the Aral Sea region. Water International, Stockholm Environ Inst Boston Center, vol 17, no 2, pp 55–67

Razakov RM (1990) Ecological activities in the Aral Sea coastal region: investigations and programme of actions. Melioratsiya-i-Vodnoe-Kozuyaistvo 1:6–8

Reed C (ed) (1977) Origins of agriculture. Mouton, La Haye

Reichelt R (1992) The extension of the Old Ogolian desert in the Sahara-Sahel and the recent reactivation and desertification. Int Symp on Evolution of desert, Ahmadabad, Feb, 1 p (Summary)

Reiners WA (1973) Carbon and the biosphere. US AEC Conf 720510, Washington, pp 317–327

Renner GT (1926) A famine zone in Africa: the Sudan. Geogr Rev 16:583–596

Retaille D (1989) Mobilité des populations sahéliennes durant la sécheresse aggravée de 1984. In: Bret B (éd) Les hommes face aux sécheresses. EST-IHEAL, Paris, no 42, pp 277–296

Rhoades JD (1982) Reclamation and management of salt-affected soils after drainage. In: Proc 1st Annu Western Provincial Conf Rationalization of water and soil res and management, Lethbridge, Alberta, Canada, 29 Nov-2 Dec, pp 123–197

Rhoades JD (1988) The problem of salt in agriculture. Yearbook of science and the future Encyclopedia Britannica. Encyclopedia Britannica, Chicago, pp 118–135

Rhoades JD (1990) Soil salinity; causes and controls. In: Goudie AS (ed) Techniques for desert reclamation. Wiley, New York, pp 108–134

Rhoades JD, Loveday J (1990) Salinity in irrigated agriculture. In: Stewart BA, Nielsen DR (eds) Irrigation of agricultural crops. ASA Monogr, USA

Robertson-Rintoul MJ (1990) A quantitative analysis of the near-surface wind flow pattern over coastal parabolic dunes. In: Nordstrom KF, Psuty N, Carter B (eds) Coastal dunes, form and process. Wiley, New York, pp 57–78

Roche MF (1986) Dictionnaire français d'hydrologie de surface. Masson, Paris, 287 pp

Rochette RM (1989) Le Sahel en lutte contre la désertification, leçons d'expériences, CILSS/PAC/GTZ, Paris. Joseph Margraf, Weikersheim, 592 pp

Rognon P (1989) Biographie d'un désert. Plon, Paris, 347 pp

Rouvillois-Brigol M (1972) Les transformations de l'oasis de Ouargla. In: Les problèmes de développement du Sahara septentrional. Coll de Ouargla, 25-26 Sept 1971, Alger, pp 35–48

Sablier E (1992) Au barrage d'Atatürk; cet immense ouvrage rend à la Turquie la haute main sur l'Orient arabe. Valeurs Actuelles 2905:24–25

Sanlaville P (1981) Réflexion sur les conditions générales de la quête de l'eau au Proche-Orient. In: L'homme et l'eau en Méditerranée et au Proche-Orient, Travaux de la maison de l'Orient. Presses Universitaires de Lyon, no 2, vol 1, pp 9–20

Sarnthein M, Walger E (1974) Der äolische Sandstrom aus der W. Sahara zur Atlantikküste. Geol Rundsch 63(3):1065–1087

Sauer CO (1952) Agriculural origins and dispersal. Am Geogr Soc, New York
Schick AP, Lekach J, Schwartz U (1987) Nahal-Yael. A research catchment representative of hyper-arid environments. In: Worksh on Erosion, transport and deposition processes with emphasis on semi-arid and arid areas. Field guidebook, Jerusalem, 169 pp
Schlumberger D (1951) La Palmyrène du Nord-Ouest. Bibliothèque Archéologique et Historique, Geuthner, Paris, 192 pp
Schneider JL (1994) Le Tchad depuis 25 000 ans. Géologie, archéologie, hydrologie. Masson, Paris, 134 pp
Schoeller H (1941) L'influence du climat sur la composition chimique des eaux souterraines vadoses. Bull Soc Géol 5(XI):267–289
Schoeller H (1948) Les modifications de la composition chimique de l'eau dans une même nappe. Association internationale d'hydrologie scientifique, Assemblée d'Oslo, pp 124–129
Schoeller H (1959) Hydrologie des régions arides. Progrès récents. UNESCO, Paris, 126 pp
Scholz F (1982) Land Verteilung und Oasen sterben. Das Beispiel des Omanischen Küstenebene "Al Batinah". Erdkunde, Bonn, pp 199–208
Seignobos C (1984) La sécheresse 1969–1974 au Tchad: la difficile interprétation des conséquences. Bull Soc Languedocienne Géogr:185–200
Shainberg I, Morin J (1987) Runoff and erosion control in cultivated fields. In: Worksh on Erosion, transport and deposition processes with emphasis on semi-arid and arid areas. Field guidebook, Jerusalem, p 26
Sharon R(1972) The spottiness of rainfall in a desert area. In: Worksh on Erosion, transport and deposition processes with emphasis on semi-arid and arid areas. Field guidebook, Jerusalem, 169 pp
Sillitoe P (1993) Losing grounds? Soil loss and erosion in the highlands of Papua New Guinea. Land Degrad Rehabil 4(3):143–166
Simaika Y (1952) Recherches sur l'hydrologie et la mécanique des fluides et l'importance des eaux souterraines dans les régions arides du nord-est africain. L'hydrologie de la zone aride. UNESCO, Paris, pp 42–58
Sircoulon J (1987) Variation des débits des cours d'eau et des niveaux des lacs en Afrique de l'Ouest depuis le début du XXe siècle. Symp AISH, Vancouver, no 168, pp 13–25
Sircoulon J (1989a) La sécheresse du point de vue climatique, hydrologique et agronomique. In: Bret B (éd) Les hommes face aux sécheresses. EST-IHEAL, Paris, no 42, pp 65–68
Sircoulon J (1989b) Bilan hydropluviométrique de la sécheresse 1968–84 au Sahel et comparaison avec les sécheresses des années 1910 à 1916 et 1940 à 1949. In: Bret B (éd) Les hommes face aux sécheresses. EST-IHEAL, Paris, no 42, pp 107–114
Sircoulon J (1992) Caractéristiques des ressources en eau de surface en zones arides de l'Afrique de l'Ouest. Variabilité et évolution actuelle in L'Aridité – une contrainte au développement. ORSTOM, Paris, pp 53–66
Skarpe C (1991) Impact of grazing in savanna ecosystems. Ambio XX:351–356
Spooner B (1974) Irrigation and society: the Iranian Plateau. In: Downing TE, Gibson McG (eds) Irrigation's impact on society. Antropological Papers of the University of Arizona 25, Tucson, pp 43–58
Sprigg RC (1959) Stranded sea beaches and associated sand accumulations of the Upper Southeast. Trans R Soc S Aust 82:182–193
Stamp LD (1940) The southern margin of the Sahara; comments on some recent studies on the question of desiccation in West Africa. Geogr Rev 30:297–300
Stone WJ (1986) Natural recharge in southwestern landscapes, examples from New Mexico. Proc Conf on Southwestern ground water issues, Tempe, Arizona. Natl Water Well Assoc, Dublin, Ohio, pp 595–602
Sturt CH (1849) Narrative of an expedition into central Australia during the years 1844, 5 and 6, vol 1. TW Boone, London, 426 pp
Susuki H (1981) The transcendent and environments. Addis Abeba Sha,Yokohama, Japan, 134 pp
Szymczak K (1997) The perspectives of studying an early Holocene occupation of the Kyzylkum. In: Proc Congr of the Deserts in central Asia, protection and development, Warszawa, 3–5 Nov
Tamura S (1992) Toward the establishment of grasses and crops on the loess plateau in China. Int Symp on Land degradation and its biological and technological rehabilitation in drylands, Tottori, Japan, pp 113–122
Teilhard de Chardin P (1956) La place de l'Homme dans la nature; le groupe zoologique humain. Albin Michel, Paris, 188 pp
The Club of Rome (1972) Limits to growth. Rome
Thomas DSG, Middleton N (1994) Desertification: exploding the myth. Wiley, New York, 194 pp

Thomas DSG, Shaw PA (1991) The Kalahari environment. Cambridge University Press, Cambridge, 284 pp
Thornes JB (1994) Catchment and channel hydrology. In: Abrahams AD, Parsons AJ (eds) 1994. Geomorphology of desert environments, vol 11. Chapman & Hall, London, pp 257–287
Thornthwaite CW (1948) An approach towards a rational classification of climate. Geogr Rev 38:55–94
Tilho J (1928) Variations et disparition possible du Tchad. Ann Géogr 37:238–260
Timberlake L (1985) L'Afrique en crise; la banqueroute de l'environnement. L'Harmattan – Earthscan, Paris, 294 pp
Tirouvallouvar (1988) Tiroukkoural. 1er siècle av JC. Editions de l'Océan Indien, Pondichéry, translated from Tamul, 269 pp
Toulmin C (1988) Smiling in the Sahel. New Sci 12 Nov:69
Toupet C (1972) Les variations interannuelles des précipitations en Mauritanien centrale. CR Soc Biogéogr 416–421:39–47
Toupet C (1984) Signification de la récente sécheresse qui a sévi dans le Sahel. Bull Soc Languedocienne Géogr 18(3–4):138–146
Toupet C (1989) Comparaison des sécheresses historiques et de la sécheresse actuelle au Sahel. Essai de définition de la sécheresse et de l'aridification. In: Bret B (éd) Les hommes face aux sécheresses. EST-IHEAL, Paris, no 42, pp 77–84
Toupet C (1992) Le Sahel. Nathan, Paris, 192 pp
Tricart J (1994) Ecogéographie des espaces ruraux. Nathan, Paris, 187 pp
Tsoar H, Møller JT (1986) The role of vegetation in the formation of linear sand dunes. In: Nickling WG (ed) Aeolian geomorphology. Allen & Unwin, Boston, pp 75–95
Tucker CJ, Fung IY, Keeling CD (1986) Satellite-derived vegetation index. Nature 319:1905
Twidale CR (1978) On the origin of Ayers Rock, central Australia. Z Geomorphol (Suppl Band) 31: 177–206
UNCED (1992) Report of the UN Conf on Environment and development, vol 1. UNCED, New York, 486 pp
UNCOD (United Nations Conference on Desertification) (1977) Desertification: its causes and consequences. Secretariat of United Nations. Conf on Desertification. Pergamon Press, Nairobi, Kenya
UNCOD (United Nations Conference on Desertification) (1978) Round-up, plan of action and resolutions, 29 Aug–9 Sept 1977, New York, 43 pp
UNEP (United Nations Environment Programme) (1991) Status of desertification and implementation of the UN plan of action to combat desertification. UNEP, Nairobi, 94 pp
UNEP (1992) The state of the global environment. Our Planet 4(2):4–9
UNEP and Commonwealth of Australia (1987) Drylands dilemna: a solution to the problem. Australian Government Publishing Service, Canberra
UNEP-GEMS – IUCC (no date) Fact Sheet 3
UNESCO (1990) The state-of-the-art of hydrology and hydrogeology in the arid and semi-arid areas of Africa. Proc Sahel Forum, Ouagadougou, Burkina Faso, 18–23 Feb 1989. International Water Resources Association, Urbana, Illinois, 990 pp
United States Salinity Laboratory Staff (1954) Diagnosis and improvement of saline and alkali soils. US Dept Agriculture Handbook 60, Washington, DC
Urvoy Y (1935) Terrasses et changements de climat quaternaires à l'est du Niger. Ann Géogr 44: 254–263
Valenza J (1970) Etude dynamique de différents types de pâturages naturels en Rép. du Sénégal (Survey of different types on natural pasture lands in Senegal Republic). Proc XI Int Grassland Congr, 1969, Dakar, pp 78–82
Valenza J, Diallo AK (1972) Etude des pâturages naturels du Nord Sénégal. IEMVT, Dakar-Maisons-Alfort. Et Agrostologique (Paris) 34:311 pp
Valenza J, Fayolle F (1965) Note sur les essais de charge de pâturages en République du Sénégal. Rev Elev Méd Vét Pays Trop 18(3):321–327
Van de Graaff WJE, Browe RWA, Buntin JA, Mackson NJ (1977) Relict early Cainozoic drainages in arid western Australia. Z Geomorphol 21:379–400
Verhagen B, Mazor E, Sellschop J (1974) Radiocarbon and tritium evidence for direct rain recharge to groundwaters in the northern Kalahari. Nature 249:643–4
Verlet B (1962) Le Sahara. Que sais-je?, 3e édn. PUF, Paris, 127 pp
Viers G (1968) Elément de climatologie. Nathan, Paris, 224 pp
Vinogradov VA, Mamedov ED (1972) Stratigrafia chetvertichnyukh otlozhnii nizoviev Zeravshana i yugo-zapadnykh Kyzyl-kumov v zvietie noveyshikh geologitfcheskikh i arkheologitcheskikh issledovanii. Biulletin Komiteta Izuchenia Chetvertitchnogo Perioda 38:142–148

Vivian RG (1974) Conservation and diversion: water-control systems in the Anasazi southwest. In: Downing TE, Gibson McG (eds) Irrigation's impact on society. Antropological Papers of the University of Arizona 25, Tucson, pp 95–112
Walker BH (1987) Determinants of tropical savannas. IUBS Monogr Ser, no 3, IUBS, Paris
Walker BH, Ludwig D, Holling CS, Peterman RM (1981) Stability of semi-arid savanna grazing systems. J Ecol 69:473–498
Warren A, Agnew C (1988) An assessment of desertification and land degradation in arid and semi-arid areas. International Institute for Environment and Development, Drylands Programme, University College, London, 72 pp
Warren A, Khogali M (1992) Assessment of desertification and drought in the Sudano-Sahelian region, 1985–1991. United Nations Sudano-Sahelian Organization (UNSO), New York, 102 pp
Watson A (1983) Gypsum crusts. In: Goudie A, Pye K (eds) Chemical sediments and geomorphology: precipitates and residua in the near surface environment. Academic Press, New York, pp 133–136
Watson A (1990) The control of blowing sand and mobile desert dunes. In: Goudie A (ed) Techniques for desert reclamation. Wiley, Chichester, pp 35–85
Weis H (1964) Murzuck – Blüte und Verfall einer Saharametropole. Bustan 5:22–36
Weiss H, Courty MA, Wetterstrom W, Guichard F, Senior L, Meadow R, Curnow A (1993) The genesis and collapse of Third Millenium North Mesopotamia civilization. Science 261:995–1004
White GF (1988) The environmental effects of the high dam at Aswan. Environment (Wash) 30(7):5–39
White R (1992) Livestock development and pastoral production on communal rangeland in Botswana. In: Worksh on New directions in African range management policy, Matopos, Zimbabwe, 13–17 Jan, Food Production and Rural Development Division, Commonwealth Secretariat, London
Williams OB, Calaby JH (1985) The hot deserts of Australia. In: Evenari M, Noy-Meir I, Goodall DW (eds) Ecosystems of the world, vol 12A. Hot deserts and arid shrublands. Elsevier, Amsterdam, pp 269–312
Winner I (1963) Some problems of nomadism and social organization among the recently settled Kazakhs. Cent Asien Rev XI:246–267, 355–373
Wispeleare de G (1980) Les photographies aériennes, témoins de la dégradation du couvert ligneux dans un géosystème sahélien sénégalais. Cahiers ORSTOM, Série Sciences Hum XVII(3–4):155–166
Wittfogel KA (1957) Oriental despotism: a comparative study of total power. Yale Univ Press, New Haven
WMO (World Meteorological Organization) (1983) Meteorological aspects of certain processes affecting soil degradation, especially erosion. Tech Note no 178, WMO no 591, Genève, Switzerland, 149 pp
Wollny E (1877) Untersuchungen über den Einfluss der Pflanzen und der Beschattung auf die physikalischen Eigenschaften des Bodens. Berlin
World Commission on Environment and Development (1987) Our common future: report of the World Commission on Environment and Development (the Bruntland Report). Oxford University Press, Oxford
Yaalon DH (1981) Pedogenic carbonate in aridic soils: magnitude of the pool and annual fluxes. Abstr, Int Conf on Aridic soils, Jerusalem
Yair A (1990) Runoff generation in a sandy area – the Nizzana Sands, western Negev, Israel. Earth Surf Proc Landforms 15:597–609
Yeager R (1989) Demographic pluralism and ecological crisis in Botswana. J Dev Areas 23:385–404
Zaletaev VS, Novikova NM (1990) Changes in biota of the Aral region as result of anthropogenic impacts in the period between 1950 and 1990. Nukus Int Conf on The Aral crisis: causes, consequences and ways of solutions, Nukus, Karakalpak, USSR, 2–5 Oct 1990
Zhu Zhenda (1990) Distribution of sand dunes in China. The principles and measures of sand dunes stabilization in China. Chinese Academy of Sciences, The Institute of Desert Research, Lanzhou, pp 1–4
Zhu Zhenda, Liu Shu (1983) Combating desertification in arid land semi-arid zones of China. Inst of Desert Research of Academia Sinica, Lanzhou, China, 69 pp

Geographic Index

A
Aarhus 118
Abéché 70
Abu Simbel 190
Addis Ababa 152
Aden 170
Adrar des Ifoghas 53, 61, 135
Afghanistan 18, 88, 161, 163, 216, 230, 231
Africa 8, 14, 23-27, 29-38, 40-42, 46-48, 54, 61, 69, 72, 77, 88, 100, 104, 111, 112, 139, 142, 146, 147, 150-152, 155, 156, 162, 176, 179, 185, 192, 204, 207, 233, 254, 257-263, 270-273
Agdz 179
Aïr 61, 135
Akcha Daria 164, 243
Aksu 246-248
Alar 248
Alaska 9
Algeria 49, 55, 106, 123, 128, 143, 149, 185, 213, 214, 216, 218, 221-225, 227, 232, 233, 264
Ali Kosh 163, 167
Alice Springs 103
Altyn Depe 163, 164
Amazonia 131
America 2, 6, 8, 9, 17, 23, 26, 69, 100, 122, 160, 177, 179, 207, 218, 239, 253, 260, 270, 271
Amezrou 178, 196
Aminabad 164
Amman 211
Amu Daria 22, 34, 88, 157, 158, 163-165, 181, 230, 236, 238, 240-244
Anatolia 23, 87, 172
Ancient Assyria 166
Andes 6, 16, 69, 142, 160, 177
Angola 14, 258
Antarctica 6, 9, 13, 170
Antofagasta 6, 14
Aoudia 227
Aoulef 11, 214, 225
Ar Rima 12
Arabia 5, 6, 20, 24, 26, 37, 83, 106, 108, 131, 145, 151, 161, 170, 185, 215, 216, 264
Arabian Peninsula 5, 12, 24, 30, 37, 211
Arad 171
Aragan 248
Aral 17-23, 51, 60, 88, 94, 102, 104, 139, 160-166, 182, 236, 238-245, 257, 258, 264, 265

Arctic 7, 9, 10
Arizona 209, 214, 220, 233
Ashkabad 18, 21
Asia 2, 5, 6, 11, 13, 17, 20, 23-25, 37, 38, 47, 56, 60, 61, 69, 73, 77, 83, 87, 88, 100, 136, 142, 147, 151, 154-157, 161, 165, 166, 173, 179, 192, 197, 207, 209, 212-218, 230, 233, 246, 254, 259, 260, 270-274
Asia Minor 60, 87, 215
Assad 87
Assiut 104, 189, 219
Assyria 166
Atacama 8, 14, 82
Ataiba 226
Atakor 53
Atbara 35
Atlantic 12, 18, 24, 132, 134, 162, 169, 250, 261
Atlantic coast 132, 134, 250, 261
Atlas 18, 105, 135, 143
Atrek 164
Australia 6, 10, 12, 13, 23, 54, 100, 103, 106, 110, 111, 131, 146, 152, 184, 218, 234, 254, 260, 262, 265, 270
Austria 157
Avdat 173, 215
Awash 187
Ayers Rock 14
Azamad 13

B
Bacho Kiro 154
Badan Jilin 266
Bafing 187
Baghdad 88, 240
Baidan Jaran 126
Bakel 33, 90, 91
Balkh 88
Baltic Sea 165, 236
Bam 85
Bamako 186
Bangkok 1
Bangladesh 236
Barada 225, 226
Bardagué 97
Bardai 97
Barsouki 17
Basra 87
Beer Sheva 173
Belgrade 185

Ben Gardane 228
Bendebal 163
Bengo 45
Benguela 14
Bilma 11, 210
Bindisi 44
Black Sea 18, 19, 165
Bogoria 30
Bokspit 34
Bol 85, 101
Bollène 1
Borborema 17
Borgou 264
Botswana 34, 107, 256
Boulder 83
Brahmaputra 160
Brava 199
Brik 12
Burkina Faso 30, 54, 85, 98, 99, 111, 135, 171, 175–177, 259

C
Caatinga 142, 150
Cairo 87, 103, 188–191, 234, 267
California 1, 14, 211, 218, 237
Cameroon 41, 145, 264
Canada 10, 99, 245
Cape Verde 171, 176, 185
Caribbean 207
Caruaru 111, 112
Casamance 141
Caspian Sea 17–19, 165, 181, 211, 217
Caucasus 18, 20
Ceara 30
central America 100, 160, 218
central Asia 5, 11, 13, 17, 20, 23, 56, 60, 61, 83, 100, 151, 154, 166, 209, 212, 213, 216, 218, 230, 246, 259
central Australia 10–13, 146
Chaco 23
Chalbi 26, 77
Champagne 68
Chardzu 21
Chari 54, 88, 90, 91, 101–103, 250
Chech 132, 214
Chihuahua 82
Chile 6, 14, 83, 170
China 5, 23–25, 30, 37, 52, 56–58, 60, 70, 83, 88, 93, 111, 112, 131, 132, 156, 160, 163, 180, 186, 191, 195–197, 214, 218, 240, 245–248, 256–259, 266, 271
Chobe 107
Choga Mami 163, 167
Choga Mish 163
Chott Fejej 223
CIS 88
Coachella Valley 181
Colorado 1, 83, 101, 181, 218, 264
Columbia 220
Copenhagen 267
crescent 25, 100, 124, 155–157, 166, 176
Crete 157, 183

D
Dadès 227
Dakar 61, 141
Dakhla 83, 218
Dakota 47
Dallols 43
Damascus 88, 211, 219, 225, 226
Darfour 11
Dashehaiz 247, 248
Dawasar 12
Dead Sea 155
Deh Luran 163
Dekkan 23
Denver 1, 218
Diama 186, 187
Diébégou 70
Diré 90, 148, 186
Diyala 167
Djaffarabad 163
Djanet 179
Djebel Marra 101, 132
Djeddi 215
Djerba 214
Djérid 223
Draa 55, 59, 178, 179, 196, 198, 209, 212, 218, 227–229
Dublin 2

E
East Africa 23, 26, 30, 42, 61, 77, 146, 152, 155, 259
Ecbatana 216
Eglab 61
Egypt 12, 53, 85, 88, 100, 104, 122, 131, 157, 158, 160, 161, 168, 170, 172, 186, 188, 218, 234, 236, 239
El Beid 101
El Kebir 93
El Krair 227
El Oued 49, 213
Elburz 211
Eleimentata 30
Ennedi 56, 93, 96, 97, 101
Er Raoui 12
Erdis 56
Essaouira 134, 193
Ethiopia 24, 26, 29, 31, 37, 54, 61, 69, 148, 187
Euphrates 87, 166–168, 239, 254, 264
Eurasia 154, 155
Eurasian plains 155
Europe 24, 29, 54, 56, 155, 186, 207, 215, 223, 239, 260, 270
Eyre 100, 131

F
Faim Steppe 21
Falo 28
Farkhad 165
Fayum 100, 162
Fderik 83
Feija 55
Fergana 165, 194, 209
Ferlo 13, 68, 73, 77, 146, 150, 258
Ferran 95

Geographic Index

Fezouata 227, 229
Fezzan 178, 179
France 1, 27, 68, 106, 122, 211

G

Galapagos 6
Ganges 160, 236
Gangir 168
Gavkor 164
Gawra 163
Geoksyour 163
Gezira 54, 264
Ghardaia 214
Gharsa 223
Ghât 179
Gila 209
Girzou 239
Gobi 161, 247
Gorouol 98
Gourara 214, 225
Gourma 69
Great Artesian Basin 106, 218
Great Sand Sea 13
Greece 2, 157, 172
Greenland 9
Guatemala 160
Gueskérou 98
Guinea 76, 156, 260
Gujarat 148

H

Haihé 245
Hama 218
Hannabou 227
Hassan Addakhil 229
Himalaya 17, 134
Himalayan chain 5
Hindu Kush 163
Hoggar 53, 61, 101, 106, 132, 135, 157
Hombori 41
Honduras 160
Hotan 197, 231, 246
Huaihé 245
Hula 99, 100
Humboldt 14

I

Igharghar 12, 215
Iguidi 12
Imperial Valley 181, 220
In Salah 214
India 23, 24, 30, 37, 38, 54, 70, 147, 148, 161, 168, 173, 180, 186, 215, 218, 234, 271
Indira Gandhi Canal 181
Indonesia 245
Indus 157, 160-163, 234, 238, 264
Iran 18, 23, 100, 128, 147, 151, 163, 167, 185, 186, 194, 197, 211, 217, 219, 225, 264
Iraq 2, 87, 155, 157, 166, 167, 234
Israel 2, 55, 82, 88, 99, 166, 184, 266
Italy 186

J

Jaisalmer 173
Japan 158, 270
Jericho 99, 220
Jerid 223
Jerusalem 173
Jordan 2, 88, 99, 100, 131, 166, 215
Jordan River 88
Jos 101
Jutland 128

K

Kabgar 74
Kalahari 6, 23, 26, 34, 73, 100, 107, 146
Kano 103
Kaouar 134, 198, 210
Kara 248
Karakalpaks 165, 232
Karakalpakstan 21, 236, 240
Karakorum 246, 247
Karakum 17, 18, 21-23, 52, 59-61, 88, 158, 181, 242
Karakum Canal 181, 242
Karchi 181
Kat 164
Kaya 175
Kayes 83, 187
Kazakhstan 17, 21, 243, 259
Kébili 230
Kelif Daria 181
Kenya 26, 30, 31, 41-43, 146, 150, 152, 153, 185, 250, 259
Kerki 21, 165, 181
Kharga 83, 193, 218
Khartoum 61
Khodjend 165
Khorezm 34, 158, 164, 165, 181
Khoulm 230, 231
Khuzistan 163
Kineret 99, 100
Kokand 194
Komadougou 98, 101, 145
Kopet Dag 18, 181
Kordofan 13, 256
Kori Teloua 97
Koriziena 98
Koufra 83, 170
Koulikoro 33, 90, 186
Koutous 145, 264
Kuiseb 15
Kunmalic 248
Kurdistan 163
Kurulik 248
Kyzylkum 17, 22, 23, 52, 55, 61, 88, 158, 231, 232

L

Labador 9
Laghouat 11
Laikipia 152
Lake Eyre 100, 131
Lake Hula 99, 100
Lake Nasser 189, 190
Lake Victoria 30, 41

Lanzhou 52, 256
Largeau 101
Larsa 239
Latin America 2, 160, 179, 207, 253, 270, 271
Layyoune 134, 264
Lebanon 155, 166, 225
Leh 214
Lesotho 100
Liaohé 245
Libya 101, 170, 172, 214, 218, 236
Lima 14
Linga 100
Lingoghin 111, 112
Lo Mustang 214
Logone 88, 90, 101, 146
Lund 35
Lut 17, 128

M

Magadi 30
Maghreb 34, 143, 212, 213, 218, 221–228, 259
Mainé Soroa 36
Makgadikgadi 107
Mal 182
Malawi 100
Mali 34, 38, 40, 41, 45, 65–70, 76, 131, 141, 148, 149, 186, 187, 225, 256, 259, 262, 264
Manantali 186, 187
Mandali 167, 168
Mandara 101
Manga 13
Manitoba 47
Mansour Eddahbi 227
Maradi 168, 263
Maritime Alps 59
Marsabit 41
Massawa 103
Mauritania 94
Médénine 209
Mediterranean 3, 5, 10, 11, 18, 19, 23, 24, 46, 48, 54, 59, 69, 82, 87, 143, 147, 155, 167, 169, 172, 177, 185, 186, 189, 213, 215, 217–219
Mediterranean steppe 48
Mediterranean zone 24, 172, 213
Meghna 160
Melrir 215
Merca 108
Merouane ben Djeloud 49
Merv 165, 181
Merzouga 227
Mesopotamia 87, 157, 160–163, 166–168, 238, 239, 254
Mesopotamian foothills of the Zagros 185
Mexico 1, 34, 69, 110, 160, 177
Mexico City 1
Middle East 2, 5, 25, 54, 69, 87, 155, 156, 162, 163, 166, 207, 219, 239, 240, 259, 272
Middle Empire 158
Mohave 82
Moldavia 155
Mombasa 42
Mongolia 60, 233, 259

Mopti 40, 61, 90, 186
Morocco 14, 54, 55, 59, 123, 134, 155, 178, 179, 185, 193, 196, 198, 209, 212, 216–224, 227, 229, 232, 264
Moscow 165
Mouinak 244
Mourdi 13
Mourdiah 261
Mourgab 88, 163, 164, 165
Mozambique 258, 261
Murray 172, 234
Mushâ 219
Mya 215

N

Nahrvan 240
Nairobi 250
Naivasha 30
Nakuru 30
Namarel 74
Namib 6, 8, 14–16, 23, 26, 55, 83
Namibia 6, 14
Nara 260, 261
N'Djamena 36, 61, 90, 101
Nefud 37
Nefzaoua 179
Negev 56, 60, 71, 93, 106, 171, 173, 215, 218, 265
New Guinea 156
New Zealand 270
Niamey 28, 36, 61, 90, 91, 186
Niger 11–13, 23, 28, 32–36, 38, 43, 53, 54, 61–64, 66, 72, 88, 90, 91, 94, 95, 111, 112, 116, 131, 141, 145, 146, 148, 151, 168, 169, 186, 187, 198, 225, 259, 263, 264
Nigeria 34, 36, 103, 144, 145, 148
Nile 12, 37, 54, 61, 85, 86, 88, 103, 155, 157, 160, 162, 166, 187–190, 234, 236, 264
Nioko 149
Nordeste 17, 23, 28, 29, 32, 36, 111, 112, 139–142, 150, 153, 173, 182, 183, 238
North Africa 23, 34, 48, 139, 179, 207
North America 6, 9, 17, 122, 260, 270
North Atlantic 18, 162
Norway 9
Nouadhibou 83
Nouakchott 264
Nubariya 189
Nukus 18, 21, 22, 181, 236

O

Obeid 163, 167, 263
Ogan 248
Olgas 14
Oman 21, 185, 229
Orontes 218
Otrar 164, 165
Ouagadougou 177
Ouargla 216, 233
Ouarzazate 179, 227
Oursi 85, 94, 99
Oxus 164, 165

Geographic Index

P
Pacific 19
Paipol 185
Pakistan 69, 161, 234, 235, 238
Palmyra 185, 211
Pamir 18, 50, 163
Patagonia 6, 26
Pernambuco 111
Persia 163, 185, 216, 218
Peru 6, 14, 82, 83, 160, 210, 218
Petra 215
Phoenix 209, 210, 214, 220, 233
Punjab 181, 211

Q
Qattara 193
Qum 186

R
Rajastan 5, 36, 37, 128
Red Sea 103, 170, 236
Repetek 18, 22
Rhéris 179
Rhir 216
Rhodesia 109
Rio 2, 252
Rio de Janeiro 2, 252
Rocky Mountains 16
Russia 47, 155, 165

S
Sabz 163
Sahara 5, 6, 10–13, 18, 24, 26, 28, 32–37, 40, 46, 47, 52, 53, 56–61, 72, 80–83, 88, 93, 100–105, 112, 122, 131–135, 142, 146, 147, 155–157, 161, 162, 169–172, 178, 210, 212–218, 221–225, 227, 249, 257, 260, 264
Sahel 10, 12, 13, 17, 18, 23, 24, 27–37, 40, 43, 53, 56, 59, 61–63, 65, 68–75, 83, 88, 90, 91, 98, 105, 111, 112, 122, 123, 131, 132, 135, 137, 139, 144, 146–148, 153, 158, 168, 169, 172, 179, 208, 233, 250, 258–263, 273
Samara 167
Samarkand 17
San Joaquin 1
Sankarani 186
Santorin 158
Sao Francisco 150, 238
Saoura 12
Sary Poul 88
Saudi Arabia 131, 170, 264
Sefid 163
Ségou 186
Sélima 13
Sélingué 186
Senegal 28, 32–34, 36, 54, 61, 63–67, 73, 88, 90, 91, 138, 141, 142, 146, 150, 182, 186, 187, 237, 259
Serdeles 179
Sergipe 173
Sertao 112, 182, 264
Shangyou 247
Shirintagar 88
Siberia 9, 18, 82, 155
Simpson Desert 131
Sinai 11, 95, 111
Sinkiang 34, 88, 162, 163
Somalia 26, 54, 61, 108, 146, 149–152, 185, 199, 255, 258
Sonora 82
South Africa 34, 261
South America 6, 8, 23, 26, 260
South Sahel 62, 63
Spain 23, 110, 186, 218
Sudan 13, 26, 35, 37, 44, 54, 86, 88, 98, 101, 122, 170, 188, 250, 256, 263, 264
Sumé 112
Surham Daria 164
Swakopmund 15
Syr Daria 17, 21, 88, 157, 164, 165, 194, 212, 240–244
Syria 2, 87, 166, 185, 215, 216, 225
Syrte 170

T
Tadémaït 214
Tadjikistan 240, 243, 245, 259
Tafassasset 12
Tafilalt 179, 212, 214, 227, 229
Taghouzi 178
Taif 185
Taitmar 246, 248
Taklamakan 23, 34, 53, 88, 111, 126, 132, 161, 197, 231, 246–248, 266
Talas 88
Tamanrasset 11, 12, 96
Tamdi 231, 232
Tamentit 214, 225
Tamgroute 59
Tanzania 146
Taouat 225
Tarfaya 132
Tarim 88, 214, 238, 245–248
Tashkent 21, 22, 165, 241
Tassili 61, 106, 132, 179
Tassili des Ajjer 179
Taurus 20, 87, 166
Tazabagjab 158
Tchou 249
Tedjen 88, 163, 181
Telak 13
Ténéré 23, 134
Tengger 93, 195, 196, 266
Termit 131
Thailand 42, 157
Thar 5, 34, 83, 131, 161, 233
Tibati 145
Tibesti 61, 97, 101, 132, 135, 150, 157, 169
Tibet 17, 37
Tidikelt 214, 225
Tigris 87, 166, 167, 239, 254
Tikanlik 248
Tillabéry 36
Timkent 164
Timmimoun 214
Touat 214, 215

Touggourt 49
Tozeur 179, 223, 227
Trujillo 210
Tsabong 34
Tunisia 55, 62, 93, 116, 123, 172, 206, 209, 212–214, 221–224, 230, 232
Turan 5, 17, 18, 21, 22, 47–56, 59, 70, 90, 162, 177, 183, 205
Turkana 30
Turkestan 47, 50, 164, 165
Turkey 2, 87, 166
Turkmenistan 34, 163, 165, 235, 241, 243
Turtkul 21

U
Ukraine 47
Urals 18, 26, 164
Uruk 157, 163
USA 6, 34, 54, 83, 181, 186, 220, 233, 237, 245, 254, 262, 270
USSR 48, 56, 122, 151, 170, 181, 194, 207, 212, 231, 234, 240, 243, 245, 248, 264, 270
Uzbekistan 194, 240–243, 259

V
Valréas 1
Venice 1

Victoria 30, 41, 100

W
Wadi Halfa 191
Wadi Natrun 103
Waksh 165
West Africa 14, 32, 34, 35, 40, 41, 61, 111, 112
western Asia 5, 87, 156, 157
Wola 156

X
Xiaohaiz 247
Xinjiang 245, 246, 247, 248

Y
Yanfolila 260
Yani Daria 164, 165
Yarkant 246, 247, 248
Yatenga 176, 177
Yedseram 101
Yellow River 51, 52, 157, 245, 264
Yemen 176, 185
Yobe 145

Z
Zagora 227

Subject Index

A
aborigines 146
Abraham 161
Abu Simbel 190
achaba 143
Achemenid 185, 239
action
　abrasive 120, 196
　aeolian 12, 53, 100, 113, 137
　activity, faunal 31
aeolian circulation, global 19
aeolian competence 122, 125
aerosols 52, 65
agricultural tools 155
agriculture
　hydraulic 158, 160
　irrigated 53, 141, 157, 161–163, 165, 168, 176, 182, 188, 215, 219, 226, 234, 237, 241, 244
agrostologists 77
agrosystems 213
air cushion 19
albedo 33–35, 235
Alexander the Great 239
allochthonous 14, 16, 28, 55, 64, 85, 88, 90, 91, 135, 150, 157, 163, 166, 180, 214
anemochorous 68
anticyclone 8, 13, 18, 82
aquifer 49, 58, 96, 103–105, 145, 170, 175, 178, 184, 190, 213, 216, 217, 223, 235, 258
area
　of aeolian transport 134
　accumulation 134, 201
　catchment 97, 173
　loess (cf. loess) 51, 237
　source 131–134, 193, 200
　transportation 200
aridification 32, 34, 38, 45, 157, 158, 160–162, 168, 185, 204, 216, 235, 240, 271
　Holocene 185
　increasing 34, 38
　postglacial 32, 34, 162
aridisation 158
attenuation 98
Azvesta 164

B
badlands 108, 253

bank 137, 141, 144
banquetas 176
barley 54, 139, 161–163, 167, 168, 177, 212, 221, 239, 265
barren ground 57, 260
barrier 15, 20, 101, 127, 197, 199, 227, 248
　green 197, 227, 248
basketry 212, 214
bilharziosis 190
biogas 70
biomass 17, 56, 62, 63, 66–69, 110, 143, 144, 265, 271
　herbaceous 62, 66, 67
　vegetal 143
biotechnology 265
biotope 4, 162, 167
bioturbation 107
boiling point 106
bone industry 155
Borana 146, 152
boreholes 150, 209, 216, 218, 223, 224, 258, 259
　deep 150
boundary layer 114, 115
breeders 98, 100, 142, 151, 153, 210, 226, 232, 233, 256, 260
breeders (c.f. pastoralists) 41, 146, 153, 163, 187, 272
breeding
　Borana 146, 152
　cattle 142
　sheep 156
building materials 189
Bulinus haematobium 190
bush fires 71

C
Caatinga 142, 150
camel caravans 231
camels 97, 174, 179, 218, 265
Canaan 161
canal, Nahrvan 240
canopy 57, 65, 71, 115
capacity
　infiltration 72
　retention 50, 52, 63, 68, 106, 122, 182
　storage 106, 172, 182
cartonville 150
castor oil 140
casts 143

catchment 97, 123, 173–175, 180, 184, 215
   water 173
Celts 215
ceramics 155, 157
cereals 42, 44, 138–141, 155, 162, 177, 212, 221, 225, 239, 261, 266
   dry 141
chadouf 168, 218
check dams 174
chergui 123, 227
chott 216
   artificial 216
cisterns 60, 150, 172, 173, 215
civilization
   Egyptian 86, 158
   hydraulic (cf. hydraulic society) 4, 137, 154, 157, 158, 160–163, 212
   Iranian 185
   Minoan 157
   Namazga 163
   oasis 212
   urban 269
climatic changes 122, 154, 168, 252, 258
climatic regulator 18, 243
coefficient of variation 83
community
   vegetal 57
cone, alluvial 50, 54, 216, 217, 218, 226, 230, 231
constraints
   aeolian 79
   climatic 42
   demographic 154
   ecological 214
   hydrological 154, 263
   physical 79
construction 67, 70, 71, 87, 150, 152, 157, 160, 165, 173, 174, 177, 179, 184–187, 191, 217, 219, 227, 229, 245, 274
Continental Intercalaire 80, 104, 223, 225
continentality 8, 26, 80
cotton 22, 54, 56, 102, 139–142, 148, 164, 165, 180, 191, 212, 214, 226, 231, 236, 240–244, 259
cover
   dead 140
   sandy 53, 64, 95, 135, 201
   vegetal 73, 110, 124
cradle of agriculture 156, 162
creep 118, 119
creeping 118, 119
cultural individuality 146
culture
   Aterian 155
   forage 223
   irrigated 28, 61, 138, 142, 212
   mechanized 139
   Mundigak 161
   subsidence 264
currents
   aeolian (cf. wind) 114, 132, 136, 193, 201
   ocean 15
cycles, cosmic 32

D

dam
   arched 186
   earth 172, 175
   reservoir 185, 192, 237
date palms 167, 178, 179, 210–213, 225
decadence 4, 203, 204, 234, 240
decline 3, 4, 147, 158, 185, 221, 224, 229, 231, 240, 245, 252, 258, 269, 270
deficit
   evapotranspirational 85
   pluviometric 24, 25, 28, 29, 36, 41
   water 25, 77, 81, 107
deflation 13, 21, 23, 35, 52, 75, 100, 113, 116, 117, 121, 124, 128–132, 135, 194, 199, 200, 210
   aeolian 35
deforestation 1, 30, 44, 69, 70, 77, 148, 255
deglet nour 222
degree of aridity 5, 6, 177
delta 10, 18, 34, 64, 87, 95, 100, 141, 146, 162–164, 186–190, 236–238, 243, 244
demographic explosion 80, 188, 190, 208, 222, 231, 259
   pressure 139, 149, 190, 206, 220, 221
deposits 16, 21, 35, 50, 51, 54, 60, 77, 85, 94, 102–106, 119–122, 131, 183, 185, 188, 189, 201, 221
   aeolian 12, 13, 16, 18, 19, 31, 35, 46, 47, 51, 53, 55, 67, 70, 71, 73, 75, 79, 95, 98–100, 107, 112–125, 128, 132, 134, 136, 137, 162, 172, 189, 192–194, 200, 201, 228, 236, 243, 256
   silty 119, 122, 180, 239
depression
   aeolian 31, 98, 99, 100
   continental 94
   lacustrine 168
desalination 1
desert
   advance of the 34, 35, 157, 255, 256
   Australian 23, 146
   Chinese 11, 51, 112, 128
   coastal 11, 26
   cold polar 3
   continental 3, 5, 6
   littoral 6, 14
   Ogolian 135
   subtropical 17
   warm tropical 3
desert coating (patina) 55
desertification 2, 4, 35, 51, 52, 61, 108, 157, 161, 204, 208, 216, 220, 223, 232, 249, 250–263, 273
desertion 216, 222, 224, 270
determinism 160, 204, 219
   environmental 160, 219
development
   of oases 179
   global 266
   sectorial 205
   sustainable 79, 201, 208, 221, 233, 267, 269, 271
diagonals
   arid 5
   dry 6
diaspores 75

Subject Index

diffusor 120
discharge of water courses 32
disequilibrium
    ecological 32, 43
    economic 43
djebars 211
Djeitun 163
domain, hydroclimatic 184
domestication of cattle 168
drainage 12, 48, 50, 85, 105, 107, 190, 191, 209, 216, 230, 233–237, 239, 243, 247, 255, 265, 274
    relict 12
drought
    agricultural 31
    edaphic 31
    meteorological 31
drying, climatic 156
dunes
    active 35, 53, 111, 122, 192, 195
    barchan 117, 119, 123–128, 132, 198, 201, 210, 227
    classification of 113, 123
    deflation 128
    elongate 128
    erosive 130
    fixed 34–37, 108, 111, 122, 231, 263
    genesis of transporting 117
    linear (sifs) 16, 123, 124, 227
    longitudinal 13, 124, 127, 130, 131, 200, 201
    longitudinal red 45
    mobile 131, 193, 195
    overgrown 37
    ovoid dome 124
    parabolic 100, 123, 129, 200
    pyramidal 13, 124
    sand 62, 63, 108
    transverse 23, 124, 126, 201
duricrusts 13, 54
dust storm 18, 19, 55, 72

E

easterlies 115
ecological catastrophe 113, 166, 182
    refugees 113, 253, 263, 273
ecosystem 9, 24, 59, 79, 80, 85, 154, 163–166, 175, 177, 188, 193, 210, 212, 244, 245, 248, 252–254, 256, 261, 272
    dry 79, 154, 166, 244, 252–254, 272
    Sudanian 24, 261
edible 66, 71, 74
El Niño 14
encrustation 81, 93
    organic 93
endorheic basin 95
energy
    kinetic 93, 108–110
    solar 22, 71, 80, 184
environment, dry 77, 83, 109, 154
epeirogenic 100
ergs (cf. SGAE) 197
erodability 31, 76, 108, 109, 117
erosion
    aeolian 31, 67, 71, 73, 107, 112, 113, 116–118, 122, 128, 189, 192, 200, 228
    by water 35, 107, 108, 111, 136
ethnic groups 155
Eucalyptus 174
evaporation
    potential 18, 85, 101, 242, 246
evapotranspiration 6, 24, 25, 27, 62, 85, 104, 105, 107, 115, 180, 210, 235, 237, 264
    potential (PET) 6, 25, 26, 105, 180, 263, 264
    real 27, 264
expansion, Neolithic 167

F

famine 4, 32, 36, 37, 41, 42, 44, 65, 68, 143, 234, 249, 250, 253, 254, 258, 261, 262, 273
FAO 25, 46, 54, 56, 69, 109, 248, 252
fertility 23, 29, 31, 48, 50, 52, 54, 79, 116, 122, 139, 245, 257, 272
fertilizers 174, 184, 189–191, 199, 206, 229, 231, 238, 244, 265
    nitrogen 48, 56, 57, 66, 67, 71, 191, 265
fixation 93, 181, 192–195, 198, 199
    biological 9, 24, 46, 57, 96, 192–199, 235, 250, 252, 254
    of mobile sand 192, 193
floods, spring 166, 217
floristic heritage 98
fluctuations, climatic 2, 34, 37, 59
föhn 15, 17, 18
force
    buoyancy 117, 118
    Coriolis 114, 132
    friction 114
forests
    gallery 110, 212, 247
    grazed 143
French ONG 187
friction (cf. wind) 114, 115, 119, 120
frost, night 243
fructification 67, 76, 176

G

GAAS 53, 131–135, 200, 201
    closed 132
    open 132
    Saharan 132
gallery (cf. irrigation)
    underground (cf. foggara) 213, 216, 217, 224, 225
gaylak 232
Gengis Khan 230
geohistory 3, 154, 168
germination 28, 43, 58, 68, 74, 76, 116, 199, 254
ghanat 216, 217, 218
ghourds (cf. dunes) 123, 124, 128
gigantomania 165, 166
gilgai (cf. soils) 54
glacial maximum 38, 106
glaciers 34, 50, 88, 99, 104, 136, 211, 231, 241, 245, 246, 267
glazing (cf. battering crust, scalding) 238
goats 42, 66, 72, 142, 147, 150, 174, 223

Gobi (cf. reg) 52, 55, 59, 115, 124, 247
goz 13
Grande Bere 36
grasses
    perennial 28, 62, 71, 76, 111, 144
    xerophytic 13
gravity constant 117, 120
grazeable (cf. grazing, pastoralism) 41, 43, 56, 58, 61, 64, 66–68, 72–77, 100, 102, 110, 113, 143–146, 152, 153, 187, 206, 209, 221, 232, 237
groins 175
gueltas 97
gullies (cf. run-off) 1, 12–14, 17, 28, 30, 31, 50, 51, 55, 58, 62, 72, 83, 85, 88, 91–98, 102–105, 107, 134, 140, 161, 167, 171–176, 185, 199, 211, 217, 245, 247, 265
gullying 31
Güneydogu Anadolu Project (GAP) 87

H
Haloxylon aphyllum 60
hamada 52, 53, 60
Harappa culture 161
harmattan 53, 132, 134
heliophytes 64
herbicides 206, 239, 244
herds 41, 43, 64, 66, 72, 73, 77, 108, 138, 143–147, 150, 151, 153, 182, 205, 207, 221–223, 232, 234, 273
Holocene 12, 30, 37, 38, 100, 101, 106, 135, 156, 158, 185
    lower 30
hope 203, 233, 266, 267
horizon (cf. soils) 31, 46, 48, 49, 71, 80, 103, 104, 107, 116
humus (cf. soils) 46, 50, 55, 56, 71, 140, 156
hunters 146, 156, 162, 164
hydrology 2, 24, 27, 28, 30, 79, 88, 102, 104–106, 149, 152, 154, 186, 240, 246, 263
hydromorphism 190
Hyksos 158, 161, 239
hyperarid conditions 155
hyperhalophilic 47

I
indebtedness 206
index, bioclimatic 6
individual pumping 219
infiltrability 30
infiltration 30, 31, 49, 55, 57, 72, 76, 81, 93–95, 102, 103–108, 111, 116, 140, 171–176, 183, 190, 243, 254
inselbergs, inselgebirge 14
insolation 9–11, 24, 66, 80, 81, 115, 137, 166, 184, 210
International Institute for Environment and Development 266
intertropical convergence 7, 12
irrigated soil (cf. soil) 234, 235, 247
irrigation
    by gravity 213
    drop 237
    modern 168, 177, 188, 239
    underground 184
isohyets 36–41, 56, 61, 180, 256

J
jet stream 82, 114

K
kanayet (cf. irritation) 216
karakul sheep 231, 232, 265
karez (cf. irrigation) 216
Kazakhs 232
Kernwüste (cf. desert) 6
kewir 100
khadin 173
kichlak (cf. breeding) 232
King Menes Marmer 157
kirat 225
ksar 225

L
lacustrine period 215
lake
    artificial 188, 220
    L. Nasser 189, 190
land
    marginally arable 43
    pastoral 43
land reform 229
Lavlakyan 164
law, Bernoulli's 120
layer
    alluvial 103
    phreatic 99, 106
level
    piezometric 49, 96, 218
    water 30, 34, 58, 91, 101, 166, 188, 191, 210, 216, 219, 220, 226, 237, 240
liman 171
litter 81
livestock 13, 42, 43, 61, 62, 64–66, 68, 71–75, 97, 98, 100, 107, 138–148, 150–153, 164, 172, 177, 187, 205, 212, 221, 232, 233, 238, 256, 260, 265
loess 21, 23, 50–52, 56, 60, 106, 119, 121, 122, 237
lucerne 212, 223, 225, 233
Lugalzaggisi, king of Uruk 157

M
maize 27, 41, 42, 48, 111, 112, 139–141, 189, 191, 212, 264, 265
man
    archaic 154
    modern 4, 154, 247, 269
manure 176, 220
margins
    Sahelian 59
    subhumid dry 3
market gardening belts 223
matter
    dry 62–67, 71, 76, 176, 180
    organic 46–48, 55, 57, 63, 67, 74, 76, 117, 139, 140, 156, 235, 238, 257
medicinal plants 215
migration 7, 13, 52, 64, 127, 130, 146, 149, 150, 161, 188, 193, 195, 198, 270
milk 41, 43, 74, 150–153, 223, 232

Subject Index

goat milk 223
millet 27, 41, 54, 56, 65, 70, 138-141, 162, 168, 175, 176, 195, 264
mixed 119, 174, 186
mobilizable material 122
monsoon 30, 37, 38, 123, 155
moradores 140
morbidity 244
mortality 222, 244
motor pumps (cf. irrigation) 179, 218, 219, 224-226, 229
mu 248
mull (cf. soils) 49

N

Nabateans 215, 216
nadi 171, 172
nebkas 116, 123
Neolithic 155-158, 164, 167, 265, 269
  miracle 155
night dew 210
Nile 12, 37, 54, 61, 85, 86, 88, 103, 155, 160, 162, 166, 187, 188, 189, 190, 234, 236, 264
  discharge 155
nitrates 56, 238
nitrogen 48, 56, 57, 66, 67, 71, 191, 265
nitrophilic 74
nodules (cf. nitrogen) 48, 71, 235
noria (cf. irrigation) 164
Nubians 191
nutrients 31, 71, 73, 98, 100, 110, 116, 139, 144, 238, 265
nycthermal 209

O

oasis
  natural 210
  urban 211, 225
Ogolian 135, 155
olives 166, 228
optimum
  climatic 155-157
  Holocene 38
overexploitation 13, 30, 51, 72, 111, 122, 139, 144, 161, 205, 219, 224, 225, 244, 262, 263
overgrazing 31, 34, 41, 43, 55, 56, 71-76, 77, 102, 108, 111, 143, 145, 161, 205, 231, 255, 256, 259

P

palisades 67, 193-195, 198
palm fronds 196
pans (cf. ponds) 100, 107
parasites 190, 210, 223, 244
particles
  sandy 75, 112, 119
pastoral 41, 43, 57, 61, 65, 74, 78, 137, 143-147, 150, 153, 154, 168, 180, 221, 222, 228, 232, 260, 269
pastoralists (cf. breeders) 41, 146, 153, 163, 187, 272
  pastoralism 41, 143, 152, 209, 221
pastures, grazing 28, 41, 43, 61-64, 66, 68, 71-73, 75, 77, 78, 100, 107, 137-140, 143-146, 149, 152, 153, 171, 181, 188, 192, 208, 221, 222, 237, 245, 248,

257, 261, 265
  aerial 64
pedogenesis (cf. soils) 46, 77, 116, 135
perforated 183
permeability (cf. porosity, soils) 48, 50, 51, 53, 63, 93, 96, 106, 107, 121, 235
pesticides 44, 206, 223, 239, 244, 271
Peul 36, 145-151
  Woodabé 151
phosphates 191
phosphorus 56, 238
piedmont 17, 21, 60, 82, 97, 168, 230
pingo 10
pinnules 76
plant roof (cf. canopy) 57, 65, 71, 115
plants
  domestic 156
  perennial 35, 210
  relict 60
  woody 9, 69, 116
pluviometry 25
polar ice 1, 51, 170
political hierarchy 160
pollen 116, 216
population, world 257, 270, 272
porosity (cf. permeability) 48, 50-53, 63, 93, 96, 106, 107, 121, 235
pottery 158, 161, 214
prairie 99
  dry 99
precipitation
  annual 6, 10, 14, 21, 24, 25, 32, 54, 55, 58, 92, 93, 96, 109, 111, 139, 233, 242, 253
  excessive 29
  mean 13, 17, 47, 73, 94, 101, 173, 260, 263
pressure
  demographic 139, 149, 190, 206, 220, 221
  of moving air 117
  static 118
production
  primary 32
  vegetal 41, 66
productivity 29, 44, 53, 66, 67, 73-75, 111, 116, 142, 144, 224, 250, 252, 257, 261, 265
prophylaxis 190
proteins 64-66, 143, 265
protohistory 163
pumping station 107

R

rains
  summer 23, 59, 167
  winter 23, 27
Ramses II 190
rate of evaporation 97, 101, 106
recharge 94, 99, 104-107
  of aquifers 106
recurrent 29, 250, 273
reflectance 215, 233
reg (cf. Gobi, soils) 52, 55, 59, 115, 124, 247
regeneration 66, 71, 72, 75, 194-197, 204
  natural 196, 197

regime
  bidirectional 123, 124
  monodirectional 123, 124, 128, 130
  pluviometric 63, 96, 265
  seasonal 105
  wind 123, 124, 128
regulating structures (cf. dams) 167
rehabilitation
  artificial 200
  natural 113, 200
relative humidity 242
remote sensing 33, 35, 118
resilience 45, 77, 208, 256, 270
resources 2, 28, 31, 35, 43, 56, 78–81, 83, 85, 91, 100, 104, 106, 107, 136–138, 145, 153, 168, 170, 177–180, 192, 204, 205, 208, 210, 212, 219–224, 228, 231, 233, 241, 245, 248, 252, 262–273
  food 222
  human 80, 81, 245, 267, 272
  permanent 80
  renewable 80
reverse osmosis 170
rhettara (cf. irrigation) 216
rice 24, 56, 102, 138, 141, 142, 162, 177, 180, 186–189, 191, 206, 212, 265
ripple marks 116, 122
river navigation 188
rock paintings 156
roots, underground part of plants 46, 48, 54, 58, 67, 106, 107, 139, 176, 238, 240, 241, 248, 265
rotation 72, 114, 120, 220
run-off
  diffuse (rill wash) 31, 50, 71, 77, 83, 91–93, 96, 98, 105–108, 112, 161, 171–173
  laminar 114, 125
  lines of 107
  sheet flow 92, 112

S

Sahel 10, 12, 13, 17, 18, 23, 24, 27–37, 40, 43, 53, 56, 59–63, 65, 68–75, 83, 88, 90, 91, 98, 105, 111, 112, 122, 123, 131, 132, 135, 137, 139, 144, 146–148, 153, 158, 168, 169, 172, 179, 208, 233, 250, 258, 259, 260–263, 273
saheli 123, 227
Sahelian (cf. zone) 6, 10, 13, 24, 25, 30, 34, 36, 59, 61–67, 70, 75, 76, 88, 94, 96–98, 101, 105, 107, 131, 153, 158, 176, 177, 186, 249, 250, 256, 258, 260, 261, 264
salinas 100
saline concentrations, efflorescences 49, 54, 235
salinity 1, 18, 47, 60, 100, 181, 186, 190, 210, 223, 229, 236, 238, 243, 246, 247
salinization 52, 56, 99, 113, 140, 169, 189, 190, 203, 223, 226, 228, 229, 230–240, 243, 244, 247, 248, 255, 257, 263, 274
  secondary 52, 229, 230, 235, 236
saltation 31, 116–120, 132, 134, 193, 194, 198
salts, soluble 47, 49, 235, 238
sand invasion 196, 200, 224, 227, 228
sand ridges 95, 131
Sargon 157

savannah 26, 32, 37, 61, 71, 144, 250, 256, 260, 261
  bush 261
  tree 260, 261
scalding (cf. battering crust) 184, 238
scale, regional 29, 104, 106, 240
Schistosoma 190
seasonality 63
sedimentary balance 23, 123–126, 128, 131, 135, 201
  negative 23, 128, 131
  positive 201
sedimentation 50, 239, 240, 243
seeds 28, 58, 64, 67, 68, 73–75, 176, 200, 211, 212, 223
  date 223
self sufficiency 140
Semitic dynasty of Akkad 157
services
  sanitary 153
  social 148
  veterinary 151
sheep 42, 72, 142, 143, 147, 156, 174, 221, 231, 232, 265
sheet flood
  wash 31, 58, 71, 77, 83, 91–93, 96, 98, 105–108, 112, 161, 171–174, 215, 216, 259, 265
Silk Road 214, 247
slavery 157, 158
snow 9, 21, 87, 99, 104, 211, 217, 241, 246
society
  feudal 218
  hydraulic (hydraulic system) 160, 212
soil degradation 110, 234, 235, 252, 259, 261
soil texture 116, 180
soils
  alluvial 54, 227
  desert 46–48, 50
  ferruginous 35, 46, 47
  hydromorphic 64
  leached 47, 49
  relict 46
  saline 48
  salty 60
  sandy 60, 62, 63, 75, 106, 180, 189, 206, 228
  tropical 47
solonets (cf. soils) 50, 56
sor 49
sorghum 27, 41, 54, 56, 138, 140, 141, 162, 174, 264
source
  glacial or periglacial 197
  sand 198
space, interdunal 62
species
  annual 63, 66
  hardy 158
  perennial 34, 58, 256
spineless cacti 140
splash 31, 76, 93
sprays (cf. irrigation) 183
stalks 56, 70, 175, 195
State of the World Population 270
steppe
  Algerian 63, 221
  environment 154, 215
  Sahelian 131

Subject Index

thorn 61
wooded 41, 263
stomata 58, 115
subhumid dry 3, 10, 12, 24, 25, 27, 30, 47, 69, 71, 72, 77, 93, 94, 138, 144, 248, 252, 254, 260
subsidence 1, 17, 32, 33, 43, 72, 82, 115, 237, 264
subtropical countries 11
Sumerian empire 157
supply, water 72, 73, 104, 145, 176, 180, 183, 191, 211
system
    breeding 222
    customary 147
    dry ecological 3, 253
    hydraulic 160, 165, 216, 217, 240
    irrigation 54, 86, 137, 158, 160, 163, 165, 167, 178, 184, 258, 274
    root 58, 63, 116, 139, 174, 191, 196, 211, 228, 237, 238

T

takyr 50
tamarind 65, 212
tanka 173
tent 148
tenure 2, 68, 148, 153, 179
terraces 34, 54, 87, 110, 161, 175, 176, 177, 182, 215, 216, 217, 247
thermal inversion 14
thermonuclear 208
thmen 225
tît 212
tolerance threshold 107, 204, 218
tomatoes 72, 141, 212, 222
tourism 214, 220, 222
traction 118, 119, 148, 179, 218
    animal 148, 179
transhumance (cf. nomad, nomadism, pastoralist) 56, 80, 143–146, 148, 153, 209, 232, 258
transport
    aeolian 18, 51, 118–121, 134
    dust 121
trees
    deciduous 64
    evergreen 64
    forage 64, 65, 143
    pioneer 199
tribe 146
    nomadic 146
tugai 231, 247
tundra 9, 10
typology 178, 201, 219, 220

U

underflow 61, 82, 83, 85, 95, 103, 104, 106, 184, 185
UNDP 56
unemployment 245
UNEP 2, 56, 109, 234, 248, 250, 252, 254, 257, 260, 261
UNESCO 6, 25, 46, 153, 190, 250, 253
UNICEF 207, 271
United Nations 2, 56, 88, 250, 252, 257, 270, 273
United Nations Population Fund 270

upwelling 14
USLE (Universal Soil Loss Equation) 109

V

variability, interannual 29, 81–83, 158, 210
varnish 53, 55, 93
vegetation
    contracted 61, 97
    halophilic 60
    tree 71
veterinary medicine 206, 260
viscosity 93, 118
vulnerability 31, 76, 77, 108, 113, 119, 139, 188, 192, 204, 208, 232

W

wanda wassu 36
warming 30, 32, 155, 156, 161, 170
    postglacial 156, 161
wars
    civil 68
    tribal 143
water
    artesian 104, 214, 224
    fossil 224
    run-off (cf. run-off) 171, 173
    surface 10, 58, 73, 81–83, 85, 102–105, 137, 150, 158, 165, 180, 214, 216, 223, 243
water retention 50, 52, 63, 68, 96, 116, 122, 182, 254
water right 179, 224
water supply 72, 73, 104, 145, 176, 180, 183, 191, 211
wells 43, 67, 72, 74, 106, 107, 145, 150–152, 170, 175, 179, 183, 198, 209, 213, 216–219, 223–226, 257, 258
    Arab 87, 88, 100, 146, 147, 166, 167, 212, 213, 216
    artesian 145, 179
    pulley 218
wheat 41, 54, 139, 141, 161–164, 167, 168, 177, 212, 222, 225, 231, 239, 264
wild ungulates 73
wind (cf. trade winds) 4, 12–14, 18–20, 31, 46, 51, 53, 55, 66, 68, 71, 72, 76, 79, 80, 100, 110–137, 176, 192–201, 210, 218, 227, 228, 231, 234, 248, 252, 254, 256, 257, 259, 263, 264, 269
    aeolian circulation 19
    aeolian currents 114, 132, 136, 193, 201
    harmattan 53, 132, 134
    monsoon 30, 37, 38, 123, 155
    surface 19, 20
    westerlies 7, 16, 115
windbreak 121, 174, 197
winnowing 116, 117, 122, 134, 200
winter season 102
WMO (World Meteorological Organisation) 252
wood, firewood 69, 70, 75, 174, 222
woody 9, 17, 62, 63, 65, 69, 116, 221
World Bank 29, 43, 177, 187, 208, 248
World Food Programme 151
WWR (World Weather Records) 83

X

xerophytic, bush (cf. vegetation) 13, 32, 77

Y

years, dry 28, 36, 64, 65, 76, 98, 145, 269
yellow fever 205
yields 62, 137, 139, 168, 175, 180, 189, 191, 231, 252

Z

zai 176
zay 176
Zénète 212
zone
   arid 7, 9, 26, 29, 32, 45, 85, 92, 95, 97, 98, 101, 103, 140, 145, 185, 250
   climatic 3, 27, 46, 113, 269
   Guinean 25, 92
   hyperarid 26
   Saharan 25
   Sahelian 25, 63, 88, 96–98, 101, 107, 158, 260, 261
   Sudanian 25, 36, 61, 92, 158, 260, 261
zoochorous (cf. anemochorous) 68, 75
Zoroastrian epics 164

Printing: Mercedesdruck, Berlin
Binding: Buchbinderei Lüderitz & Bauer, Berlin